D0907063

Form and Space Vision

YVES LE GRAND

Form and Space Vision

Revised Edition

Translated by Michel Millodot and Gordon G. Heath

INDIANA UNIVERSITY PRESS
BLOOMINGTON AND LONDON

Library of Congress catalog card number: 66–12728

Manufactured in the United States of America

Translators' Preface

Professor Le Grand has, like Helmholtz, written three volumes of *Physiological Optics* over the same ten-year period of this century. His first volume, published in 1946 with second and third revised editions in 1951 and 1965, has not yet been translated. His second volume, published in 1948, was translated in 1957 under the title *Light, Colour and Vision.* This is the third volume, published in France in 1956 under the title *L'Espace visuel*; Professor Le Grand has revised and updated it for the English translation.

This project was made possible by a grant from the American Optometric Foundation, which is gratefully acknowledged. We wish to thank Dr. Ronald W. Everson for his help and advice on Chapters 16 and 17, and Mrs. Susan Millodot who has assisted us throughout this long task which has been completed because we remain convinced that Professor Le Grand's uniquely comprehensive views of *L'Espace Visuel* belong in the English literature of physiological optics.

Michel Millodot
Gordon G. Heath

Montreal and
Bloomington, 1966

Author's Acknowledgment

It is a great pleasure for me to thank Michel Millodot and Gordon G. Heath for undertaking the difficult task of translating my book. As far as my knowledge of English allows me to say, it seems an excellent work and I am happy to welcome the new friends they have enabled me to make.

Yves le Grand

Paris, 1965

Contents

Introduction

As the title implies, this text is devoted to problems regarding the vision of details, forms, movements, and depth; in short, to the essential elements of our knowledge of the world through our vision. This text is not, however, a treatise on visual perception. We will utilize only the minimum amount of psychological concepts. Our goal consists of gathering and coordinating the greatest possible number of visual phenomena (physical and physiological) as they appear to normal subjects.

To interpret the mechanisms of vision it is necessary to begin analyzing the structure of the retinal image. We shall discuss very briefly the theory of ocular dioptrics as it is covered in all optometry texts. Gaussian or paraxial optics is the basis of calculations of ophthalmic lenses; it neglects in particular the complexity of light rays, which, after refraction through the transparent media of the eye (cornea, aqueous humor, crystalline lens, vitreous humor), form the retinal image.

But the hypothesis that each medium of the eye possesses a well-defined refractive index is correct only for monochromatic light of a given wavelength. If this wavelength changes or if the light that enters the eye is complex, the phenomena become complicated because of the *chromatism* of the eye, which is due to the variation of refractive indices with wavelength. The resulting *chromatic aberration* is studied in Chapter 1. Then we consider *spherical aberration*, i.e., the deviation from Gaussian optics in the case of a point source situated along the axis of the eye. With the knowledge of these experimental results the wave theory of light can be applied to study the structure of the retinal image of a point of light. The case of several points, lines, or details of a more complicated geometric figure will naturally lead to the problems of *visual acuity* and vision of forms.

It is well known that the keenest vision is obtained when images are on the *fovea*, that region of the retina which immediately surrounds the fixation point. But the rest of the *visual field*, in spite of its lesser efficiency in the resolution of details, plays an essential role that we will study in its various aspects. Then we shall enlarge the static point of view by introducing the notions of *movement*. Following this we shall recall that a normal subject has two eyes, and show that binocular vision plays a preponderant role in *stereopsis* and the structure of visual space.

As was pointed out above, the text will reflect essentially the point of view of the physicist. We shall attempt, however, to give the reader some concepts of the physiology of the *visual centers*, which in spite of the limitations of our knowledge on this subject enable us to explain certain remarkable facts.

Finally the last few chapters (Section B) will be devoted to some applications: vision through the atmosphere, artificial illumination and visual tasks, stereoscopy. These are difficult and important techniques for which knowledge of physiological optics seems to us fundamental.

The bibliography is given in two parts. "References" lists those works primarily contributing to the drafting of this volume and cited in the text by numbers. "General References" are those cited by author and date and comprise volumes and review articles valuable to readers wishing to extend their investigation of a particular topic. I hope thus to have fulfilled my goal: to provide a tool for the use of those who are interested in the problems of vision.

Yves le Grand

Form and Space Vision

Section A

FACTS AND THEORIES

CHAPTER ONE

Chromatic Aberration

The formation of the image of a point of light on the retina of the human eye is described in an elementary fashion in many texts. In the simple schema referred to as paraxial or Gaussian optics, the eye is assumed to be a centered optical system: All the rays of light coming from a point source, after refraction through the system, intersect at one point, which is the image. If the source is at infinity, or practically so, this image is at the *second focal point* of the system; in an unaccommodated emmetropic eye this focal point is situated on the retina; the image is therefore in focus. But, strictly speaking, this only happens with monochromatic light, that is to say, a single radiation of a perfectly defined wavelength λ. In effect, the refractive index n of each of the media of the eye varies with λ and the position of the image is affected; if the eye is adjusted for one particular wavelength it is not adjusted for any other, and the image is focused either in front of or behind the retina. We shall see that if the eye is emmetropic for yellow, it is highly myopic for blue and slightly hyperopic for red. It is this important phenomenon that we shall study in this chapter, conforming to paraxial optics.

Dispersion

The variation of the refractive index n with the wavelength λ is a general phenomenon except in a vacuum, where the velocity of electromagnetic waves is a universal constant. All substances show dispersion, which means that the velocity of the waves in a particular medium depends upon λ; the same is true of the index n, since n is equal to the ratio of the velocity in vacuum to the velocity in that particular medium. As a rule, except in the absorption bands, n increases when λ decreases. It follows that the propagation of light in a transparent medium is not characterized by only one number, but by a function n of the variable λ. This function does not have a strict theoretical representation, and therefore the dispersion is defined only by a table of numerical values of n for values of λ sufficiently close together. For substances that do not show appreciable absorption in the visible spectrum (between 400 and 700 mμ wavelengths), an old custom, from the days of Fraunhofer, assigns values of the index for some of the lines of the solar spectrum, such as the lines A (768.1 mμ), C (656.3 mμ), D (589.3 mμ), F (486.1 mμ), and G (430.8 mμ). The indices are designated n_A, n_C, etc.

Sometimes a dispersing medium is characterized by two parameters only, the index of refraction n_D for the sodium D line (which is actually a doublet 589.0 and 589.6 mμ) and the *coefficient of dispersion* or *constringence* defined by the ratio (always positive)

$$\nu = \frac{n_D - 1}{n_F - n_C} \tag{1}$$

It is easy to show that combining lenses of different constringences helps to resolve, at least approximately, the problem of *achromatism*. Let us suppose, for example, two thin lenses, the powers of which are, respectively (in air),

$$D = (n - 1)a, \qquad D' = (n' - 1)a' \tag{2}$$

where n and n' are the indices of refraction of the lenses for the same wavelength λ and a and a' functions of the radii of curvature of the surfaces. If the lenses are combined, the power of the system is the sum $D + D'$ and if this power is to remain constant when the wavelength varies by $\delta\lambda$, the following condition applies:

$$\delta D + \delta D' = 0$$

or

$$a\delta n + a'\delta n' = 0$$

Replacing a and a' by their values from (2), we have

$$D\frac{\delta n}{n - 1} + D'\frac{\delta n'}{n' - 1} = 0$$

By choosing for λ the line F, and for $\lambda + \delta\lambda$ the line C, this gives approximately

$$\frac{D}{\nu} + \frac{D'}{\nu'} = 0 \tag{3}$$

If the powers D and D' are of opposite sign we will be able to satisfy this condition with two different types of glass, a crown glass, which has a ν of the order of 60, and a flint glass, for which ν' can be lower than 40. Such *achromatic doublets* are commonly used in optics.

Cornu's Formula

When very high accuracy is not required, we can represent the dispersion of a transparent medium in the visible spectrum by a formula proposed by Cornu (see Chrétien [11]):

$$n = n_\infty + \frac{K}{\lambda - \Lambda} \qquad (4)$$

This formula does not have any theoretical basis but it utilizes only three positive constants and its approximation is sufficient in many cases.

As an example, Table 1 contains experimental values of the refractive index of pure water according to Dorsey (1940) and of sea water of a coefficient of salinity of 37.4, according to Bein (1935), both at the temperature 20°C. The table also contains calculated values using formula (4), showing exact concordance for the wavelengths printed in boldface.

Application to the Ocular Media

The dispersion of the aqueous humor, which is a liquid, is easy to measure experimentally using a small sample. The measurement is not

TABLE 1

Indices of Refraction of Pure Water and Salt Water at 20°C

	PURE WATER		SALT WATER	
λ, mμ	*Measured*	*Calculated*	*Measured*	*Calculated*
360.2	1.347 75	1.347 32		
404.7	1.342 84	1.342 66		
435.8	1.340 21	1.340 21	1.347 45	1.347 45
486.1	1.337 12	1.337 15	1.344 21	1.344 24
546.1	1.334 46	1.334 46	1.341 43	1.341 43
589.3	1.332 99	1.332 94	1.339 89	1.339 85
656.3	1.331 15	1.331 12	1.337 98	1.337 95
706.5	1.330 02	1.330 02	1.336 81	1.336 81
808.0	1.328 15	1.328 29		
n_∞		1.318 48		1.324 92
K		6.6620		6.8153
Λ, mμ		129.2		133.3
ν	55.8		54.6	

so easy for the vitreous body, which is not a very homogeneous gel. It is even more difficult for the cornea, which consists of several layers, and nearly impossible for the very heterogeneous crystalline lens. However, several authors have attempted it on animals' eyes or enucleated human eyes. Let us mention the somewhat inaccurate measurements of Kunst (1895), who gave the following results for $n_F - n_D$: cornea, 0.00375; aqueous humor, 0.0041; lens cortex, 0.00535; lens nucleus, 0.0060. These early results indicated, in comparison with water, an excess in the dispersion of the lens and a deficiency in the dispersion of the cornea, since the corresponding value for water is 0.00413.

The more recent results reported by Polack [54] confirm this indication. According to his measurements, the constringence of the aqueous humor would be 53.3 at 20°C. The value for the vitreous body would be of a similar order. The measurements were made by the method of deviation in hollow prisms. Using the method of total reflection (refractometer with a half sphere), Polack obtained for the cornea and lens the values 56 and 52, with some reservation (particularly for the latter figure). Tagawa (1928) confirmed these results.

The theorists of ocular dioptrics agree that the heterogeneous nature of the lens prevents any significance being attached to results derived from calculations of the path of light passing through the eye. The best solution is to use an empirical *general index* for the crystalline lens, which assumes that the lens is homogeneous and has the same thickness and radii of curvature as the actual organ. But it is evidently hazardous to evaluate the dispersion relative to this general index.

Considering these experimental and theoretical uncertainties the simplest solution seems to be to assume that all the transparent media of the eye obey Cornu's formula (4) in the visible spectrum, taking the same value for the constant Λ, let us say 130 mμ (which is halfway between pure water and salt water; see Table 1). We will also take the following values for the constringence: cornea, 56; aqueous humor and vitreous body, 53; lens, 50 (this last value is the average of the values estimated by Polack and Kunst). Kunst found 49 for the cortex and 47 for the nucleus. Assuming the classical values for line D, the indices of refraction calculated for some of the lines of the solar spectrum are given in Table 2.

These indices apply for a temperature of about 18°C, which is the standard laboratory temperature; the temperature of the human eye is less than that of the body by only 1 or 2°C. It is of the order of 36°C, and therefore all the indices in situ must be at least 0.001 less than in Table 2, but this hardly alters the result. We will therefore keep the values given in Table 2 for our calculations.

TABLE 2

Refractive Indices of Ocular Media

	n_A	n_C	n_D	n_F	n_G	n_∞	K
Cornea	1.3726	1.3751	1.3771	1.3818	1.3857	1.3610	7.4147
Aqueous humor	1.3331	1.3354	1.3374	1.3418	1.3454	1.3221	7.0096
Lens (general index)	1.4144	1.4175	1.4200	1.4259	1.4307	1.3999	9.2492
Vitreous body	1.3317	1.3341	1.3360	1.3404	1.3440	1.3208	6.9806

Chromatism of the Schematic Eye

The propagation of paraxial rays in the eye can be determined by calculation from the preceding data. Values for the radii of curvature of the four optical surfaces of the eye (anterior and posterior surfaces of the cornea and lens), as well as their positions, have to be assumed for calculation. These values are given in Table 3. All positions are given with respect to the apex of the cornea and all distances are in millimeters (for the definition of principal points, focal lengths, power, etc., see basic geometrical optics texts).

TABLE 3

Schematic Eye Values

Position of the	
anterior surface of cornea	0
posterior surface of cornea	0.55
anterior surface of the lens	3.6
posterior surface of the lens	7.6
Radius of curvature of the	
anterior surface of the cornea	7.8
posterior surface of the cornea	6.5
anterior surface of the lens	10.2
posterior surface of the lens	−6

	SPECTRAL LINE				
	A	C	D	F	G
Position of the					
first principal point	1.596	1.595	1.595	1.596	1.596
second principal point	1.906	1.907	1.908	1.913	1.917
first focal point	−15.307	−15.187	−15.089	−14.860	−14.680
second focal point	24.416	24.296	24.196	23.970	23.791
Displacement of second focal point	0.219	0.100	0	−0.227	−0.406
Anterior focal length	−16.903	−16.783	−16.683	−16.455	−16.275
Posterior focal length	22.510	22.390	22.289	22.057	21.874
Power, diopters	59.161	59.585	59.940	60.771	61.442

Regarding Table 3 we will note:

1. When the wavelength changes, the first principal point remains stationary and the second principal point barely moves, as Polack demonstrated.

2. As λ decreases, the second focal point moves regularly in the negative direction, that is, toward the cornea.

3. The value $\delta x'$ of the displacement of the second focal point (as a convention let us assume $\delta x' = 0$ for line D) can be calculated by the following empirical formula:

$$\delta x' = -0.3621y + 0.0065y^3 \tag{5}$$

where $\delta x'$ is in millimeters and the variable y is defined as a function of λ (in mμ) in the following equation:

$$y = \frac{1000}{\lambda - 130} - \frac{1000}{589.3 - 130} = \frac{1000}{\lambda - 130} - 2.1772$$

4. Table 3 concerns the unaccommodated eye. When the eye accommodates, the general index of the lens increases, and supposing the constringences remain the same, the calculation arrives at values of $\delta x'$ which increase approximately 2% per diopter of accommodation [25]. This variation could be neglected and we can assume that chromatism of the schematic eye is independent of accommodation.

Longitudinal Chromatic Aberration

Let us call D the power of the eye for a specific wavelength λ, R its *refraction* measured from the first principal point (that is, its ametropia; $R = 0$ represents the emmetropic eye, $R < 0$ a myopic eye, and $R > 0$ a hyperopic eye). Call n' the index of refraction of the vitreous body and l' the distance between the second principal point and the retina. The classical relationship between these conjugate foci is

$$\frac{n'}{l'} = R + D \tag{6}$$

Let us write the same equation for another wavelength λ_0 (for instance, line D):

$$\frac{n'_0}{l'_0} = R_0 + D_0 \tag{7}$$

By subtraction it becomes

$$R - R_0 = (D_0 - D) - \left(\frac{n'_0}{l'_0} - \frac{n'}{l'}\right) \tag{8}$$

$R - R_0$ is called the *longitudinal chromatic aberration* of the eye because it represents, in diopters, the separation of two sources of light situated on the axis of the eye and emitting the wavelengths λ and λ_0, respectively, the images of which are both situated on the retina within the approximation of paraxial optics. We note that this longitudinal chromatic aberration is smaller than the difference of powers, $D_0 - D$, because the two terms in parentheses have the same sign. The index and the power both increase when λ decreases.

We can modify equation (8) by replacing l' by l'_0; the error is negligible because of the very small displacement of the first principal point with the change in wavelength. Therefore, l'_0 can be replaced by its value in equation (7) and it follows that

$$R - R_0 = -D + D_0 \frac{n'}{n'_0} + R_0\left(\frac{n'}{n'_0} - 1\right) \tag{9}$$

With this equation, using Tables 2 and 3, the longitudinal chromatic aberration of the schematic eye can be calculated for any ametropia (assuming, of course, that we are dealing with *axial* ametropia; i.e., the power is the same as the emmetropic schematic eye and only the length of the eye has changed). Some results are given in Table 4.

It can be noted that the aberration of the myope is greater than that of the emmetrope and that of the hyperope is less, but the differences are insignificant.

If the ametropic eye is corrected by a lens which exactly corrects its ametropia for the reference wavelength λ_0, the power of this lens is more or less R_0 for this wavelength if we neglect the distance between the lens and the eye, whereas for another wavelength λ it becomes $R_0(n - 1)/(n_0 - 1)$, where n_0 and n are the indices of the lens for λ and λ_0, respectively. The longitudinal chromatic aberration will be equal to the difference between the ametropia R corresponding to wavelength

TABLE 4

Longitudinal Chromatic Aberration of the Schematic Eye (in diopters; reference wavelength, line D)

	R_0	LINE A	C	F	G
	0	0.59	0.27	−0.63	−1.14
Without correction	+10	0.56	0.26	−0.60	−0.98
	−10	0.62	0.28	−0.66	−1.20
With correction	+ 5	0.63	0.29	−0.67	−1.22
	− 5	0.55	0.25	−0.59	−0.96

λ and the power of the correcting lens for this same wavelength, and using equation (9) it will be equal to

$$R - R_0 \frac{n-1}{n_0-1} = -D + D_0 \frac{n'}{n'_0} - R_0 \left(\frac{n-1}{n_0-1} - \frac{n'}{n'_0} \right)$$

Take as an example a crown glass widely utilized in spectacles, for which $n_A = 1.5143$, $n_C = 1.5174$, $n_D = n_0 = 1.5200$, $n_F = 1.5261$, $n_G = 1.5310$. The calculations arrive at the values found at the bottom of Table 4. The situation is reversed because the dispersion of the lens is greater than that of the eye. It is now the myope that has the least aberration and the hyperope the greatest, and the differences are about twice as marked as previously.

Comparison with Experiment

The calculations developed above reveal the considerable importance of longitudinal chromatic aberration: a subject emmetropic for yellow-orange (line D) would become myopic by 1 diopter for violet (line G) and hyperopic by nearly 0.6 diopter for a deep red (line A), according to Table 4.

One can wonder at the fact that chromatic aberration was denied by great scientists, Euler (1747), for instance. Yet Newton (1704) described the colored fringes which appear on the edge of a white patch on a black field when one covers half of the pupil of the eye and correctly explained the chromatism of the eye (see the end of this chapter). Newton rightly concluded that the eye is affected by chromatic aberration, but he believed wrongly that it was an unavoidable defect of optical systems because he assumed that the constringence was the same for all transparent media. On the contrary, the false assumption introduced by Euler (achromatism of the eye) led him to a correct deduction: that Newton had made a mistake in his theory of dispersion and that an achromatic combination was possible according to equation (3). He actually was unaware that experiment had already proved him right, and that as early as 1733 an English amateur named Hall had had built by George Bart the first achromatic telescope, of which the objective was a crown-flint doublet [11]. This discovery remained unknown until Dollond, the famous optician of London, had it patented under his name in 1758. Dollond later demonstrated (1789) that the eye could not be achromatic, because the power of all its optical elements is positive. This conclusion is not strictly true: The posterior surface of the cornea has a negative power, although small indeed; moreover the external layers of the lens are divergent menisci. A slight correction of the chromatism should not be impossible, especially if the constringences of the media of the eye are sufficiently different from one another.

The direct measurement of longitudinal chromatic aberration consists of determining the position of the *punctum remotum* (i.e., the position of the light source which produces a retinal image in focus in the unaccommodated eye) for all the wavelengths. Although the first trials were by Maskelyne (1789), it was not until Fraunhofer (1814) that satisfactory results were obtained. He looked at a spectrum through an eyepiece with a reticule, the other eye being open and looking into the distance in order to avoid accommodating. The movement between eyepiece and reticule in the violet measured the aberration. An emmetropic eye for line C became myopic by 1.5 diopters for line G. These results were confirmed by Matthiessen (1847), then by Helmholtz [20]. Helmholtz illuminated a small hole with monochromatic light and determined his punctum remotum by obtaining the clearest retinal image at the farthest possible distance. He showed that by assuming the dispersion of water for the eye, calculations gave more or less the same values as experimental results, with, however, slightly larger empirical values, which he attributed to an excess dispersion of the lens.

The direct method was also utilized by Nutting (1914). He presented to a subject two monochromatic point sources of different wavelenghts, one seen by reflection from a semisilvered mirror and the other through it. The distance of the sources was adjusted until they appeared equally clear. This somewhat inaccurate method was improved by Ames and Proctor (1921). The criterion of optimum sharpness of a grating was also utilized by Polack [54], Biot (1946), and Hartridge (1947), but with only one observer in each case. The objective methods of refraction (retinoscopy, for instance) are not easy to use, because the choroid and the retina scatter wavelengths shorter than 570 mμ very little; however, Lau et al. (1955) obtained very good results on three subjects with a refractometer using interference filters. The most complete measurements are those of Pinegin (1941) on 9 subjects and Wald and Griffin (1947) on 14 subjects. Wald and Griffin had their subject look through a semisilvered mirror at a white and black target situated at 10 meters, where the luminance of the white parts was 0.3 cd/m^2; this test prevented the subject from accommodating. By reflection in the mirror placed in front of his left eye the observer saw a monochromatic point source superimposed on a black area of the test. The point source was a small hole illuminated by a mercury lamp with filters which isolated lines or narrow bands of the spectrum. Between this hole and the eye an achromatic objective was placed so that its focal point would be on the cornea of the observer. The distance between the hole and the objective directly measured the distance of the punctum remotum when the point was seen clearly without accommodation. (Strictly speaking the focal point of the objective should have coincided with the first principal point of the eye, but the difference is negligible.)

The results are represented in Figure 1. We have assumed conventionally that $R_o = 0$ for 589 mμ (line D); the authors had adopted the mercury 578-mμ line and we have corrected their values by -0.05 diopter. There are important variations among individuals which are perhaps significant, but which also result from the inaccuracy of the criterion of optimum sharpness and from the disturbing effect of slight accommodation. At these low levels accommodation is not very constant, as we shall see later; in fact, individual results varied over a range of the order of 0.5 diopter. From one observer to the other, the variation of power D_o (and D, to which it is proportional) gives us according to equation (9) a very slight variation of the aberration. A technique similar to that of Wald and Griffin was used by Bedford and Wyzecki (1957) on 12 subjects with similar results.

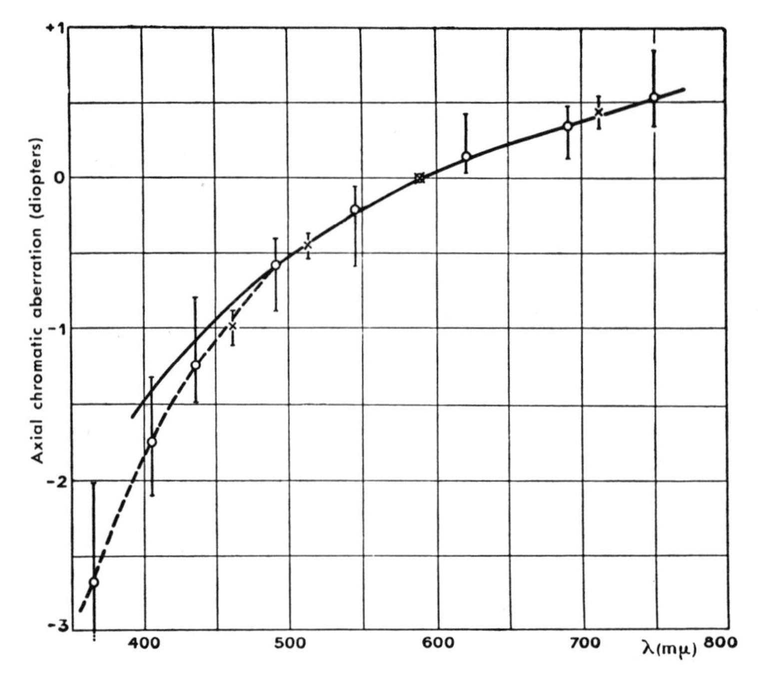

Fig. 1. Longitudinal chromatic aberration of the human eye, as a function of the wavelength. o, measurements of Wald and Griffin; x, measurements of Ivanoff. The dashed curve represents the experimental aberration; the solid curve, the calculated aberration. The vertical lines indicate the variations of results among subjects.

Chromatic Parallax

We can obtain an indirect measurement of longitudinal chromatic aberration by employing an old phenomenon, that of chromatic parallax. Let us suppose that we have a distant point source of light emitting two monochromatic radiations—red and blue, for example. (This is approximately what is obtained when a cobalt filter, which transmits only the two extremities of the visible spectrum, is placed in front of an incandescent light.) A pinhole disk *E* is placed in front of the eye (Fig. 2). If the hole is shifted from the axis of the eye so that the beam of light penetrates the eye through the edge of the pupil, the dispersion of the eye separates the red radiation and the blue, in the same manner as a prism, because of the oblique incidence of the beams, and consequently two separated luminous spots are formed on the retina, one red *R*, the other blue *B*. This phenomenon is evidently linked to the longitudinal chromatic aberration of the eye, because if the source of light is at infinity and if the eye is emmetropic for red, the image *R* coincides with the second focal point for the red radiation, whereas the focal point for the blue radiation is at *B′*, in front of the retina.

When the disk *E* is moved, the images *R* and *B* are superimposed when the pinhole is on the axis of the eye, and their relative positions are reversed when the pinhole is on the other side of the axis. In this type of chromatic parallax, which Guild (1917) called *internal*, the images are displaced in the same way whether the eye or the pinhole is moved, and only their relative displacement matters. But two sources, one red and one blue, could be placed at different distances from the eye so that their images would both be in focus on the retina. These images will be superimposed when the sources are aligned along the axis of the eye, and in this case the movement of the screen in front of the eye will not produce any parallax. On the other hand, parallax would appear if the

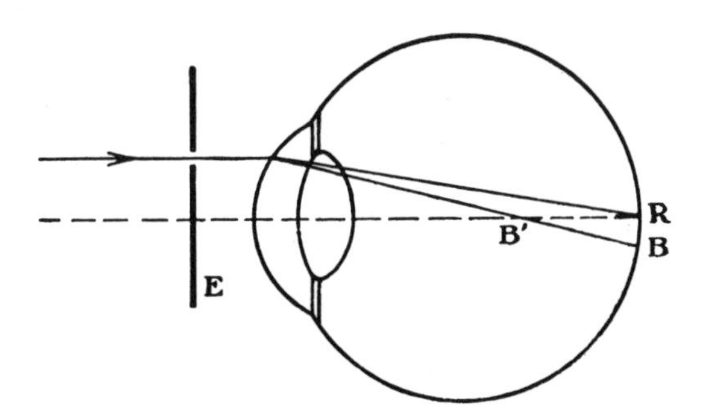

FIG. 2. Chromatic parallax.

eye moved because the sources are no longer aligned on the axis of the eye. In this *external* chromatic parallax the movements of the pinhole and the eye play different roles. When the longitudinal separation of the two monochromatic sources of different colors is gradually increased, the parallax that is observed as the pinhole is moved in front of the immobile eye disappears for a specific separation of the sources; but the parallax that is observed by moving the eye behind a fixed pinhole remains constant, because the sum of the internal and external parallax remains the same. This is what is observed, for instance, with a spectroscope when the magnification of the telescope is 10 to 20 times. The exit pupil of the eyepiece, which is then of the order of 1 millimeter in diameter, plays the part of the fixed pinhole. If a blue radiation of the spectrum is made to coincide with the reticule which is illuminated by diffused light of another color, some parallax is noted by displacement of the eye. This does not change as the adjustment is varied, and hence an accurate determination is impossible. The only solution is to illuminate the reticule with the same wavelength as the line observed.

This method of parallax was used by Young (1802) to measure the chromatic aberration of the eye by a variation of Scheiner's principle (which consists of viewing through two small holes in a screen held very close to the eye and noting the doubling of objects which are not in focus), but this method is not very accurate, because the retinal images lack sharpness when viewed through a pinhole. Sheard (1926) replaced the double pinhole by a shadow of a small object placed in front of the pupil, but the difficulty of keeping accommodation constant accounts for the inacurracy of the results. Sheard observed a plateau between 520 and 620 mμ.

The method of Ivanoff [25] avoids all these drawbacks and excellent accuracy can be obtained: two point sources S_0 and S (Fig. 3) emit the radiation λ_0 and λ (λ_0 was the sodium D line and λ a narrow band produced by a monochromator in the spectrum of an incandescent lamp). An achromatic lens L gives a single image S' of these two sources, one directly and the other by reflection from a plane mirror M. The image S' is formed in the plane of the pupil of the subject, whose head is held in a fixed position (by a dental impression). The objective L has a long focal length, so that the distances LS' is several meters. The subject thus sees the lens as two semicircles, the lower illuminated by S_0 and the upper by S. A. vertical reticule is fixed on the lower part of the lens; an adjustable vertical reticule is on the upper half. If the two reticules are in alignment they appear so when S' is on the axis of the eye (Fig. 3*a*); otherwise chromatic parallax shifts them (*b*) and the subject moves

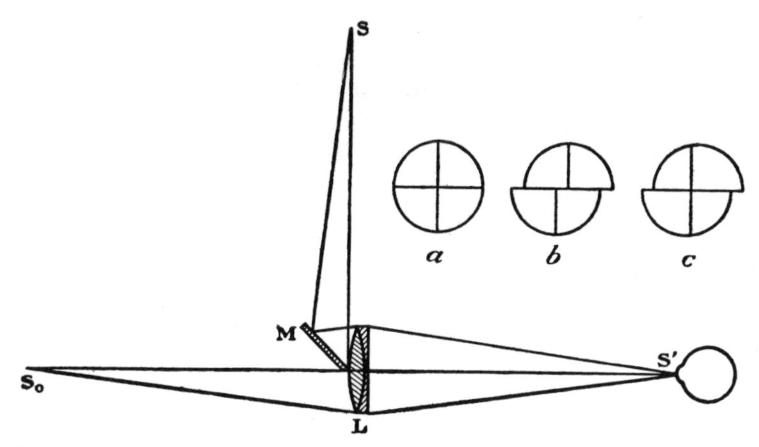

FIG. 3. Measurement of chromatic aberration (Ivanoff). *a*, *b*, and *c* are appearances of the lens *L*.

them back into alignment *(c)* by adjusting the reticule. This method is accurate (of the order of 0.1 diopter) because first, accommodation scarcely influences the alignment of the reticules and, second, the criterion of vernier alignment is much more objective and sensitive than the one of optimum sharpness. The longitudinal chromatic aberration is deduced directly from the displacement of the reticules. Measurements were made on 11 eyes (six subjects); there were very slight differences between the eyes of the same subject, more marked differences from one subject to another. The results have been reproduced in Figure 1.

The results of Wald and Griffin and those of Ivanoff are in good accord with theory, for $\lambda > 500$ mμ. For shorter wavelengths, the experimental aberration is greater than was anticipated in theory, and this difference is even more marked in the ultraviolet region (at 300 mμ Pinegin found more than 6 diopters of longitudinal chromatic aberration). This difference is due to the lens, because a measurement made by Ivanoff shows that the dispersion of an aphakic eye is notably reduced for short wavelengths.

We can relate this high dispersion of the lens in the violet and ultraviolet to the absorption of these wavelengths by the organic constituents of the lens, whereas water remains transparent. However, this absorption varies appreciably with age, whereas we have not noted any correlation between age and longitudinal chromatic aberration in the short wavelengths.

Transverse Chromatic Aberration

The longitudinal chromatic aberration that we have just studied is an aberration of *position*; i.e., the chromatic images of an object are located in different planes. Along with it there is an aberration of *dimension*; the size of the retinal image varies with the color. These two effects are usually combined; only one of the monochromatic images is in focus on the retina, the others are each surrounded by a blurred border. But we can isolate the transverse chromatic aberration by means of an extremely small pupil; this is what we will utilize in this section.

Strictly speaking, the positions of the entrance and exit pupils of the eye, which are conjugate to the iris through the media anterior to it (cornea and aqueous humor) and posterior to it (lens and vitreous body), respectively, vary slightly with the wavelength, but calculation shows that these variations are quite negligible. Let u denote the visual angle subtended by an object PM, seen from the center C of the entrance pupil AB (Fig. 4). The conjugate ray $C'M'$ leaving the center C' of the exit pupil $A'B'$ makes an angle u' with the axis. According to the Lagrange-Helmholtz relation

$$n\overline{AB}u = n'\overline{A'B'}u'$$

When the wavelength λ changes, only n' and u' change in this relation, and therefore the dimension OQ of the retinal image is proportional to u' and inversely proportional to n' for each value of λ. In white light, where all wavelengths are present, the images of long wavelengths overlap those of short wavelengths because n' increases when λ decreases; for example, if M is a point source of white light, the pseudo-image O will resemble a small spectrum with the red toward the outer end and violet toward the inner, the variation of angle u' for any wavelength being such that

$$\frac{\delta u'}{u'} = -\frac{\delta n'}{n'}$$

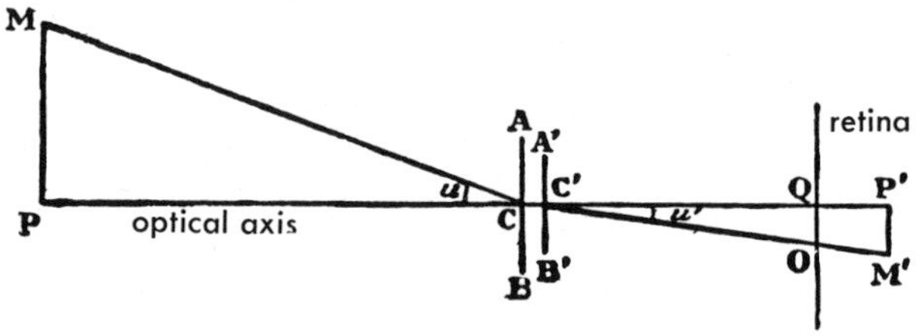

FIG. 4. Calculation of transverse chromatic aberration.

This spectrum resembles the image (if n' were constant) of a real spectrum in object space, with the relation

$$\frac{\delta u}{u} = \frac{\delta u'}{u'}$$

Suppose, for instance, that the limits of the spectrum are lines A and G. According to Table 2, $\frac{\delta n'}{n'} = 0.0092$ and, for $u = 5°$, the apparent width δu of the spectrum would be less than 3′, which would be difficult to perceive at this distance from the fixation point (this angle would be even less if we took into consideration the changes of the pupil size due to chromatic variations). It could be argued that the existence of angle α (between visual axis and optical axis), which is of the order of 5° and which is due to the fact that the fovea is at a slight eccentricity from the optical axis of the eye, makes foveal vision correspond to an angle $u = 5°$; this little spectrum would thus be visible. But in reality there is no reason why the optical axis of the eye, actually ill defined, should be such that all the rays, whatever their wavelengths, be superimposed onto it, because the eye is not a centered system. Such an *achromatic axis* can be determined experimentally by Ivanoff's method (see Fig. 3); its intersection with the cornea is the point of entrance of rays such that the apparent separation of the reticules is nil. But this point of intersection varies a great deal from one person to another and sometimes it is quite off the center of the pupil.

If, instead of a luminous point as a source, a white patch on a black background is used, we should theoretically see the image surrounded by a yellowish fringe, since the images of longer wavelengths overlap, but the width of this fringe is significant only for high values of u; that is, in peripheral vision; details are then blurred and the fringe remains invisible.

Colored Fringes

We have just seen that fringes produced by transverse chromatic aberration do not play any visual role. On the other hand, one can perceive other fringes caused by circles of diffusion resulting from longitudinal chromatic aberration. When the eye is in focus, with an average-sized pupil, these fringes are not noticed; we will study the reason for this in Chapter 3. If these two conditions are not fulfilled the fringes become visible.

Thus Jurin (1738) observed that a white surface seems surrounded by either a bluish fringe or a yellowish fringe, depending upon whether the surface was farther or closer than the plane conjugate to the retina. In the first instance, all the images are in front of the retina but those of shorter wavelengths are the farthest away and consequently these images form the greatest circle of diffusion on the retina; it is the opposite in the second instance. These effects are hardly noticeable, and it is unlikely that they play any part in the appreciation of depth as Polack (1900) maintained.

One can see the fringes more readily when, with correct focus to produce a sharp image, the pupil is partially covered; Newton (1704) described the blue fringe which overlaps a white patch on a black background when the pupil is partially covered on the same side as the white patch; the fringe is yellowish if the pupil is covered on the other side. Mollweide (1805) gave the exact interpretation of this phenomenon; we can account for its colorimetry as follows.

Suppose B is a white sheet of paper (Fig. 5) on a black ground. M is a point along the edge of the paper at the intersection with the optical axis. C is the center of the pupil of which half is covered by an opaque screen E. The retina is conjugate with B for the wavelength λ_0, and M'_0 is the image of M for this wavelength. N is a point on the retina close to the edge of the image so that $M'_0N = y'$. If N is outside of the image $B'M'_0$ for λ_0 $(y' > 0)$, only radiations of wavelengths shorter than λ_0 will reach N. Let us call one of these shorter wavelengths λ, so that the image of M for λ is at M', and $M'_0M = \delta x'$. The point N will be illuminated by radiation of wavelength λ emerging from points on the screen B close to M which have passed through the pupil in the shaded area on the diagram. If we call the radius of the pupil r and let α denote an angle (positive and less than 90°) such that

$$CD = r \sin \alpha$$

the ratio of the area of the shaded segment to the area of the half-pupil is

$$1 - \frac{2\alpha}{\pi} - \frac{\sin 2\,\alpha}{\pi} \tag{10}$$

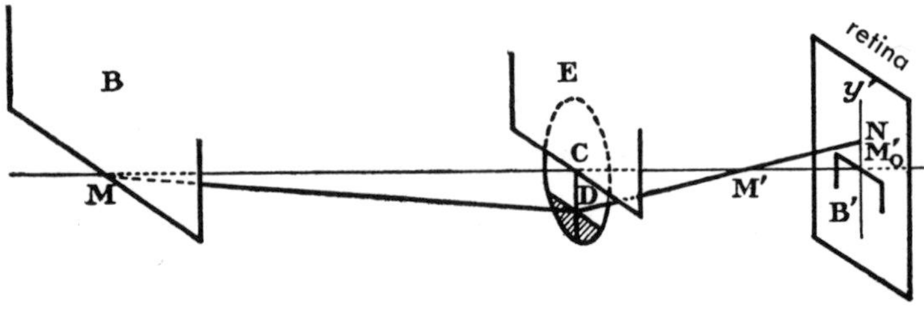

Fig. 5. Diagram of the colored fringes with partial covering of the pupil.

Taking into consideration the fact that M'_0M' is small compared to CM'_0, we have

$$-\frac{y'}{\delta x'} = \frac{CD}{CM'_0} = \frac{r \sin \alpha}{CM'_0}$$

In the schematic eye, the distance CM'_0 between the exit pupil and the retina is 20.51 mm. If we assume an entrance pupil 4 mm in diameter, the calculated exit pupil is $r = 1.84$ mm. Finally, between the ordinate y' on the retina and the visual angle there is the relationship

$$y' = 4.85\, u \tag{11}$$

when y' is in microns and u in minutes of arc. Consequently, by evaluating $\delta x'$ in microns, one arrives at

$$-\sin \alpha = \frac{20.51 \times 4.85}{1.84} \frac{u}{\delta x'} \tag{12}$$

This problem of colorimetry is easy to resolve by the method of selected ordinates well known in colorimetry. For a given value of the visual angle u with λ_0, one first determines $\delta x'$ for each of the selected wavelengths, using equation (5), then the angle α using equation (12), and from equation (10) one can deduce the factor which gives the area of the pupil utilized. The summation of all these factors establishes the chromaticity coordinates of the light which illuminates the point N of the retina.

If the white paper were the other way around, then since the superimposition of this case and the one preceding would obviously give a uniform white, the colors of the fringes must be complementary to the

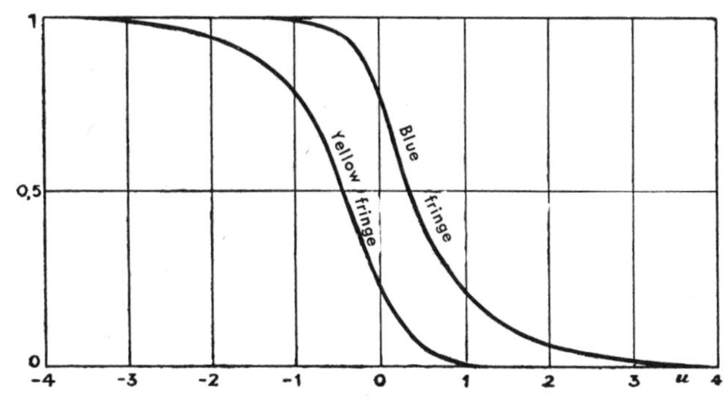

FIG. 6. Photometry of colored fringes. Abscissa: angles in minutes of arc, starting from the image which is in focus (line D); positive values are outside of the image. Ordinate: retinal illumination in relative values or the apparent luminosity of the image.

preceding ones and the values of u are of opposite sign. This is not strictly true (we are neglecting spherical aberration and diffraction) but is sufficient to given an idea of reality. As an example we have calculated the case in which the focus is made on line D ($\lambda_0 = 589$ mμ) and the paper emits standard C of colorimetry. Figures 6 and 7 show the photometry and the colorimetry of the fringes, respectively. Note that most of the blue fringe spreads beyond the edge of the image and most of the yellow fringe spreads inside the image which is in focus. This is due, of course, to the choice of λ_0, which is yellowish orange.

One can also observe colored fringes when looking through the edge of a correcting lens, which produces a prismatic effect. For example, if a

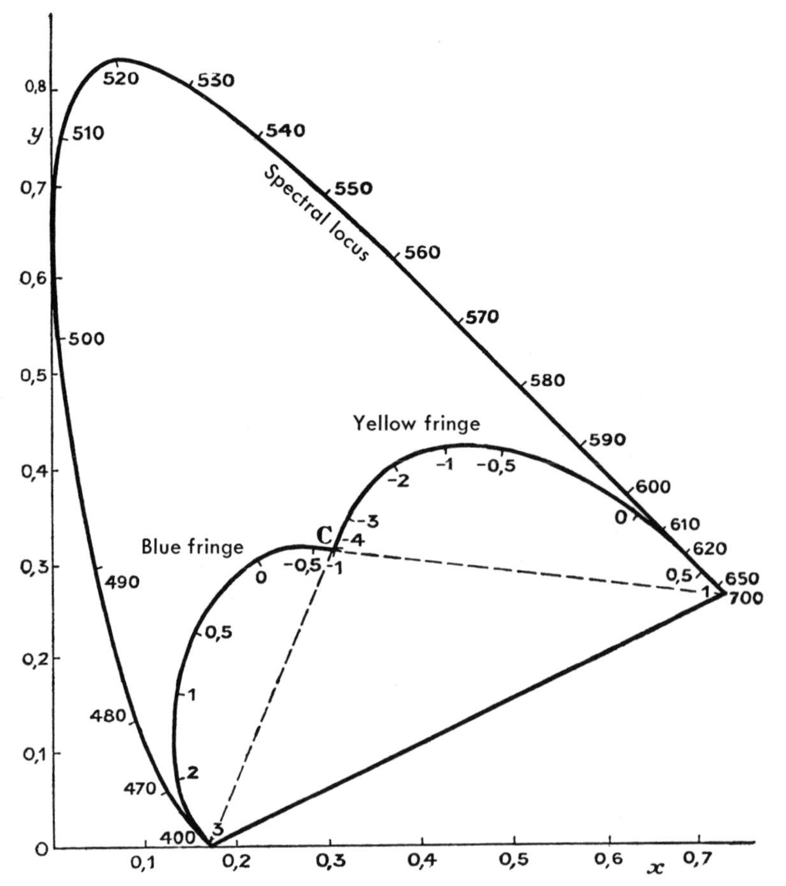

FIG. 7. Colorimetry of colored fringes on the CIE diagram. The spectral locus is calibrated in wavelengths (in mμ). The fringes are expressed in angles with the same convention as in Figure 6.

subject wearing prescription lenses observes through the top of them a horizontal mercury lamp which is far away, he easily sees the separation of the blue, green, and yellow images of the lamp as in a spectroscope having weak dispersion. But chromatism of the eye, of course, does not play a part in this experiment; the effect is due solely to the spectacle lens.

CHAPTER TWO

Spherical Aberration

When an optical system has an axis of revolution, any ray of light emerging from a point M on this axis will remain in the plane of incidence which contains the axis and, consequently, will again intersect the axis at a conjugate point M' (real or virtual). The distance CI between the optical axis and the point of intersection I of the incident ray with the entrance pupil of the system is the *incident height*, designated by h (Fig. 8). By symmetry, all the rays from M with the same h intersect the axis at the same point M', but the position of this point is a function of h that we can express as follows:

$$x' = a_0 + a_1h + a_2h^2 + a_3h^3 + \cdots$$

The coefficients a are functions of both the distance x of the source and the wavelength λ. It is evident that the odd coefficients are zero, because if h is changed in sign it corresponds to a symmetrical ray with respect to the axis, and the refracted rays all converge on the same point M'. Therefore it follows that

$$x' = a_0 + a_2h^2 + a_4h^4 + \cdots = a_0 + \delta x' \tag{13}$$

The term a_0 is the distance of the *paraxial image* and is equal to x' for very small values of h. Classical optometry utilizes only this value, which is a function of x alone (λ is assumed to be constant). In Chapter 1 we studied the variation of a_0 with λ (longitudinal chromatic aberration).

In equation (13), $\delta x'$, the sum of all the terms except a_0, is called the *spherical aberration.* It is an expression of the distance between the paraxial image and the point where all the rays of light actually intersect the axis; it is a function of h, x, and λ. We can design optical systems, referred to as *corrected,* which are free of spherical aberration for specific values of x and λ. The system is then *stigmatic* for a given point

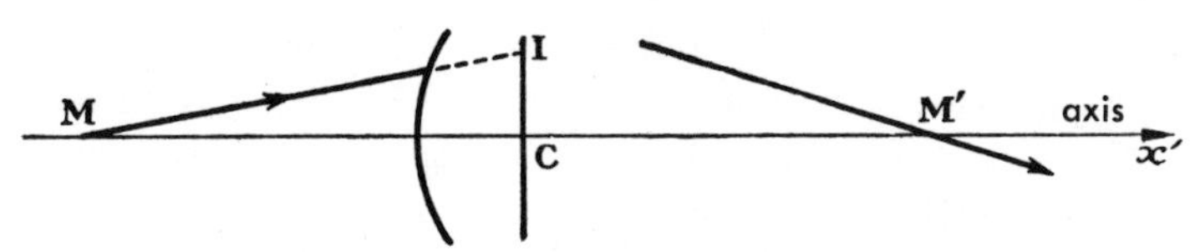

Fig. 8. Ray tracing in a centered system.

object and wavelength. A spherical optical surface is not usually corrected, and this is perhaps where the name spherical aberration comes from.

In an uncorrected system $\delta x'$ is either positive or negative. When h increases M' moves either in the direction of the light or against it and the system is referred to as *overcorrected* or *undercorrected*, respectively. The surface enveloping the light, to which all rays are tangent, is called the *caustic*, and the point P' of the caustic will point one way or the other, depending on whether the system is overcorrected or undercorrected (Fig. 9).

It can be noted that P' is the focus of the *paraxial* rays (h small), whereas the other extremity of the caustic is formed by the *marginal* rays (h maximum) which pass at the edge of the diaphragm. If the shape of the diaphragm is circular and centered on the axis, the caustic is a surface of revolution.

Case of the Eye

As far as the eye is concerned, the existence of an axis of revolution is only an approximation. When this approximation is reasonably good, which assumes the astigmatism of the cornea and crystalline lens to be slight, one can speak of spherical aberration for the foveal image, since the fovea is approximately on the axis. But in general the eye is not a

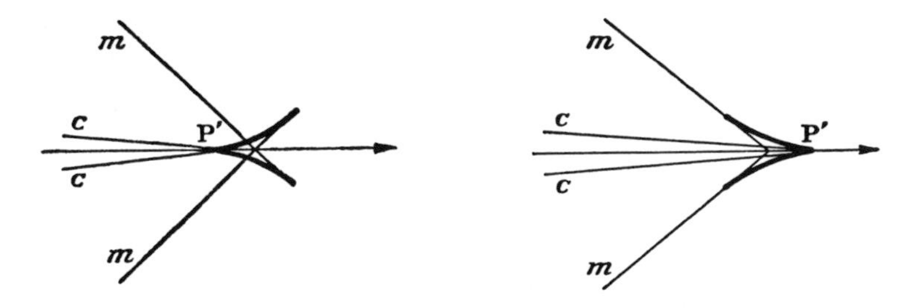

Fig. 9. Overcorrected (left) and undercorrected (right) spherical aberration. m, marginal rays, c, paraxial rays.

surface of revolution, and in each plane of incidence one can define a different spherical aberration which can even involve odd powers of h. If the astigmatism is high, one can speak of spherical aberration only in the principal meridians, since in the other meridians the rays of light do not remain in the plane of the axis. We shall omit these complicated cases, which are actually not very interesting, for imagery in high astigmats is so poor that aberrations cannot possibly make it any worse. Henceforth we will therefore assume that the astigmatism is small enough to be neglected.

The Schematic Eye

If we were to assume that the four optical surfaces of the eye (two of the cornea, two of the lens, assuming the lens to be homogeneous) were spherical, the trigonometric calculation of the exact path of a ray could be applied. If the incident ray is parallel to the axis and intersects the entrance pupil at the height h in millimeters, we have

$$\delta x' = -0.1375h^2 \tag{14}$$

the terms of the fourth power or higher being negligible. It follows that the eye would be quite undercorrected: for $h = 2$ mm, the rays would intersect 0.55 mm in front of the paraxial image, which corresponds to 1.5 diopters of myopia if the eye were emmetropic for the paraxial rays. It has been calculated that each diopter of ametropia displaces the image approximately 0.37 mm. As the eye accommodates, this undercorrection becomes greater.

But this hypothesis is fictitious, because there is no spherical surface in the eye, except for the aphakic eye wearing a contact lens. As h increases, the surface of the cornea becomes flatter, and the problem is further complicated by the heterogeneous nature of the lens. We shall see later that equation (14) is absolutely erroneous. For paraxial approximations, as in calculating chromatic aberration, the schematic eye can be quite satisfactory, because only the axial radii of curvature of the surfaces are used and not their real shapes, but it is useless in the case of spherical aberration; only by experiment can valuable information be obtained.

Early Measurements

It was Jurin (1738) who first observed the spherical aberration of the human eye. Young (1801) attempted some measurements with a four-slit optometer using Scheiner's principle. For most subjects there was undercorrection in the absence of accommodation and overcorrection above a certain amount of accommodation. Young also noted that the shadow of a wire lattice projected on the circle of diffusion of a near object appeared rectilinear without accommodation, but convex toward the center when he accommodated. When the cornea was immersed in water, the effect remained unchanged, which proved that the lens was causing this effect. Volkmann (1846) also repeated Scheiner's experiment by looking at a needle through four small holes aligned in an occluder placed in front of the eye. His results were inconsistent from one subject to another. Matthiessen (1847) thought that spherical aberration of the eye is negligible on the average.

Besides these attempts at direct measurements, some information had been obtained by looking at the image of a point source on a dark background, a star, for example; this image is a cross section of the caustic by the retina. Some subjects see it as circular, but most observers perceive an irregular shape, which explains the antiquated representation of a star by a polygon "star." Sometimes the image is so irregular that several may seem present: de la Hire (1660) observed multiple images of a star. Some people see the crescent of the moon with several horns. Donders (1864) attributed these phenomena to the heterogeneity of the crystalline lens (such effects are absent in aphakics). These phenomena are accentuated with age; sometimes there is *monocular polyopia* or multiplication of the image, which is an early symptom of cataract.

Some temporary irregularities of the cornea also alter the retinal image of a point. Meyer (1853) describes striae which can be seen when partially closing the eyelids for a certain length of time. They would be caused by small prisms of lacrimal fluid deposited on the cornea. Bull (1891) pointed out that after looking monocularly into a microscope for a long period of time the inactive eye remained blurred for several minutes and saw horizontal lines double. This would be due to horizontal folds on the cornea caused by wrinkling of the epithelium after prolonged lid closure.

When the caustic curve is very irregular, the notion of spherical aberration, although of theoretical importance, loses a great deal of interest; therefore we shall avoid these cases and deal with subjects in whom the image is approximately one of revolution.

The Research of Tscherning and Gullstrand

Tscherning [66] devoted some research to spherical aberration. On the basis of Young's experiment of the shadow of a wire lattice, he built an *aberroscope*, a planoconvex lens with a grid pattern on the plane surface. The lens is placed against the eye and the subject looks at a point source farther away than the focal point; the shadow of the grid pattern is seen projected on the circle of diffusion, the lines usually appear convex toward the center when the accommodation is relaxed and concave above a certain amount of accommodation. The opposite would happen if the point source were nearer than the focal point, as in Young's experiment. Tscherning concluded from this that the eye is undercorrected at rest and overcorrected when accommodating. This conclusion is often correct, as we shall see later, but Tscherning's experiment does not prove anything.

In fact, if the lens of the aberroscope were stigmatic and produced an image S′ of the source S, if the eye were also stigmatic and produced an image S″ of S′, and, finally, if it could be assumed that the only optical surface of the reduced eye were a plane, the image of the grid pattern on the retina in this case would be exactly similar in shape to the original. If with the same hypotheses the eye were undercorrected, the marginal rays would be more inclined, and as they intersect the retina after having crossed the axis (in the case of a point source further away than the focal point), their intersection with the retina would be farther away—hence the convexity toward the axis of the image of the grid.

But this reasoning is erroneous for several reasons. First, a grid pattern is perceived as a rectangle when it is so in reality and not when its retinal image is so; the *distortion* of the image produced by the eye would have to be known, although this has very little significance since the retina is not plane. Furthermore, the lens of the aberroscope has its own aberration. Last, but not least, the curvature of the optical surface of the reduced eye which we considered alters the results a great deal.

Gullstrand [20], a great antagonist of Tscherning, did not hide his criticism of his rival's aberroscope, but the crude lucubrations of Gullstrand were worth no more. He attempted to study subjectively the form of the caustic to which he attributed the appearance shown in Figure 10. A is the paraxial image; the eye would be undercorrected up to about $h = 2$ mm (points B and C), the distance between A and BC could be as high as 4 diopters. Above this value of h, the aberration is reversed; for $h = 4$ mm, the eye is undercorrected (DE); the maximum concentration of light would be in a plane M, so AM is about 1.5 diopters and the paraxial image of an emmetropic eye would therefore correspond to a hyperopia of that order. We shall see later that sometimes the caustic curve resembles this one, but the number of diopters is exaggerated, probably because Gullstrand's method, which consisted of shifting the retina with respect to the caustic curve by means of lenses, was impaired by the fluctuation of accommodation.

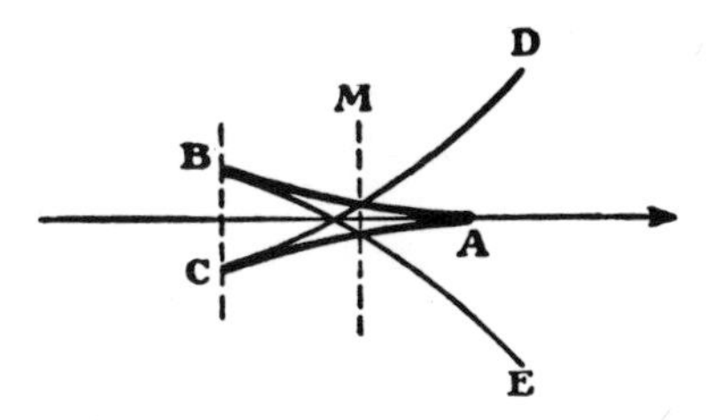

FIG. 10. Caustic curve of the eye, according to Gullstrand.

Recent Research

Volkmann's method, which utilizes annular pupils and, therefore, admits only those rays having a particular height of incidence h, was employed by Ames and Proctor (1921) on three eyes; by Arnulf et al. on four eyes (1941, published in 1948); and by von Bahr on 30 eyes (1945). As a rule the eye was unaccommodated. The results varied a great deal from one subject to another. On the whole there was a slight undercorrection in daylight vision (diameter of the pupil less than 4 mm), but in nocturnal vision, when the pupil was larger, the aberration was as high as 1 diopter. The aberration often became overcorrected for high values of h, as Gullstrand had supposed.

Ivanoff's measurements (1953, 1956) are the most accurate for diurnal vision. He utilized the same parallax method as for chromatic aberration (see Chapter 1), but instead of two rays of different wavelengths entering the pupil at the same point, he used two rays of the same wavelength, one entering the pupil through the center $h = 0$ (along the achromatic axis) and the other through a point of the pupil at a given eccentricity h. The results varied from one subject to another and even with the same subject, depending upon the state of accommodation. As an example we have reproduced the curves obtained by Ivanoff on two typical subjects in Figure 11; when h increases an undercorrection is shown by the curve shifting to the left. The most commonly observed curves are of the type drawn in a continuous line. There is undercorrection when the eye is unaccommodated and a tendency toward overcorrection when the subject accommodates 3 diopters or more, which confirms Tscherning's opinion, but sometimes the eye is overcorrected when unaccommodated (dashed curves). Ivanoff's method does not give any information regarding $h = 0$, as Westheimer (1955) pointed out, but in some subjects, at least, there seems to be in the central region some rapid variation in the aberration which may be accounted for by a local heterogeneity of the crystalline lens, probably due to its embryonic nucleus.

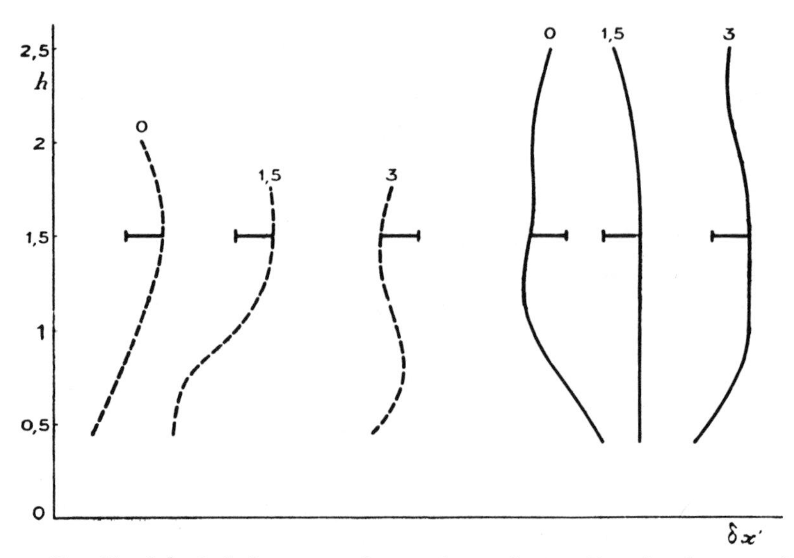

FIG. 11. Spherical aberration of two subjects for 0, 1.5, and 3 diopters of accommodation. Abscissa: The aberration (overcorrected to the right and undercorrected to the left); the segment represents 0.1 diopter. Ordinate: The departure from the achromatic axis (Ivanoff).

The results were nearly always different along the horizontal meridian from one side of the pupil to the other, which demonstrates an asymmetry of the eye. The results also varied from one meridian to the other and even from one part of the pupil to the other (Smirnov, 1961; van den Brink, 1962; Krauskopf, 1964). For all these reasons, it seems illusory to speak of an "average" spherical aberration of the human eye; the only general conclusion is that in diurnal vision spherical aberration reaches at most a few tenths of a diopter, which is much less than the values found with equation (14).

Objective Methods

In the early days of retinoscopy, Jackson (1888) described certain appearances due to spherical aberration. Close to the neutral point the shadow movement is often in opposite directions in the center and at the periphery of the pupil (scissors movement). Several authors attempted to measure spherical aberration by this technique (Pi, 1925; Stine, 1930; Coates, 1940–1941), but with very limited accuracy. It could be done more easily by using an objective optometer which utilizes only a small pupillary area, such as Fincham's optometer; but unless the patient is under atropine the constant fluctuation of the accommodation makes the measurements difficult.

A better method consists of forming the image of a point source on the retina and utilizing this image as a source to study the aberrations of the eye like those of any objective lens, as for instance by Foucault's method. The scattering of light by the retina eliminates the distribution of phase due to light entering the eye, and consequently only the reflected light plays a part in the results. However, the sensitivity of the method is reduced by the fact that the incident light on the retina produces a large image. One of the difficulties is that of maintaining a wide enough pupil without mydriatics in spite of the necessarily high intensity of light. Arnulf et al. (1954) ingeniously avoided this problem by superimposing on the same photographic plate the impressions taken by successive flashes of an intense source, while the subject kept his fixation steady; this method displays the spherical aberration and the defects caused by the lack of homogeneity. With a similar method, replacing photography by an electronic method, Berny (1965) investigated the eyes of two subjects in different accommodative states.

Scattering of Light

Besides the light refracted according to Descartes' laws and producing an image, including spherical aberration, there exists an appreciable amount of light that is scattered within the eye by small heterogeneities of the transparent media and produces a luminous veil on the retina. Helmholtz wrote, "When we observe a bright light on a dark background we notice a whitish haze with its maximum intensity near the light. As we cover the luminous source, the background resumes its blackness." This proves the origin of this phenomenon to be visual, not atmospheric. If the source of light has a small apparent diameter, some local concentration of scattered light is perceived in the shape of colored halos of which the center is the source, and the red color is on the periphery of this halo. With a very small bright source, a pattern is seen made up of numerous bright, thin radiating lines referred to as *ciliary coronas* (Fig. 12). Any normal eye perceives this pattern; it becomes more distinct with age but exists in youth, since Tscherning [67] showed it to a child of 7.

All these phenomena can be easily interpreted by the classical theory of diffraction of light by small heterogeneities situated in the medium through which the light is propagated. If these were of very small dimension compared to the wavelength λ of the light and were scattered haphazardly, the intensity of the scattered light in a specific direction

FIG. 12. Diagram of a small source of light seen on a dark background. The lines radiate over a distance of some 10°.

would vary as λ^{-4} (Rayleigh's law). This accounts for the blue of the sky, in which the molecules of air are the scattering particles. In liquids this molecular scattering also exists but is usually hidden by the diffusion of bigger particles in suspension; this *Tyndall's effect* is not very selective; i.e., its intensity varies only slightly with λ. Furthermore, instead of the intensity decreasing regularly from the geometric image, there may be maxima in certain directions. For example, round particles of radius a, floating in the eye at a distance ξ in front of the retina, cause a circular corona, with maximum relative intensity at a distance from the image given by the following theoretical expression:

$$\frac{1.63\lambda\xi}{2an'}$$

where λ is the wavelength of the radiation in air and n' is the refractive index of the vitreous body. By replacing the distances on the retina by visual angles using equation (11), we arrive at the angular radius,

$$u = \frac{0.34\lambda\xi}{2an'}$$

where the distances are in microns and the angle in minutes of arc.

This phenomenon is interpreted as a ring for which u is equal to a little over 2°, and it seems to be of corneal origin (Druault, 1923). On myself, for example, I find $u = 130'$ for $\lambda = 0.546\ \mu$. Assuming a distance of 22 mm from the retina to the image of the cornea through the lens we have $\xi = 2.2 \times 10^4$ and consequently $2a = 23\ \mu$; this is actually the order of dimension given anatomically for the diameter of the epithelial and endothelial cells of the cornea. In glaucoma, for example, pathological halos may be seen. With certain types of contact lenses the lack of respiration of the cornea produces temporary halos.

When the pupil is wide enough, some subjects may see a second halo, nearly twice as large as the first, for which a crystalline origin is indicated by the variations in appearance that occur when the pupil is partially covered [29]. The periodic variations in indices of the radial fibers of the lens probably act in the same manner as a diffraction grating. If c is the grating interval (the width of one fiber and one interspace), the radius of the ring on the retina corresponding to the first-order spectrum is

$$\frac{\lambda\xi}{cn'}$$

Hence the angular radius with the same conventions as above becomes

$$u = \frac{\lambda\xi}{4.85cn'}$$

On myself, I found $u = 195'$ for $\lambda = 0.546\ \mu$; with $\xi = 17$ mm, the grating interval c is 7.3 μ, which is feasible for the average diameter of the fibers of the crystalline lens.

It is probable that the corona is due to the fibers of the lens, since the continuous movements that are noted in the corona are actually reduced (Raman, 1919) when the accommodation is stabilized by fixing a target, but Ronchi and Zoli (1955) believe that the cornea plays a part in the formation of this corona.

When one wears glasses in the rain the drops of water on the lenses produce beautiful diffraction effects that are easily seen at night when looking at a distant source. The edges of the diffusion image of each drop are surrounded by *caustic rings* which are beautiful to observe, and which compensate for—alas, only slightly—the well-known drawbacks of ophthalmic lenses.

Photometry of Scattered Light

The phenomena described above do not lend themselves easily to theoretical calculations of the distribution of scattered light on the retina. Stiles (1929) attempted it, assuming Rayleigh's law, and he found a retinal illumination proportional to $u^{-1} - 0.005$ for a medium value of u (in degrees), but this hypothesis is not very likely, as proved by the white (but not blue) appearance of the scattered light of a white source.

There are in fact, in addition to scattering, other causes for the parasitic light. First, as the sclera and choroid are not quite opaque, one can easily perceive the red color of shadows when the sun illuminates the eye laterally. In albinos in whom the pigmentation of the choroid is very light, the red light which filters through can be easily seen through the pupil and causes discomfort to the subject. Second, the choroid does not absorb all the light reaching the retina; part of it is scattered laterally and contributes to a general veil. Some measurements of this scattering of light by the retina have been made by Toraldo di Francia and Ronchi (1952) on animals. Another part of the light is reflected back to the anterior segment and out of the eye (this is used in ophthalmoscopy and retinoscopy), but the cornea reflects a certain amount back onto the retina and a veil is produced which is not negligible. The contribution of the different media to the scattering of light in the eye was studied by Vos et al. (1963, 1964) and by Boynton and Clarke (1964).

The experimental measurement of this parasitic light reaching the retina besides the image is a very interesting problem, but difficult; it has been attempted in several ways.

1. The illumination in the eye can be measured through the choroid and sclera of albinos, or through holes pierced in the tunics of enucleated animal eyes. This method was attempted by Bartley [3], Boynton (1954), Ranke (1954), and de Mott (1959) but is not very accurate. Furthermore, the enucleation alters the clarity of the eye. With an ophthalmoscope we could also measure the brightness of the retina of a subject around the image, but scattering in the instrument itself may diminish the value of this technique. However, close to the image, photographic methods (Flamant, 1955) can be employed, as we shall see later.

2. A *threshold* (absolute, differential, or critical fusion frequency) could be measured with a small test on a black background using an intense source at an angular distance u from the test; the measurement is then repeated without the intense source but with a uniform artificial veil of which the luminance can be adjusted so that the same value for the threshold is obtained. If only the parasitic light played a part, its value would be measured as a function of u. Unfortunately, the study of glare proves that this is not so and that *inhibition* caused by the intense source changes the threshold appreciably. Therefore, such measurements should be considered with caution. They have been made by numerous authors, such as Pokrowski (1926), Bartley and Fry (1934), Toraldo di Francia and Ronchi (1953), and Berger (1954).

3. The true luminances can be matched directly or successively with the brightness caused by the parasitic light. This method is very delicate, and the same criticism as in the previous method can apply. However, it seems that inhibition has less effect on such a matching than on a threshold. On three young subjects, I have in this way determined [29] a curve (Fig. 13) which represents for the dark-adapted eye the brightness L of the background at a distance u from the source ($1° \leqslant u \leqslant 30°$) or rather the ratio L/E of this luminance to the illumination produced by the source in the plane of the pupil ($0.1 \leqslant E \leqslant 10$ lux). Variation with wavelength λ is slight and becomes nil for $u = 30°$; for higher values of the angle, blue light would be slightly more scattered (Tyndall effect), whereas for lower values of the angle it would be the red, perhaps because the retina itself scatters long wavelengths to a greater extent.

Compared to the measurements of glare, these results show that ac-

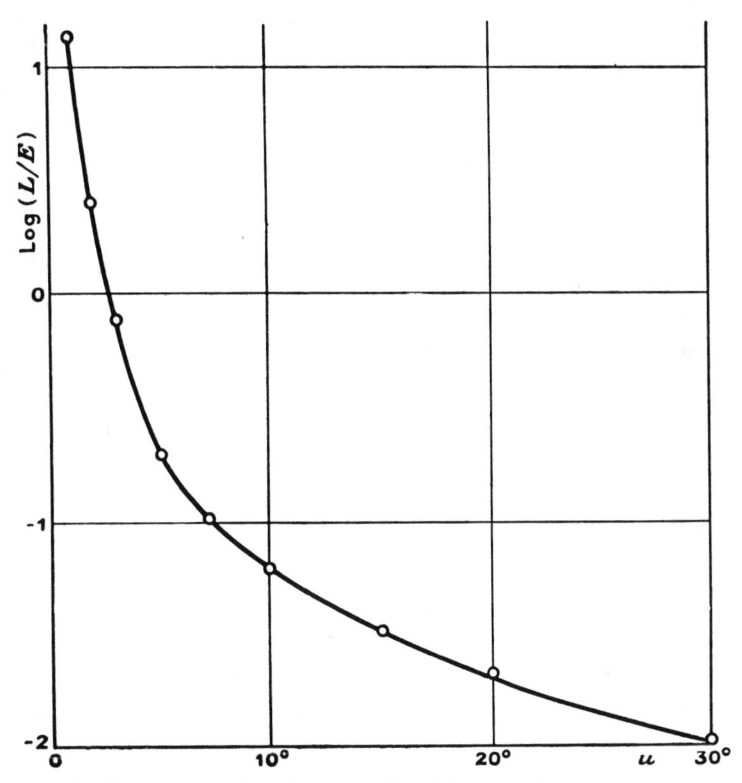

FIG. 13. Brightness L (in cd/m^2) of the diffusion veil at angular distance u of a small source producing on the pupil the illumination E (in lux). Ordinate: log (L/E).

cording to the position of the source, scattered light accounts for nearly all the glare or down to less than half, the remainder being due to inhibition. This predominance of the scattered light (especially near the source) was formerly verified by Fry (1953) and Alpern (1953), who utilized transillumination through the sclera, and by Boynton et al. (1954), who varied spatial conditions of glare.

From Figure 13 it is easy to evaluate approximately the luminous flux incident around the main part of the image. For angles between 1 and 30°, we obtain 6.3%. It is therefore reasonable to assume that approximately 10% of the incident flux is scattered around the image in young subjects with a highly pigmented choroid. For older people with greater scattering from the choroid and physiological defects in the ocular media, one can expect values much higher than this minimum.

4. Fry (1953) proposed to measure scattered light by means of the pupillary constriction, since in some conditions it would measure the total flux falling on the retina and not only that of the image. On this principle Fugate (1954) devised a technique by which superimposition of large and small images would enable one to differentiate between the image and the scattering. But besides the uncertainty of the measurements of the pupillary constriction, the phenomenon depends upon too many factors for this method to obtain high accuracy.

5. Finally, it was proposed that scattered light should be studied by means of *electroretinograms*, i.e., the differences in potentials produced in the eye when the retina is exposed to light. These potential changes can be recorded and measured easily, even on man. Fry and Bartley (1935) assumed that scattered light could play a more important part in the electroretinogram than would appear from the relative smallness of the diffused flux, because of the large surface of action which would compensate, and more, for the weak illumination. Several investigations have confirmed this hypothesis, particularly those of Marg (1953) and Boynton (1953). Boynton believes he has isolated in the *b* wave of the electroretinogram two successive components very close to each other, one of which would be caused by the image and the other by scattered light. But analysis of these experiments is difficult and so far has not enabled measurement of scattered light as such. An experiment of Wirth and Zetterström (1954) would seem to contradict these ideas. They placed, through the vitreous body in front of the retina of a cat's eye, small plastic cones blackened on their lateral surfaces, through which light is conducted without scattering. The electroretinogram that they recorded hardly differed from its usual appearance (see also Armington, 1955).

CHAPTER THREE

Photopic Vision of a Point of Light

The knowledge of chromatic and spherical aberration is sufficient to cover thoroughly the problem of the image of a luminous point situated on the optical axis. This is almost the case in foveal vision of a point, because the fovea is close to the optical axis of the eye. But for a point source to be seen by the fovea it must be of a sufficiently high level of luminance and this is what is meant by *photopic* in this chapter.

Image through a Stigmatic Instrument

In a geometrically perfect instrument the image of a point would be a point if the wave nature of light did not intervene. All the luminous flux would be concentrated in the smallest possible point image and the illumination on a screen would be maximum at this point and nil around it.

This is, of course, only fictitious, and after Fresnel established the mathematical basis of the wave theory of light, the astronomer Airy (1834) calculated the exact distribution of illumination within the image. His calculations can be found in any text on optics and we shall only recall the principle. Suppose a point object on the optical axis emits a monochromatic radiation of wavelength λ (in vacuum or in air) which in the image space of index of refraction n' becomes

$$\lambda' = \lambda/n' \tag{15}$$

Assume the exit pupil to be a circle of diameter d' centered on the axis and situated in the plane $A'B'$. Since the instrument is stigmatic the wave front passing the edges of this pupil is a sphere of radius $x' = S'C'$ and center C' (Fig. 14). A screen E is placed perpendicular to the axis at C'. To calculate the illumination at a point M' on the screen E, we will consider any point P' on the surface of the wave front and we will evaluate the *difference in phase* defined by the angles

$$\varphi = 2\pi \frac{P'M' - x'}{\lambda'} \tag{16}$$

If ds is an element of the wave front around P' the following two sums can be calculated:

$$A = \iint k \cos \varphi \, ds \qquad B = \iint k \sin \varphi \, ds \tag{17}$$

where the integrals apply to the entire wave front and k refers to a factor which maintains a constant value k_0 when the radii $P'M'$ are close to the normal to the wave and decreases rapidly when they deviate from it. The illumination at M' is equal to $A^2 + B^2$.

On the screen E, the image is one of revolution around point C', where the illumination has its maximum value

$$E_m = \pi^2 d'^4 E' / 16 x'^2 \lambda'^2 \tag{18}$$

where E' is the illumination in the plane of the exit pupil. Away from C' the illumination decreases until it becomes nil at a distance

$$r = 1.22\, \lambda' x' / d' \tag{19}$$

One finds [13] that 84% of the total luminous flux is spread in this

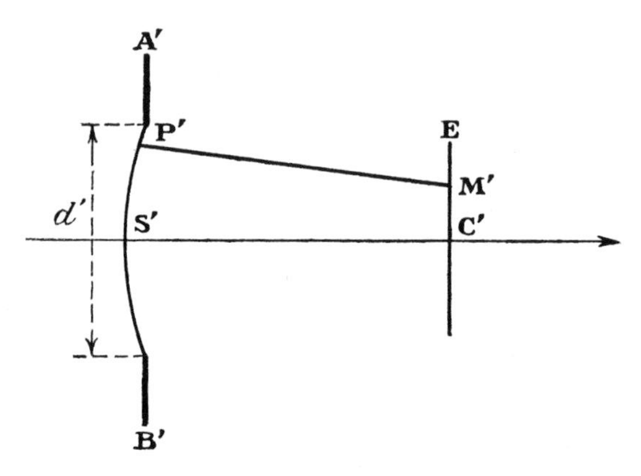

Fig. 14. Calculation of the illumination of an image in a stigmatic instrument.

circle of center C' and radius r, which is referred to as *Airy's disk*. This disk is surrounded by rings, the first of these containing 7% of the total flux.

Let us apply equation (19) to the schematic eye; the distance between the exit pupil and the second focal plane is $x' = 20.5$ mm. Moreover, $d' = 0.92d$, where d is the diameter of the entrance pupil; finally, $n' = 1.336$. Therefore, using equation (15) we arrive at

$$r = 20.3\ \lambda/d \tag{20}$$

r and λ being expressed in the same units and d in millimeters. For example, for $d = 4$ mm and $\lambda = 0.59\ \mu$, one obtains $r = 3.0\ \mu$. Consequently, if the eye were free of any spherical aberration, the image of a distant point of light emitting line D would be a disk 6 μ in diameter surrounded by rings. Actually it is exaggerated to consider the limit of the Airy disk to be the point where the illumination becomes exactly zero; if one uses as the limit the point where the illumination is 1% of what it is in the center, the value of r in (20) would have to be multiplied by 0.90; if one uses 10%, by 0.71.

Photometry of Point Sources

Even if the eye were perfect and if the retina had a continuous structure, the wave theory of light would limit the apparent size of small sources of light on a dark background. A source would be *punctual* as long as its image did not differ practically from the theoretical Airy disk. But on the retina of the schematic eye, an angle of 1′ corresponds to 4.85 μ; consequently the diameter of the Airy disk (6 μ) is equivalent to the size of the geometric image of a source having an apparent diameter of 1.25′. It may be concluded, therefore, that a source of apparent diameter smaller than this value must have the same appearance as one of infinitesimal size.

This assumption is verified by experiment. Le Grand and Geblewicz (1939) devised an instrument which changed the apparent diameter of a source from 15″ to a higher value u, without a change in flux; the value of u at which the subject notes a change in one out of two trials is about 1.5′. In these experiments, the source was seen on a black background and the illumination that it produced in the plane of the pupil was 10^{-5} lux.

On a light background the results are slightly different. For example, Blackwell (1946) determined, on a uniform white background of luminance L, the sources which can just be distinguished from the background in one out of two trials. The luminance of these sources depends upon the visual angle u (apparent diameter), but below a certain critical value of u the illumination produced by the source becomes independent of u, which is a photometric criterion of a point source; the results are shown in Figure 15. Evidently, if the image of the point source is not exactly in focus on the retina the thresholds are increased (Ogle, 1960).

Direct measurements of the apparent dimension of a point source on a dark background have been made by several authors, particularly Dale (1919), Tonner (1943), and Fiorentini (1950). This can be done either by moving a small luminous source vertically between two others situated on the same horizontal line, until its lower edge appears tangent to a line formed by the upper edges of the other two sources, or by moving two very thin vertical lines so that they seem tangent to both sides of the source. The results are of the same order of magnitude as the Airy disk, but the image seems to increase in size as the source becomes more intense, either because the edges of the diffused circle become perceptible above threshold, or because of the retinal structure.

As a result of this, for sources with an apparent diameter of less than 1′, the appearance depends only upon the total flux entering the eye and consequently, for a given pupil size, upon the illumination E

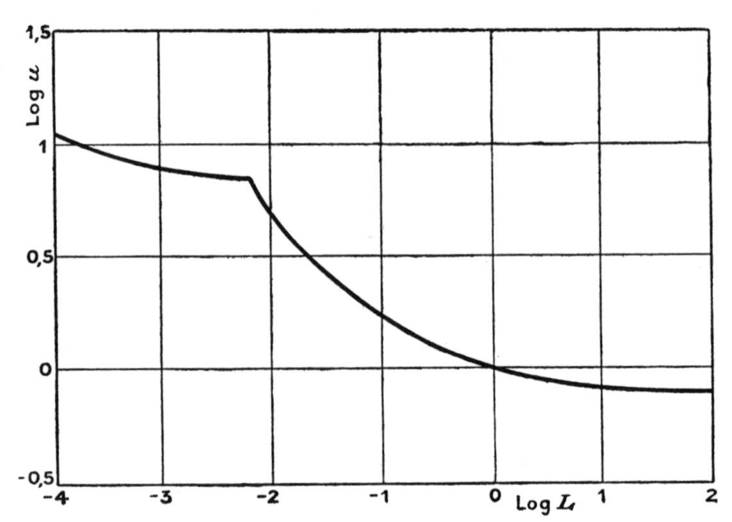

FIG. 15. Apparent diameter u below which a source is photometrically punctual, as a function of the luminance L of the background on which it is just visible, according to Blackwell. (L in cd/m^2, u in minutes of arc). The point of transition between the two curves separates the photopic and scotopic regions.

in the plane of the pupil. The term *point brilliance* is sometimes given to the illumination produced by a point of light on a surface area situated in the plane of the observer's pupil and perpendicular to the rays of light. This brilliance can be expressed in lux; it equals the intensity of the source (in candelas) divided by the square of the distance (in meters) if there is no absorption between the source and the observer.

Twenty centuries ago the astronomers of Alexandria classified the stars into six *magnitudes*, the first corresponding to the brightest and the sixth to the just visible. This terminology is probably related to the apparent variations in size which we mentioned earlier. In 1859 Pogson drew attention to the fact that brightness of stars of successive magnitude varied in a geometric progression, which was later explained by Fechner's law. The magnitude m is defined by

$$\log_{10} E = a - 0.4\,m \tag{21}$$

where a is a constant depending upon the atmospheric absorption; at sea level and for a star near zenith, in the summer, a is equal to -5.68. According to Tousey and Koomen (1953), at 3000 meters above the earth we have $a = -5.656 - 0.0508 \operatorname{cosec} \theta$, where θ is the angular height of the star above the horizon.

The methods of photometry of point sources are not as precise as those of extended sources. According to Danjon (1932) the sources should be placed 10 or 15′ from each other and they can be matched with an error of about 10% by trained observers; with 100 trials one can achieve a precision of 1%. The matching must be done quickly, using a bracketing method after each match, and the movements must be regular and progressive without stops or hesitation; otherwise it would be the hand rather than the eye which would respond. Each observer is affected by a constant error which may reach 30%; it diminishes with the brightness and it is eliminated by alternating the position of the stars every five or ten trials. Of course, all these findings apply to reasonably bright sources seen by the fovea.

The problem becomes difficult if the sources to be compared emit lights of different colors. Moreover, the values of the *relative luminous efficiency* (formerly the *relative visibility factor* $V\lambda$) of small sources probably differ from the classical values obtained with sources several degrees in diameter. Measurements by Wright [74] and Thomson (1947) suggest wide variations from one point to another in the fovea, which also depend upon the state of adaptation. Temporarily we may keep the classical values but treat them with reserve as far as point sources are concerned.

The transition from point sources, whose appearance depends only on the illumination E in the plane of the pupil, to extended sources characterized by their luminance L, i.e., the ratio of E to the solid angle Ω, is a problem not yet well known. Le Grand and Geblewicz (1939) have shown that when matching two sources having apparent diameters of 15″ and 90″, the flux of the latter source exceeds the former at the moment of apparent equality by between 3 and 13% (average 7%) for 13 subjects. When one of the sources is 120″ in diameter the comparison with 15″ becomes very uncertain.

Out-of-Focus Effects

It is interesting to study what happens to the retinal image in a stigmatic eye when the retina is no longer coincident with the second focal point in the case of a distant point source, or, more generally, when the retina is not conjugate to the object. The distribution of illumination in the *diffraction pattern* which then replaces the Airy disk has been calculated by Conrady (1919) and Buxton (1921). The reader can also consult the papers by Chrétien [11] and Lapicque [28].

For an object situated on the axis, calculation of the illumination at the center of the pattern is simple (Fig. 16): the spherical wave passing through the exit pupil $A'B'$ of diameter d' has its center at C' and so its radius is $x' = S'C'$. The retina is perpendicular to the axis at M' so that

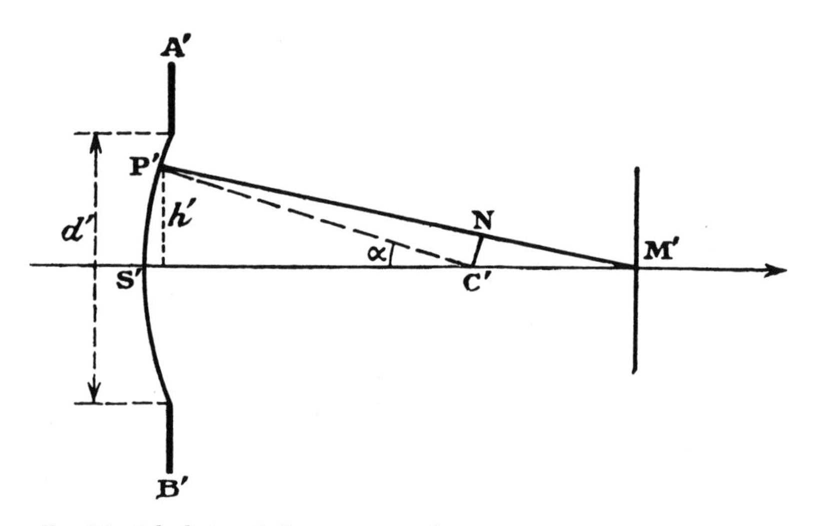

Fig. 16. Calculation of illumination in the center of the diffraction pattern when the retinal image is out of focus.

$C'M' = \delta x'$. We can assume approximately that

$$P'M' = x' + NM' = x' + \frac{\delta x'}{\cos \alpha}$$

$$\cos \alpha = (1 - \sin^2\alpha)^{1/2} = \left(1 - \frac{h'^2}{x'^2}\right)^{1/2} = 1 - \frac{h'^2}{2x'^2}$$

The difference between distances $P'M'$ and $S'M'$ is therefore equal to $\delta x'\ (h'^2/2x'^2)$ and the difference in phase becomes

$$\varphi = \frac{2\pi h'^2}{\lambda' 2x'^2}\delta x' = \Phi z \tag{22}$$

where Φ is the maximum value of φ obtained at the edge of the pupil. At that point, $h' = d'/2$, so

$$\Phi = \pi d'^2 \delta x' / 4\lambda' x'^2 \tag{23}$$

Let z be an additional variable

$$z = 4h'^2/d'^2$$

The integrals of equation (17) become

$$A = k_0 \int \cos \varphi \cdot 2\pi h' dh' = \frac{\pi k_0 d'^2}{4} \int_0^1 \cos (\Phi z)\, dz = \frac{\pi k_0 d'^2}{4} \frac{\sin \Phi}{\Phi}$$

$$B = \frac{\pi k_0 d'^2}{4} \int_0^1 \sin (\Phi z)\, dz = \frac{\pi k_0 d'^2}{4} \frac{1 - \cos \Phi}{\Phi}$$

Hence the illumination at point M on the retina is

$$E = A^2 + B^2 = \left(\frac{\pi k_0 d'^2}{4}\right)^2 \frac{2(1 - \cos \Phi)}{\Phi^2} = \frac{2(1 - \cos \Phi)}{\Phi^2} E_m \tag{24}$$

where E_m is the maximum illumination at the center of the Airy disk ($\Phi = 0$). If in equation (24) one takes $\Phi = \pi/2$, the expression for E becomes

$$E = \frac{8}{\pi^2} E_m = 0.81 E_m$$

The image is thus almost identical to the Airy disk except for a reduction of illumination in the center (<20%). This is why Lord Rayleigh (1879) stated that the difference of extreme phase of the rays forming the image could be as high as $\pi/2$, or $\lambda'/4$ in wavelength, without producing any noticeable difference of the image. Beyond that value, alteration of the image increases very rapidly: for $\Phi = \pi$, $E = 0.41\ E_m$, and for $\Phi = 2\pi$, $E = 0$, there is a dark spot on the axis.

If we introduce the wavelength in air into equation (23), then Rayleigh's criterion can be expressed by the formula

$$\delta x' = 2.36 \frac{\lambda}{n'} \frac{x'^2}{d^2}$$

Assuming $\lambda = 0.59\ \mu$, $x' = 20.5$ mm, $d = 4$ mm, and $n' = 1.336$, we obtain

$$\delta x' = 27\ \mu = 0.027 \text{ mm}$$

As one diopter out of focus corresponds to $\delta x' = 0.37$ mm, then Rayleigh's criterion represents a depth of focus of $1/13$ diopter in front and behind the exact conjugate point. Helmholtz arrived at this value experimentally, since he could not perceive any difference in sharpness between 12 m and infinity. Other authors such as Bourdon [6] and Fincham (1937) give greater values, up to 0.25 diopter. But with very low contrast targets, Luckiesh and Moss (1940) have shown that a difference in the adjustment of 0.1 diopter altered the sharpness of the target and they devised an accurate optometric subjective method on this principle. It seems, therefore, that for the best eyes the variations of adjustment are limited by Rayleigh's criterion. According to Campbell (1954), for pupils wider than 2.5 mm in diameter the depth of focus is slightly increased by the Stiles-Crawford effect, which reduces the efficiency of light passing through the edge of the pupil and reaching the retina at a slightly oblique angle. The Stiles-Crawford effect acts as an "apodization" in the plane of the pupil and modifies the measured spread function which does not include the Stiles-Crawford effect (Metcalf, 1965).

Effect of Chromatic Aberration

So far, we have only considered a monochromatic point source. Suppose the source emits white light. The problem becomes more difficult even if the eye is assumed to be free of spherical aberration. Its chromatic aberration cannot be neglected. The eye will be exactly in focus for only one wavelength λ_0 and out of focus by a value $\delta x'$ for any other wavelength λ. How will the subject choose (unconsciously) the wavelength λ_0 which produces the best images?

The criterion of this image is not obvious, a priori. Newton thought that the subject chose λ_0 in the brightest region of a prismatic spectrum, that is, the yellow-orange band; d'Alembert (1767) believed that the eye tried to reduce colored fringes caused by superimposition of the circles of diffusion. This attempt would result in placing the images of the two extreme ends of the spectrum (red and violet) at equal distances in front and behind the retina and focusing the blue-green radiation around $\lambda_0 = 0.5\ \mu$. Helmholtz assumed that λ_0 coincided with the maximum relative luminous efficiency, that is, greenish yellow $\lambda_0 = 0.56\ \mu$.

The first attempt at a theoretical solution of this problem was made by Lapicque [28], who assumed that the selection of λ_0 is such that the brightness is maximum at the center of the image. Suppose $w_\lambda d\lambda$ to be the energy which penetrates into the eye in the form of radiation, of which the wavelength varies between λ and $\lambda + d\lambda$. The corresponding illumination of the pupil is proportional to $w_\lambda V_\lambda d\lambda$, letting V_λ denote the relative luminous efficiency of simple radiations, and consequently from equation (18) the maximum illumination at the center of the Airy disk is

$$dE_{\mathrm{m}} = \frac{C}{\lambda^2} w_\lambda V_\lambda d\lambda$$

where C is a more or less constant factor. But because of the out-of-focus effect, the illumination in the center of the diffraction pattern is given by equation (24) and becomes

$$dE = \frac{2(1 - \cos \Phi)}{\Phi^2} dE_{\mathrm{m}} = \frac{2C(1 - \cos \Phi)\, w_\lambda V_\lambda}{\lambda^2 \Phi^2} d\lambda \tag{25}$$

The value Φ is given by equation (23) and the out-of-focus effect by an expression analogous to equation (5). By integrating (25) between the extremities of the visible spectrum the illumination E in the center of the pattern is obtained. Assuming for w_λ the spectral energy distribution of the standard illuminant C (which closely resembles average daylight), one finds that the maximum of the illumination E corresponds to $\lambda_0 = 560$ mμ, which conforms to Helmholtz's assumption. With an incandescent light (source A) it would correspond to $\lambda_0 = 575$ mμ. Lapicque found higher values of wavelength λ_0, but his graphical method of calculation lacked precision.

Actually, the hypothesis of maximum illumination in the center of the image is unfounded, and only experiments can give the answer to the problem of which wavelength λ_0 is in focus. Hartridge (1918) looked at an object illuminated with white light, and a test was superimposed upon it by reflection from a semisilvered mirror and illuminated with monochromatic light; he found that when the light was yellow, the adjustment was the same. By a method of parallax, Polack [54] noted that with relaxed accommodation the eye tends to be in focus for the red extremity of the spectrum, whereas the accommodated eye tends to be in focus for shorter wavelengths. This variation is quite natural, since the eye can thus take advantage of its chromatic aberration to diminish its efforts of accommodation. Ivanoff [25] demonstrated this effect by a method of parallax, with white light on 10 observers (Fig. 17); the tendency is to focus in the red when unaccommodated, around 580 mμ (yellow) for 1 diopter of accommodation, and about 500 mμ (blue green) for 2.5 diopters of accommodation. One notes that the physical criterion of the "best image" does not play any part, and the eye adjusts itself so that it spares its accommodation. This phenomenon explains the apparent residual accommodation of aphakics in white light, which is frequently over 1 diopter; Rosenbloom (1953) even reported 3 diopters in one aphakic subject.

What should be remembered from this discussion is that contrary to a common hypothesis, the eye does not adjust itself to one wavelength λ_0, as a rule; in particular, the very common choice of the radiation $\lambda_0 = 589$ mμ (line D) has neither physiological nor physical value but actually corresponds, on the average, to an emmetropic eye accommodating 1 diopter. If the eye is ametropic the principle of the least effort of accommodation holds true, and permits one to predict which focus the subject will choose.

The Image Pattern in White Light

Away from the center of the pattern the illumination can no longer be calculated by integration. Either an approximate mathematical method must be used, as was done by Lapicque, or a mechanical inte-

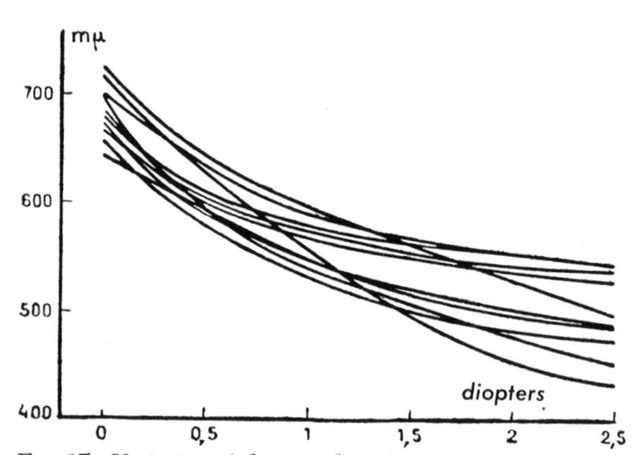

Fig. 17. Variation of the wavelength in focus on the retina for white light as a function of the state of accommodation, for 10 observers (Ivanoff).

gration (with an instrument devised by Marechal at the Institute of Optics in Paris). Figure 18 shows the distribution of retinal illumination for an entrance pupil 4 mm in diameter with a source of light emitting standard illuminant C when the eye is in focus for $\lambda_0 = 560$ mμ. The figure also shows the Airy disk that would be obtained if the same total luminous flux were produced by the radiation 560 mμ alone. It can be noted that the central part has nearly the same dimensions for both images, but the central illumination in white light is only 44% of the illumination in monochromatic light and the central disk only receives about 70% of the flux falling on the Airy disk, i.e., less than 60% of the total flux; the rest is scattered on a wide halo.

The illumination can be calculated for points further away from the axis using the approximation of geometric optics. Let dE' be the illumination in the plane of the exit pupil produced by radiation of wave-

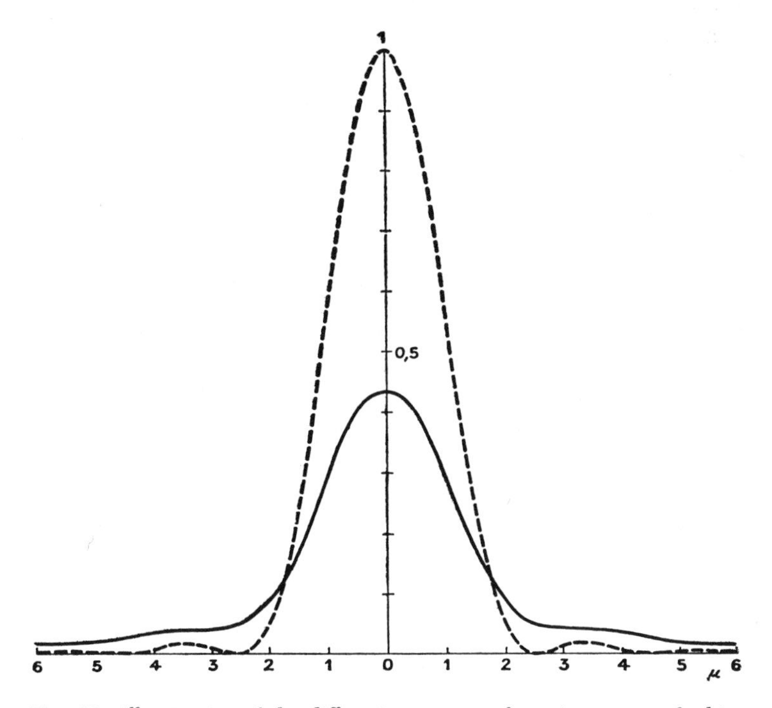

FIG. 18. Illumination of the diffraction pattern of a point source of white light C, assuming the eye to be free of spherical aberration and in focus for the radiation 560 mμ (continuous line). The abscissa is the distance along the retina starting from the center of the pattern. The dashed line represents the Airy disk, assuming that the same luminous flux is produced solely by the radiation 560 mμ.

lengths between λ and $\lambda + d\lambda$; the uniform illumination of the corresponding circle of diffusion is $dE'(x'^2/\delta x'^2)$. If this illumination dE' of wavelength λ_0 produced an illumination equal to unity in the center of the Airy disk, using equation (18) it would follow that

$$dE = 16x'^2\lambda_0'^2/\pi^2d'^4$$

Hence the value of the illumination for points away from the axis is

$$16x'^4\lambda_0'^2/\pi^2d'^4\delta x'^2$$

The results shown in Table 5 have been calculated by the above method.

Figure 19 shows the variation of color in a diffraction pattern. The center is almost pure yellow (563 mμ); then it changes to less saturated yellow-orange, pink, purple, and farther away than 12 μ to blue, of which the dominant wavelength is almost 460 mμ; the edge of the halo becomes darker and tends toward violet.

These results apply for a pupil 4 mm in diameter. With a decrease in pupil size the color becomes weaker, and when the pupil is 1 mm the pattern is almost achromatic. This is due to the fact that chromatic aberration diminishes with pupil size, whereas the spread by diffraction increases. For pupils smaller than 1 mm, the colors would be reversed; of course, pupils of this size are hardly ever encountered but may be caused by artificial means.

Comparison with Experiment

The blue halo which surrounds the image corresponding to the long wavelengths can be easily seen when looking at a white point source through a cobalt lens which transmits only the extremities of the spectrum. One sees a red point surrounded by a blue diffused circle; in near

TABLE 5

Illumination toward the Edges of the Image
(the illumination in the center of the Airy disk is unity)

Distance from the center, μ	Calculated illumination
6	0.018
10	0.005
12.5	0.0017
15	0.0007
17.5	0.00033
20	0.00016
25	0.00005
30	0.00001

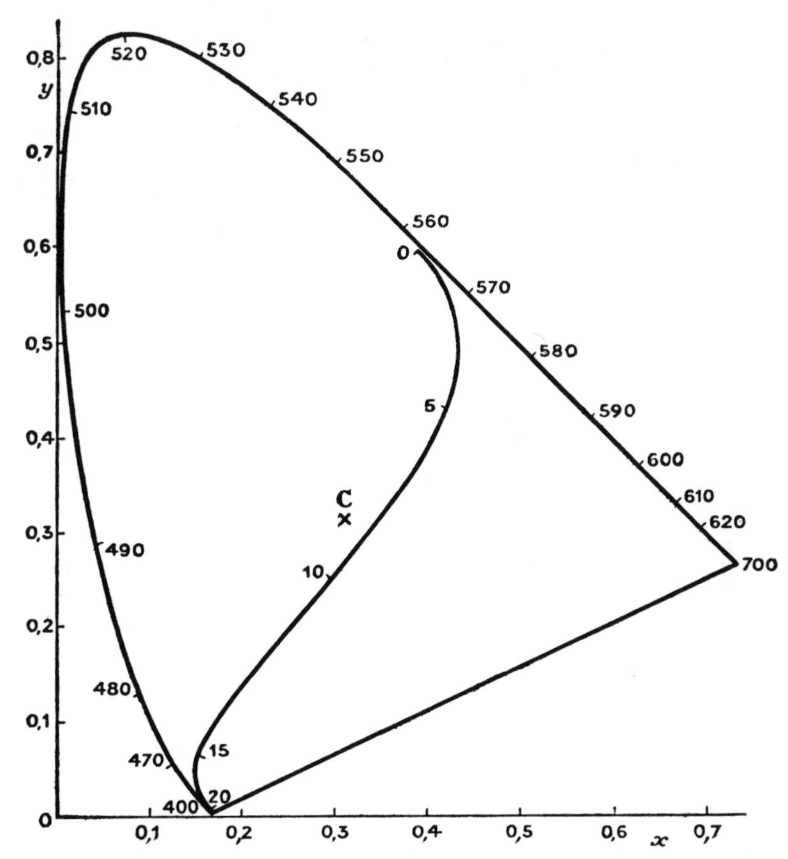

FIG. 19. Colorimetry of the diffraction pattern of a point source of white light C, assuming the eye is free of spherical aberration and in focus for the radiation 560 mμ. The curve is graduated in distances on the retina (in μ) starting from the center of the diffraction pattern.

vision the focusing is changed and one sees a blue point surrounded by a red halo.

But in white light the source is seen as a white point and not as a yellow point surrounded by a blue halo. This paradox has given rise to lengthy discussions.

Actually, the fact that a blue halo is not seen has already been explained by Helmholtz: the illumination of this halo is weak and it disappears because of the parasitic scattered light. By reducing the flux in the central part, the halo can be seen (for example, by using a cobalt glass).

It is more difficult to explain why the yellow central part seems white;

first, one must remember that colorimetry of point sources is very imprecise and when looking at a star of low intensity through yellow glass it is very difficult to distinguish white from yellow points, as Escher-Desrivières and Jonnard showed (1932). At a sufficient level the source appears yellow, perhaps because the scattered light itself (which spreads over a large area of the retina and therefore enables better chromatic sensitivity) is seen yellow. Furthermore, one must keep in mind that the retinal image is only an intermediate step in perception and not an image such as that produced by an optical instrument the eye then examines. "The eye is the only optical instrument whose image is not destined to be looked at" is a somewhat paradoxical observation which explains a great deal. In fact, it is not surprising that one should perceive as a white point, a yellow spot surrounded by a white diffused halo, because that is the way one sees distant sources which, at near, are surely identified as white. One sees as a yellow point, a yellow spot surrounded by a yellow diffused halo, and this is logical enough.

Therefore it does not seem necessary to think of an "antichromatic reflex" as did Hartridge [19d], which would eliminate chromatic appearances in the vision of point sources. This reflex plus the local fluctuations of color sensitivity along the retina was also used by Hartridge to explain the fluctuations of colors and intensity which are superimposed on the twinkling of stars and even occur (slightly) for point sources in laboratory conditions, as Holmgren (1884) noted. But this phenomenon could be explained more easily on the basis of constant fluctuations of accommodation of the order of 0.1 diopter, which Arnulf et al. (1951) demonstrated by observing the retinal image of a subject, using direct ophthalmoscopy.

However, in support of the antichromatic reflex one can mention some interesting experiments by Kohler (1951) to which we shall return in Chapter 14. He prescribed prismatic lenses to a few subjects for some weeks and even months. At first, the subjects complained of colored fringes produced by the prisms but these disappeared after awhile; if the spectacles were removed, colored fringes, probably of a subjective nature, since they were also seen with monochromatic yellow light, would appear in the opposite direction.

Colorimetry of Small Sources

It has long been known that color vision of very small sources differs greatly from that of extended sources. Charpentier (1888) noted that small blue objects on a background of equal luminance were the first to disappear foveally when decreased in size. Fick (1888) stressed the complexity of these phenomena; for example, it is possible that two small colored objects which seem colorless when separated regain their color when brought together, or, on the contrary, they become neutral if they were seen colored when separated (as yellow and blue, for example). Hering (1895) was the first to obtain colorimetric measurements on this phenomenon and demonstrated that the equation yellow = red + green was altered with smaller dimensions of the test, the mixing becoming more and more red as the angular diameter of the test diminished. It is actually a commonly observed fact that if two sources, red and green, appear of equal intensity when viewed at near, the red source seems more intense farther away. Horner and Purslow (1947) and Shaxby (1947) have confirmed this fact by measurements; but there are great individual variations. Hering inferred the cause to be macular pigmentation, but it is more likely caused by the differential sensitivity to color of the foveal receptors.

Foveal dichromatism has often been described; hues of small red, orange, and blue-green sources change only slightly but their saturation lessens; yellow, yellow-green, and green acquire blue whereas blue, violet, and purple lose some blue. Hence, possible confusions depend upon the dimensions and the luminance level of a test; for example, according to Hartridge [19d] a yellow test becomes colorless below an apparent diameter u (Table 6) as a function of luminance L of the background (which is the same as the test).

TABLE 6

Apparent Diameter u Below Which Yellow Color Disappears, as a Function of Luminance L

L, cd/m^2	0.003	0.03	0.3	3	36	300	9000
u, minutes of arc	70	35	20	12	7	2.5	1.1

Several systematic studies have been carried out on the colored appearance of point sources, particularly by Kompaneisky (1944) and Middleton and Holmes (1949). The latter observed a small source of angular diameter 1 or 2′ (sample of paper of reflection factor 0.2) on a gray background of reflection factor 0.03. The subject assessed the apparent color of this source by comparison with Munsell specimens illuminated with the same light but viewed at near through an aperture in a gray mask. One example of the results found is shown in Figure 20. It is seen that yellow-green and purple tend toward gray, whereas red and blue-green hardly change. Burnham and Newhall (1953) found results of the same order in a similar study with rectangular instead of circular tests; the shape and position of the object did not play a part, whereas the reduction of surface area reduced saturation and changed hue.

The difficult problem of reproducing spectral mixture curves relative

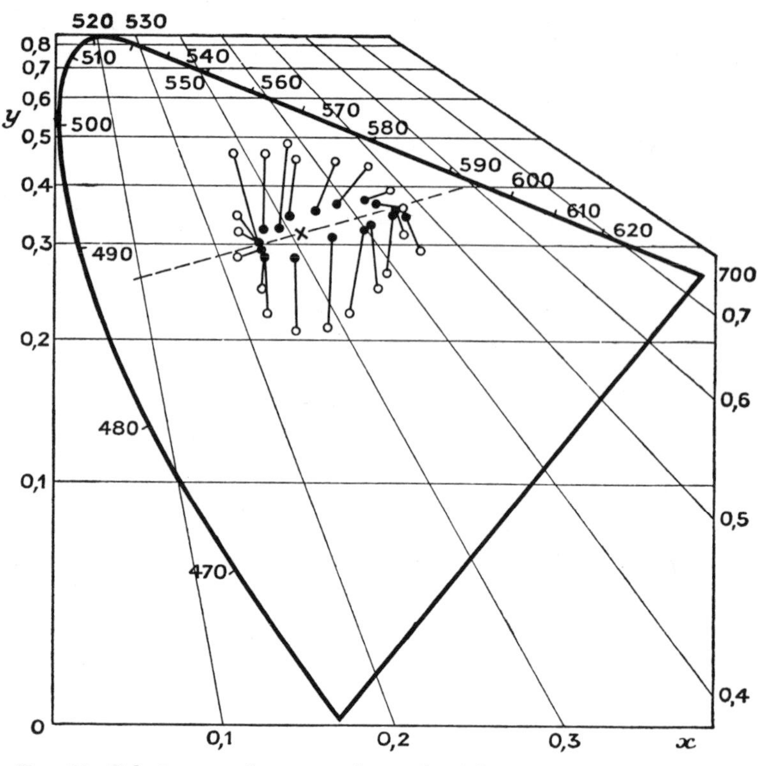

FIG. 20. Colorimetry of sources of 16 cd/m² luminance and 2′ diameter, seen on a grey background of 2 cd/m² luminance. The results are shown on a projection of the C.I.E. chromaticity chart to obtain more satisfactory spacing; the white circles represent the real chromaticities of the sources; the black points represent their apparent chromaticities (after Middleton and Holmes).

to small sources has been tackled by several researchers, particularly Willmer and Wright (1945) and Thomson (1953). The results are not sufficient to define a *standard observer* analogous to the long-established one for sources greater than 1° in diameter.

Consideration of Spherical Aberration and Scattering

So far in this study of the image of a point source we have assumed the eye to be stigmatic. What will be the appearance of the image pattern if spherical aberration is taken into consideration?

The answer depends upon the type of spherical aberration. For example, with theoretical aberration [equation (14)], an entrance pupil of 4 mm, and a monochromatic source $\lambda = 590$ mμ, maximum illumination is produced in the center of the image if the retina is situated 0.39 mm

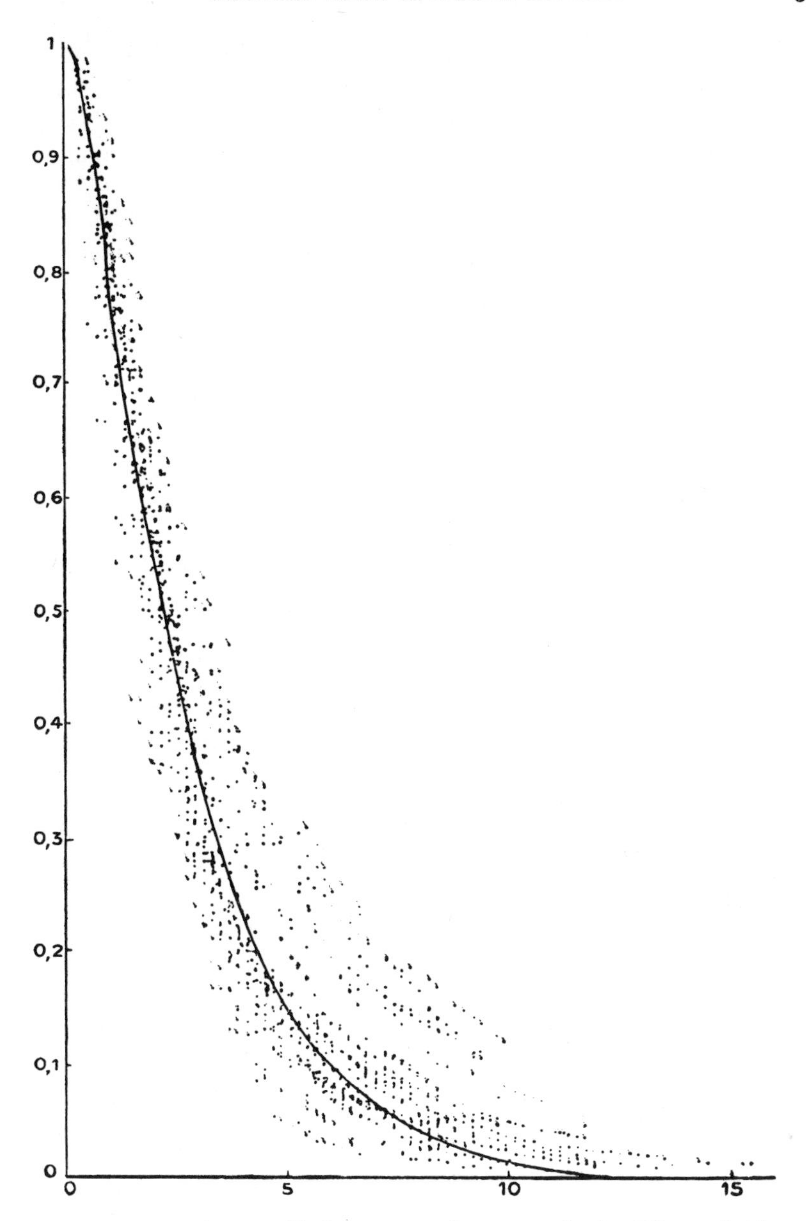

FIG. 20a. Distribution of light scatter in the image of a slit (Flamant). The dots are the measurements on seven subjects and the curve gives the experimental mean. Abscissa: the distance from the center of the image in minutes of arc.

in front of the paraxial focal point. The image then has a disk of radius 2.2 μ, *smaller* than in the case of stigmatism, but the illumination in the center is only 27% of what it would be in the center of the Airy disk; the first ring is not very well marked, but the second is more pronounced. At 6.5 μ from the center the illumination is still more than 10% of what it is in the center. In this case the distribution of light would be very different from that of a perfect eye. This is the case of a contact-lens wearer. For the aphakic there results an increased depth of focus, owing to the utilization of variable regions of the caustic surface (Le Grand, 1961).

A less unfavorable case was studied by Lapicque (1936). From the measurements of Ames, he assumed an aberration of the type represented in equation (14) but with a factor of one third this value. In this case the results remain approximately within the limits of Rayleigh's criterion; i.e., the image remains more or less of the same quality as the Airy disk. This was also the result obtained by Fry (1950) [18a].

With spherical aberrations of the eye such as those measured by Ivanoff, one might have thought that the image would be spoiled; calculation contradicts this, and Françon (1949) showed that the best image does satisfy Rayleigh's criterion. Actually this is reasonably obvious, because if a very small central surface is neglected, the emergent wave is nearly in phase for the rest of the pupil. This aberration would play a part only for very small pupils, and yet in this case the subject will unconsciously decenter his eye most of the time and the image will remain the same. However, Arnulf et al. (1954) noted a definite variation of focus for a pupil of 0.5 mm. Even with natural pupils the pupillary contraction which accompanies accommodation may serve to economize the effort of accommodation (Ivanoff). For two subjects studied thoroughly by Berny (1965), Rayleigh's criterion was satisfied for a 3 mm-pupil and slight accommodation; but although for one subject the image became blurred beyond this and the difference of phase could exceed the wavelength, for the other subject this only occurred beyond 4.5 diopters of accommodation even for a pupil of 4 mm.

Therefore one can conclude that spherical aberration does not have any practical importance, at least in daytime vision, and the quality of the image could be as good as that of a perfect instrument with monochromatic light, if it were not for the heterogeneities of the ocular media and astigmatism.

This conclusion was verified objectively by Arnulf et al. (1951), who, using a telescope of four times magnification, observed through a semi-silvered mirror the image of a luminous point on the retina of a subject.

The shape and size of the image pattern changes constantly, but rarely can an image presenting the characteristics of spherical aberration be seen. The most frequent appearance is astigmatic, then irregular halos, and more rarely, coma. These fluctuations are chiefly caused by the crystalline lens because they diminish or even disappear if the ciliary muscle is paralyzed by a cycloplegic. These microfluctuations of accommodation cause the retinal image to have a "dynamic" aspect quite different from the ordinary fixed image produced by an optical system. The amplitude of these fluctuations, measured by Ivanoff's parallax method, is of the order of 0.075 diopter for four subjects with pupils 4 mm in diameter. In the worst case of an astigmatic focal point on the retina, the image pattern is 2.4 minutes of arc, but if the retina is midway between the two focal lines this value is halved. These microfluctuations of accommodation were studied by Whiteside [73a], who photographed the reflected image from the anterior surface of the lens, and by Campbell et al. (1959) with an infrared optometer. There are large individual variations; the values sometimes exceed 0.25 diopter. The microfluctuations display two dominant frequencies, one around 0.5 cps, the second around 2 cps. In emmetropic subjects these microfluctuations disappear when fixating at infinity if there are clues in the field, whereas if the field is empty ("Ganzfeld"), only the high-frequency microfluctuations are eliminated; the others, of low frequency, increase. Moreover, Campbell (1960) noted that the microfluctuations in the two eyes are synchronous, which suggests a central origin.

To this image there is added a spread of light due to scattering. Arnulf et al. (1951) studied this matter by two ingenious methods. In the first the apparent variation of the size of a point source is measured (with a micrometer) as a function of the intensity of the source; for very small artificial pupils less than 0.5 mm in diameter the scattering is negligible in comparison to the spreading of the image by diffraction, which can be calculated. Assuming that the apparent edge of the image corresponds to a constant threshold of illumination (which one must admit is a very uncertain hypothesis) and extrapolating for a pupil of 4 mm, one obtains the values shown in Table 7.

TABLE 7

Distribution of Illumination in the Diffused Image of a Point

Distance from the center							
in minutes of arc	0	0.558	0.975	1.335	1.605	2.13	2.44
in microns	0	2.7	4.7	6.6	7.8	10.3	11.8
Illumination	1	0.66	0.42	0.28	0.20	0.13	0.07

Comparison of the results with those of Table 6 confirm Helmholtz's hypothesis: The blue halo of chromatic aberration disappears, swamped by the parasitic scattered light.

A second, more objective, method (Flamant, 1955) consists of taking a photograph of the retinal image of a slit of variable illumination (a photometric wedge over the slit). Granularity of the emulsion is effectively diminished by the artifice of a composite photograph resulting from the superimposition of 10 negatives. Enlargement of the photographic image as a function of the illumination gives a direct measurement of the luminous distribution. This method, used on seven subjects, gives a curve (Fig. 20a) wider at its base than the curve of diffraction because the scattered light varies according to $e^{-0.7x}$, where x is the distance on the retina in minutes of arc. This curve agrees very well with the measurements of scattered light made by Le Grand (see Fig. 13). These results are also in good agreement with those of Westheimer and Campbell (1962).

Thickness of the Retina

A last point to consider is the fact that the retina does not represent a surface as such, but a mass of cells occupying a certain thickness. The thickness of the photosensitive part, that is, the outer segments of the cones and the rods (maximum 40 μ), is too small to affect the image. But the receptive cells possess a greater refractive index than the vitreous body, and it is likely that this accounts for the Stiles-Crawford effect, that is, the greater efficiency of rays reaching the retina normally compared to those arriving obliquely. The laws of geometric optics no longer apply for receptors whose diameter is of the order of only a few wavelengths, and the problem of determining the concentration of energy under these conditions is like that for dielectric waveguides for radiowaves, as was shown by Toraldo di Francia (1948–1949). Jean and O'Brien (1949), using radio microwaves of 3.2 cm on model cones made of polystyrene foam, magnified to the same scale and of refractive index 1.02, were able to demonstrate the essential characteristics of the Stiles-Crawford effect. This effect would be responsible for two useful effects: diminishing the lateral scattering of light and increasing the photochemical action by funneling the energy into the receptors. It is also possible that an orientation of the photochemical molecules in the cones may play a role in the Stiles-Crawford effect (Enoch and Stiles, 1961; Ripps and Weale, 1965).

CHAPTER FOUR

Scotopic Vision of a Point of Light

A great number of psychophysical facts prove that the receptors responsible for photopic vision, which corresponds to high level of luminance, are the *cones*. As luminance is diminished, vision becomes *scotopic* and the function of the more sensitive *rods* comes into play. Rods do not perceive differences of color, which therefore disappear in scotopic vision. On the other hand, their spectral sensitivity differs markedly from that of the cones and is shifted toward the short wavelengths; this is the famous Purkinje phenomenon. The maximum of the *scotopic relative luminous efficiency* is at 507 $m\mu$, whereas the maximum photopic relative luminous efficiency is about 555 $m\mu$.

When light sources are weak the fovea becomes blind: Arago said that "to see a star you must not look at it." It is in the parafovea, about 2 or 3° from the fixation point, that two weak point sources of light can be compared more efficiently but their separation must be larger than in photopic vision. According to Wright [74], instead of 10 to 15′ the separation must be at least 30′.

Finally, in studies on scotopic vision it is essential that dark adaptation be sufficient. The International Commission for Optics (July 1950) recommended that the subject work for an hour with low artificial light (less than 20 lux at the working plane), then remain 40 minutes in total darkness. Then luminances up to 10^{-4} cd/m^2 can be utilized for research without interfering with dark adaptation. (As a matter of comparison, the levels utilized in photopic vision are of the order of 100 cd/m^2 according to the same recommendations.)

Focusing of the Eye in Scotopic Vision

The Purkinje effect must produce a modification of focus in white light; for example, for a source emitting the standard illuminant C, we saw in Chapter 3 that the best theoretical focus (obtained by the condition of maximum illumination at the center of the image spot) must be made for wavelength $\lambda_0 = 560$ mμ. Repeating exactly the same calculation but replacing the photopic relative luminous efficiency V_λ in equation (25) by the scotopic relative luminous efficiency V'_λ gives another wavelength, $\lambda'_0 = 510$ mμ. According to the values of axial chromatic aberration of the human eye, this corresponds to a value of about 0.4 diopter out of focus. Consequently, if the eye is actually in focus for $\lambda_0 = 560$ mμ at night there would appear to be a *night myopia* of 0.4 diopter.

This phenomenon was mentioned by Lord Rayleigh (1883) and confirmed by Piéron [52], but its value is usually greater than 0.4 diopter. Emmetropic people would be well advised to wear myopic lenses of 1 or 2 diopters to see best in scotopic vision. For this reason a hyperope will see better at night than an emmetrope.

As mentioned above, the Purkinje effect only accounts for 0.4 diopter, and Cabello (1945) noted that night myopia measured in monochromatic light differed by 0.3 to 0.7 diopter (mean 0.4) from night myopia determined in white light. Therefore, as Otero and his collaborators (1941–1949) have conclusively demonstrated, other processes must be present.

The first possibility was to investigate spherical aberration. Dilatation of the pupil in dim light exposes other areas of the cornea up to 4 mm from the center (pupil of 8 mm). The whole phenomenon could be explained (Le Grand, 1942) on the basis of the theoretical aberration [equation (14)], but the actual aberration accounts only for slight myopia of the order of 0.25 diopter (Cabello, 1945; Otero, Plaza, and Rios, 1948). An experiment by Wald and Griffin (1947) confirms this result: in photopic vision three subjects with naturally contracted pupils adjusted the eyepiece of a telescope, then again with their pupils dilated by homatropine, but no difference in the adjustments was found. One could, of course, object that in photopic vision the Stiles-Crawford effect diminishes the efficiency of the marginal rays.

Despite these observations, Koomen et al. (1951) still attributed an important role to spherical aberration. Otero and Aguilar (1951) have repeated measurements of night myopia with artificial pupils, proving once more the slight influence of spherical aberration. However, Bouman and van den Brink (1952) still maintain this explanation. The pros and cons regarding the theory of aberration will be found in monographs by Knoll (1952) and O'Brien (1953).

Ivanoff (1946) proposed an ingenious theory. Night vision can be improved if divergent lenses of 1 to 2 diopters are prescribed, as they stimulate accommodation and such accommodation corrects spherical aberration. The main part of night myopia would not be ametropia, but an improvement of the image obtained by a change in the shape of the lens. One could object that this practice does not improve daylight vision; but, on the one hand, the Stiles-Crawford effect applies in one case and not in the other and, on the other hand, specific blur of the image has less effect in daylight vision because of good intensity discrimination. However, Ivanoff [25] himself condemned his theory as he noted that the elimination of central rays which appreciably modify spherical aberration scarcely changes night myopia.

Biot (1950) wondered whether part of night myopia might not be due to a difference between aberrations of the eye along the axis (daylight vision, foveal) and off the axis (parafoveal vision, scotopic). But the existence of angle α can, on the contrary, convert vision along the axis to scotopic conditions. On the other hand, in daylight vision, as will be shown later, the dioptric power of the eye varies very little in the neighborhood of the axis.

The hypothesis of a difference in depth of photopic and scotopic receptors has been proved false by anatomy. The sensitive portions of the rods and cones are situated approximately in the same plane.

The only remaining possibility is the role played by the crystalline lens in night myopia. In fact, Cabello (1945) showed that night myopia was decreased a great deal when the ciliary muscle was paralyzed by atropine. Carreras (1952) noticed that aphakic and presbyopic subjects had only very slight night myopia. From this one would infer that the crystalline lens assumes a different shape at rest in the dark than in daylight. Actually, Otero et al. (1950), who photographed the Purkinje image of the anterior surface of the lens of three ametropic subjects in the dark, found that the punctum remotum was situated between -1.2 and -1.3 diopters. This result was contested by Koomen et al. (1953), who used a similar technique, except that their subjects were fixating a very weak test light instead of being in total darkness; under these conditions three of the subjects remained accommodated for the test at 5 m, although the best vision of this test was obtained with minus lenses of -1.5 to -1.75 diopters; only the fourth subject showed an increased curvature of the anterior surface of the lens corresponding to an accommodation of 0.5 to 1.5 diopters. Later measurements by Campbell and Primrose (1953) on six subjects came to an average of 0.8 diopter of accommodation in the dark. Campbell (1954) found great variations of accommodation for each subject in the dark with 13 subjects: with six photographs taken every 2 minutes the results varied between 0 and 1.1 diopters of accommodation for a given subject. The microfluctuations of accommodation, which are of the order of 0.1 diopter in daytime, increase markedly at night; on the average, Campbell found 0.64 diopter of accommodation in scotopic vision, with his 13 subjects, whereas Alpern and David (1958) found a mean of 2 diopters of night myopia on four subjects.

By infrared retinoscopy, Chin and Horn (1956) also attempted to assess the part played by spherical aberration and scotopic accommodation on nine subjects. The results seem to vary a great deal from one subject to another. In conclusion, the factors of night myopia are better known than their relative importance.

Night Presbyopia

A certain amount of night presbyopia is also associated with night myopia in scotopic vision, as was discovered by Duran (1943) and Palacios (1944) and studied by several investigators such as Wald and Griffin (1947), Kühl (1950), and Alpern and Larson (1960); the amplitude of accommodation decreases by one half at about 0.1 cd/m^2 luminance, as compared to its value in daytime, by three quarters at about 0.01 cd/m^2, and becomes practically zero between 10^{-3} and 10^{-4} cd/m^2. Of course, these values are only approximations and may vary a great deal from subject to subject.

To explain this loss of the accommodative reflex in night vision it would be necessary to know the true mechanism of this reflex in daylight, but unfortunately we know very little about it. There is not, however, a lack of hypotheses. Polack [54] supposed that the colored fringes caused by chromatic aberration when the eye is out of focus would be responsible for stimulating the ciliary muscle; Fincham (1953) attempted to verify this theory with his coincidence optometer. He compared the precision of the adjustment of 60 subjects when they were presented with either yellow monochromatic light (line D) or a complex yellow light obtained by a mixture of red and green radiations. With sodium light, 30% of the subjects did not change their adjustment when the distance to the source of light was varied, and 25% reacted abnormally; that is, they accommodated when the source was moved closer but the accommodation did not relax as the source was moved farther away. The rest of the subjects reacted normally but with less precision than with complex light. The diameter of the circle of diffusion must be between 4 and 13′ to produce the reflex, whereas a diameter of 2 to 10′ is sufficient in complex light, and for each subject the value is lower in the latter case. For dichromats there would be no difference between the two types of light. These results are interesting but, in my opinion, are not very convincing, because the mixture of red and green radiations is an artificial stimulus and their superimposition is uncertain, as all those who have done practical colorimetry will agree. The comparison of white light and monochromatic light does not give much better results (Fincham, 1951; Campbell and Westheimer, 1959). This subject will be considered again when discussing acuity. The results obtained in monochromatic light, at least as good as those obtained in white light, seem to show that the microfluctuations of accommodation must be of the same order as in white light. According to Troelstra et al. (1964), accommodation remains as precise in the absence of chromatic aberration.

If colored fringes were essential to accommodation, it would be easily understood that this function disappears at the lowest levels of luminance where vision is achromatic, but in the absence of proof supporting this role of colored fringes, the explanation must be sought elsewhere. A statement by Fincham can lead to the solution. He observed that only the fovea and the center of the fovea up to 10′ from the fixation point could, by its excitation, produce the accommodative reflex. This reflex is usually closely associated with binocular vision and convergence of the axes of the two eyes, but these functions can be dissociated by the use of monocular vision and the use of lenses which alter the image without changing the distance of the object. A study by Campbell (1954) has shown that for circular sources of apparent diameter less than 10′, the illumination E falling on the plane of the pupil is the only variable to play a role (Ricco's law of total summation). The amplitude of accommodation A (in diopters) is given by the following empirical equation:

$$A = 3 \log_{10} (E/E_0) \tag{26}$$

E_0 is the critical value of illumination at which night presbyopia is total. This equation is valid up to $E = 100\,E_0$, the corresponding value of A being 6 diopters (for the considered subject); the values by Campbell lead to a critical value of $E_0 = 8.4 \times 10^{-8}$ lux. At this level the source (white light of color temperature 3000°K) is still seen by the fovea, and disappears (foveal threshold) between 3 and 4×10^{-8} lux. Lower values of luminance are still seen by the parafovea.

The conclusion is therefore that the receptors which produce the accommodative reflex are the foveal cones, and the critical value at which night presbyopia becomes total is about twice the foveal threshold. Rod vision is insufficient to stimulate accommodation.

If night presbyopia is thus interpreted in simple terms it remains to explain why in night vision the mean position of rest of the crystalline lens often corresponds to a certain accommodation. Schober (1947) assumed that contraction of the ciliary muscle was due only to an effort to see in darkness (in young subjects concentrated attention does in fact stimulate accommodation). This phenomenon is well known in daylight vision [see for example, a study by Pheiffer (1955) on the accommodation during reading as a function of the difficulty and interest of the material] and is a source of annoyance to optometrists. It is possible that this phenomenon results in "macrofluctuations" of accommodation in scotopic vision. But, on the contrary, Whiteside (1952) noted that in the absence of any detail in the visual field (photopic and uniform), a contraction of the lens occurred (empty space myopia). This phenomenon is important for pilots flying at very high altitudes and has therefore been investigated from a practical point of view (MacLaughlin, 1954). It is particularly recommended that the instrument dials be placed at the focal point of collimators so that the pilot sees them at infinity and relaxes his accommodation. Another hypothesis which will be considered later is based on the relationship between accommodation and binocular convergence.

Threshold of Point Sources

The *absolute threshold* of a point source is the lowest amount of illumination E produced by the source in the plane of the pupil (assumed perpendicular to the rays of light) for this source to be just perceived. Astronomers believed for a long time that stars of the 6th magnitude corresponded to that threshold, but stars are seen on a background which is not black, the luminance of which is of the order of 10^{-3} cd/m^2 (in the absence of the moon). By viewing through a small aperture, thus eliminating other stars and the light from the sky, Curtis (1901) discriminated stars of the 8.5th magnitude. Russell (1917) confirmed his result, which according to equation (21) corresponds to $E = 8.3 \times 10^{-10}$ lux; this is the illumination of a source of 1 cd seen at a distance of 35 km.

The first good measurements under laboratory conditions were made by Langley (1889). While measuring the energy of a band of the solar spectrum around 555 mμ with the bolometer that he invented, he found at threshold an energy of 4×10^{-9} ergs/sec cm^2. Reeves (1917–1918) obtained 2.4×10^{-10} lux with three subjects. Buisson (1917) found 1.4×10^{-10} lux using small phosphorescent screens as sources and Löhle (1929) found a value of 3×10^{-10} lux. All these results should be considered with caution because of the difficulty with scotopic photometric units, on the one hand, and the variation of pupillary diameter, on the other hand, because it is the total flux entering the eye which determines the threshold. Taking 1.7×10^{-9} lux (that is, 1 cd at 24 km) for the mean of all these values as Buisson (1932) suggested, and a pupil 8.5 mm in diameter, the flux is slightly less than 10^{-13} lumen. Healthy emmetropic eyes were used in determining the values of the thresholds in all the above results. It must also be noted that the absolute threshold is influenced by night myopia. Ronchi (1943) noted that a greater number of stars become visible if the eye is corrected for this myopia; on four subjects, Otero et al. (1949) found values of the thresholds varying up to a factor of 2, whether night myopia was corrected or not.

It is obvious that the spectral composition of the light emitted by the source and the retinal location of the image play very important roles. The fovea is more sensitive to red than the periphery of the retina; in the extreme blue, according to Pirenne (1944), the threshold is at least 100 times lower at 4° eccentricity than in the fovea. These results are in accord with the Purkinje effect, which favors vision of short wavelengths by the rods. In white light the results depend upon the color temperature of the source and, for sufficiently low temperatures of incandescent lamps, there is more or less equivalence between the fovea and an area situated 8° away, for a source of 2.7′ diameter, according to Arden and Weale (1954). This result contradicts the astronomer's classical opinion that one ceases to see a faint star when one looks at it, but this may be due to eye movements or the color temperature of the stars. Anyhow, this slight difference of sensitivity between the fovea and the periphery for point sources is very much different for extended sources because in the periphery the sensitivity is much greater, owing to marked spatial summation. Baumgardt (1949) even assumed that a rod was no more sensitive than a cone and that scotopic adaptation was due more to a nervous process than to a photochemical one.

The variations of absolute thresholds of point sources in the retinal periphery will be considered later (Chapter 8). But there are local variations even within the fovea. According to Wright [74], the sensitivity to red is practically uniform, with perhaps a slight minimum in the center; in the green, the sensitivity is constant; in the blue, there is a marked minimum of sensitivity in the center of the fovea which is practically dichromatic, after which the sensitivity increases rapidly with distance from the center. Moreover, variations of the macular pigment play a part, particularly for short wavelengths.

Threshold of a Point Source on a Background of Finite Luminance

The threshold of point sources on a background of zero luminance was considered in the preceding section, but the case of a background of slight luminance L is much more important in practice. Measurements were made by Langmuir and Westendorp (1931), by Green (1935), and particularly by Knoll et al. (1946) on five young subjects. They used a test of 1′ diameter seen in the center of a background 20° in diameter. The test and background were illuminated with the same white light (of color temperature 2360°K) and the frequency of seeing the test was 98%. Hecht showed that their results could be represented by the empirical formula

$$E = a(1 + \sqrt{bL})^2 \tag{27}$$

provided two values are given for the constants a and b in the scotopic and photopic regions ($L > 0.01$ cd/m²). In the first instance, $a = 1.5 \times 10^{-9}$ lux and $b = 4000$ cd/m²; in the second instance $a = 7 \times 10^{-8}$ and $b = 0.4$.

Let us also cite the measurements by Blackwell (1946), who used a test of 0.6′ and color temperature 2850°K. These results have been recalculated by Tousey and Hulbert (1948) for a 98% probability of seeing. All these values are found in Figure 21; it can be noted that equation (27) represents the findings very satisfactorily.

As a matter of comparison Table 8 gives the order of magnitude of the luminance of the sky near the horizon (after Middleton [41]).

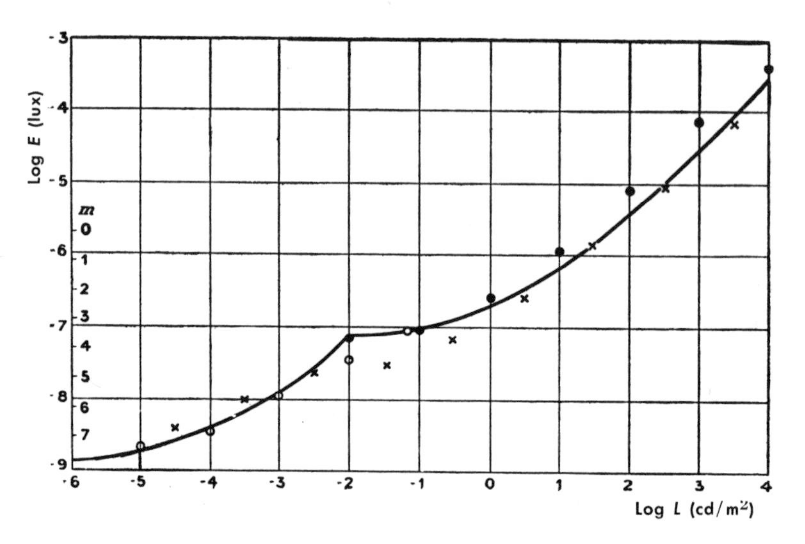

Fig. 21. Threshold E of a point source as a function of the luminance L of its background. ⊙, values of Langmuir and Westendorp. ●, values of Green. ×, values of Blackwell. The curve represents formula (27); m represents the magnitude of the stars.

TABLE 8

Luminance of the Sky Near the Horizon (in cd/m²)

Clear sky at noon	10^4
Cloudy sky at noon	10^3
Grey sky at noon	10^2
Cloudy sky at sunset	10
Clear sky, a quarter of an hour after sunset	1
Clear sky, a half hour after sunset	10^{-1}
Night sky	
full moon	10^{-2}
clear and without moon	10^{-3}
cloudy and without moon	10^{-4}

Dratz (1947) pointed out that for high values of L, the fact that the luminance of the background is or is not superimposed upon that of the source may appreciably alter the results; in the case of stars in the sky there is superimposition; this also occurred in the measurements by Knoll et al. (1946). Smith (1955) showed experimentally the fallacy of the ancient legend, dating from the time of Aristotle, that stars are visible in the daytime from the bottom of a well.

Threshold of Point Flashes of Light

Let us consider again point sources on a dark background; this time we shall assume that instead of a steady light the source is a flash of duration t. How can the threshold in the latter case be related to a steady light? Blondel and Rey (1911) established an approximate formula which in the case of the threshold of a monochromatic source can be written

$$W = P(t + t_0) \tag{28}$$

where W is the energy which enters the eye during the flash, P the flux of this same light if it is steady, and t_0 a constant of the order of 0.15 to 0.3 second. If t is smaller than a few hundredths of a second it becomes negligible compared with t_0 and W is a constant.

From the values of P that we have already mentioned (less than 10^{-13} lumen; that is, for radiation 507 mμ which produces 1746 lumens/watt, a flux less than 6×10^{-17} watt, or 6×10^{-10} erg/sec), one can expect very small values for energy W, of the order of 10^{-10} erg, when t is very small. At this level the quantum fluctuations of the light energy may come into play. The quantum theory stipulates that any receptor absorbs a certain number of photons, and that each *photon* or quantum of light of wavelength λ has an energy hc/λ, where h represents Planck's constant

(6.624×10^{-27} erg sec) and c the velocity of light in vacuum (2.998×10^{10} cm/sec). From this it is simple to deduce, expressing λ in millimicrons, the number of photons transmitted in monochromatic energy W (in ergs) which is

$$0.504 \times 10^{9} \lambda W \tag{29}$$

Assuming $\lambda = 500$ and $W = 10^{-10}$, the number of photons is 25.

The direct experimental measurement of the number of photons at threshold has given rise to many experiments. The first worthy data are those of von Kries and Eyster (1906), who found values of W between 1.3 and 2.6×10^{-10} erg, using monochromatic light of 510 mμ. From equation (29) the number of photons is therefore 33 to 67. Chariton and Lea (1929) found 17 to 30 photons for 505 mμ, and Barnes and Czerny (1932) 40 to 90 photons for 530 mμ.

These old data are difficult to interpret because the threshold is not sufficiently specified. It is a statistical value, even for one particular subject, and it must be specified by an arbitrarily chosen percentage of correct answers (flash "seen") but this percentage may modify the results to a great extent.

Very careful measurements were done by Hecht et al. (1942) on seven subjects. They used an artificial pupil 2 mm in diameter and a test field of 10′ at 20° eccentricity from the fixation point after 45 minutes of dark adaptation. The test was illuminated by monochromatic light of 510 mμ and exposed in flashes of $t = 0.001$ second duration. The threshold is defined for 60% of "seen" in the trials. The results vary considerably from one subject to another, and even for the same subject there may occur variations from day to day by a factor as high as 2. The results lie between 54 and 146 photons.

In all the above results the number of photons is that measured at the level of the cornea, but not all of these will reach the retina. It may be that half are absorbed or scattered as the light passes through the media of the eye. Moreover, not all the photons which reach the retina are effective. They have to be absorbed by the visual purple of the rods. The concentration of pigment in the rods is believed to be low, so that at the most not more than 20% of the photons would be absorbed at the wavelength 500 mμ, which is the band of maximum absorption. More recent measurements by Crescitelli and Dartnall (1954) reduce this value still more. It is therefore likely that only a few photons in each flash effectively cause the sensation of light. The sensitivity of the eye would nearly correspond to the limit imposed by the quantum properties of light.

The Poisson Equation

It has been attempted to specify more precisely the effective number of photons at the threshold of a brief flash of light by an ingenious statistical method. Let us assume that a monochromatic flash emits n photons per unit solid angle. The number n is very large even at threshold, but the probability p that each photon has, a priori, of contributing to the sensation of vision is very small. For a photon to be effective it has to enter the subject's pupil, be transmitted through the ocular media without absorption or scattering, and finally be absorbed by one molecule of visual purple in one of the rods where the source is imaged. If many identical flashes are produced, they can be classified into groups by statistical methods. In the first group are these flashes in which there is no photon absorbed, in the second group are those flashes in which only a single photon is effective, and in group $i + 1$ are those in which just i photons are effective and $n - i$ are ineffective, etc. Writing $q = 1 - p$, the probability of one of these groups is

$$p^i q^{n-i}$$

On the other hand, the possible number of the distributions of flashes in this group is the number of combinations of n objects i by i, which is

$$C_n^i = \frac{n!}{i!(n-i)!} \tag{30}$$

Therefore the probable relative frequency of a group, i.e., the ratio of the number of flashes in that group to the total number of flashes N, is

$$y_i = C_n^i p^i q^{n-i} \tag{31}$$

This is called the *binomial distribution*. If n increases while p and q remain constant, this distribution tends toward the *normal distribution* that is often encountered in biology. In the actual case, however, the conditions are different: If the number n of photons increases, the probability p that they will be absorbed decreases inversely, because the number np of absorbed photons remains constant. This is actually observed by changing the wavelength away from $\lambda = 507$ mμ, which corresponds to the minimum of n; p diminishes because the photons which reach a rod have less chance of being absorbed and the parallelism between the spectral absorption of visual purple and the scotopic relative luminous efficiency expresses the constancy of the product np. We can express this as follows:

$$np = m \tag{32}$$

where m is a constant but not necessarily an integer. Under these conditions one can demonstrate that the limit of (31), when n increases, is

$$y^i = \frac{e^{-m}m^i}{i!} \tag{33}$$

where e is the base of naperian logarithms. This is the Poisson equation (1838). The constant m represents the *arithmetical mean* of the number of photons absorbed during each flash.

Another hypothesis is usually introduced here. It is assumed that in a given flash when the number i of photons is less than the minimum number k of photons required for seeing, $i < k$, the flash is *never* seen, whereas $i \geqslant k$ the flash is *always* seen. In the latter case, the probability $\tilde{\omega}$ that a flash is seen is obtained by adding all the values of (33), so that $i \geqslant k$:

$$\tilde{\omega} = e^{-m} \sum_{i=k}^{i=\infty} m^i/i! = 1 - e^{-m} \sum_{i=0}^{i=k-1} m^i/i! \tag{34}$$

Using the same hypothesis, the value of k can be determined experimentally from equation (34), although m is unknown in absolute value. It suffices to vary m in relative value by changing the energy emitted in each flash and to measure $\tilde{\omega}$, that is, the proportion of flashes seen, in each case. If $\tilde{\omega}$ is plotted as a function of log m, a curve is obtained of which the form depends only upon k. Experimental curves can thus be compared in shape with the theoretical curve after appropriate translation along the axis of log m, since m is only known in relative value (Fig. 22). Beyond $k = 7$ or 8 the variation of the shape is too small for k to be deduced with any precision.

This method was suggested by Brumberg and Vavilov (1933) and applied by Hecht et al. (1942), who obtained values of k between 5 and 8. Van der Velden (1944) found 2, Peyrou and Piatier (1946) 2 to 5, and Baumgardt (1947) 4 to 10.

The hypothesis above assumes that there are no fluctuations due to the biological receptor (k constant) and only the quantum fluctuations of light prevent the existence of a rigorous threshold. This seems very unlikely, because when the stimulus has no quantum effect (weight, for example) there is still dispersion. With point sources the dispersion of thresholds varies from 1 to 10, whereas with extended sources it varies only between 1 and 2 or 2.5, according to Denton and Pirenne (1954); this small range indicates good retinal stability. In some subjects with good retinal stability the margin of biological variation may remain in

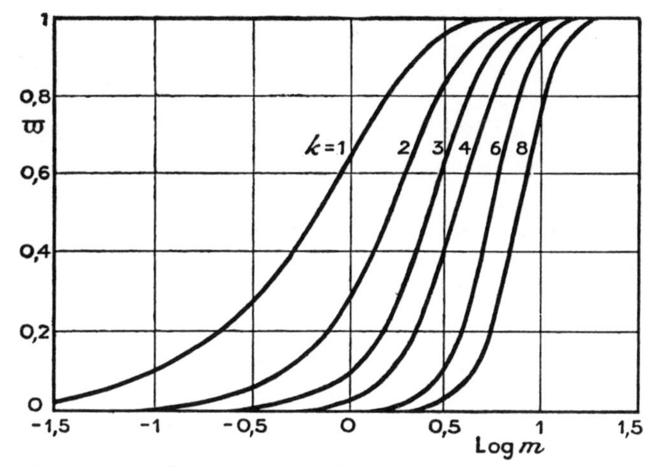

FIG. 22. Probability of seeing a flash, containing a mean number m of absorbed photons, according to the number k of photons assumed for threshold (Poisson sum).

a ratio between 1 and 1.2:1 (Crawford and Pirenne, 1954). Le Grand (1949) showed that the superimposition of these dispersions results in curves which hardly differ from those of Figure 22, but the number of photons absorbed is less than the actual number. For example, if 5 photons are necessary to cause a sensation of light, the curves $k = 2$ or $k = 3$ have the closest shape to the experimental curves. Some experiments by Blackwell (1952) also confirm the fact that the thresholds are probably never free of other fluctuations than those which arise from the quantum properties of light. It seems, therefore, that analysis of the probability of seeing curves cannot give with certainty the minimum number of photons absorbed in one flash necessary to cause a sensation of light. All that can be affirmed is that this number is very likely to be less than 10.

Other Methods

First, the hypothesis $k = 1$ can be omitted a priori, as Pirenne [53a] showed. Equation (34) would then be

$$\tilde{\omega} = 1 - e^{-m} \tag{35}$$

and the probability of "not seen" is e^{-m}. Consider now that there are N observers (or receptors) looking at the same flash. As it is sufficient for only one of them to see the flash, the total probability of "not seen" is the product of the probability for each observer, that is, e^{-Nm}, and the probability of "seen" is

$$\tilde{\omega} = 1 - e^{-Nm}$$

Compared to equation (35) it is noted that this equation shows the same probability as with one receptor and a source N times more intense. This is in contradiction with reality. Obviously it will not suffice for N sailors of a ship to come up on deck instead of only one observer in order to see a lighthouse N times dimmer!

Bouman and van der Velden (1947) have called upon the laws of temporal summation to provide a measure of the minimum number of photons k necessary to see. They assume that the flash is always seen if k photons are absorbed in a lapse of time less than a constant duration τ, whereas if $k - 1$ are absorbed in this lapse of time there is no visual sensation. From an analysis of experimental results they have deduced that $\tau = 0.02$ second and $k = 2$, but their hypothesis seems to lack support. Certainly there is a progressive diminution of summation as the time between each flash increases. Moreover, the biological dispersion is added to the quantum fluctuation, so that the only definite conclusion is that k is at least equal to 2.

The laws of spatial summation applied at threshold have also provided arguments in favor of one or another value of k. Total summation occurs for a field of 7 to 10′ (Ricco's law) in the fovea and of 30 to 45′ in peripheral vision. Beyond these values only partial summation occurs. In an area of that order there exists a considerable number of rods and the probability that two quanta of the same flash are absorbed by the same rod is negligible. Therefore it is certain that k photons are absorbed by k different rods and the effect of each alone does not cause any sensation, but the effect of them all converging on the retinal cells (bipolars and ganglions) produces an impulse along the optic nerve. Unfortunately, it seems impossible to put into an equation such a mechanism of partial summation without simplifying it to a useless caricature. Thus the arguments brought by Bouman and van der Velden (1948) or Baumgardt (1953), according to whom k must be equal to 2, do not seem justified. As will be considered later, the peripheral retina contains large sensory units which overlap one another and of which the limits are uneven. The simple hypothesis of independent and adjacent areas would tend to give a low value of k, as Pirenne and Denton (1951) showed. Independent studies by Pirenne and Marriott (1954), on the one hand, and Brindley (1954), on the other hand, have shown that k was greater than 3 and even 4, but Bouman (1955) still expresses the opinion that two quanta are sufficient. Concerning recent developments, Bouman [60b] and Pirenne [53c] should be consulted. In my opinion the most notable progress has been achieved by Crouzy (1961–1965), who clarified the difficult question of quantum effects in contrast threshold. (See also Walraven and Bouman [80].)

Supraliminal Levels

It is certain that at threshold the number of absorbed quanta is very small but cannot be more clearly specified. But it is possible that quantum fluctuations also occur at supraliminal levels, i.e., well above threshold. For example, Barnes and Czerny (1932) explained, using this hypothesis of quantum fluctuation, the slight sparkling of a small source of fixed intensity to which we referred in Chapter 3.

Guild (1944) wondered if perhaps quanta could play some part at photopic levels, because of the very slight absorption of the hypothetical pigment of cones. With 3000 measurements of differential threshold he found that the Poisson curve with $k = 17$ gives a better representation than the classical Galton's ogive, which represents the "normal" probability law. According to Mueller (1951), the shape of the curve and therefore the value of k vary with the level of intensity. Other authors, such as Bouman and Walraven (1957), Pinegin (1958), and Marriott (1959), find Poisson curves with k values between 8 and 15. On the other hand, Blackwell (1953) analyzed 27,500 measurements of the differential threshold of four subjects. For three of the subjects he found that a normal law gives the best representation, and for the fourth subject the best curve was found using a "log-normal" law, i.e., normal if the log of the luminance is the variable.

Craik (1944) objected to the quantum theory of differential threshold on the basis that an intense preadapting light which diminishes the photochemical process of the retina, therefore the probability of photon absorption, has very little effect on the differential threshold. Mueller (1950) set forth the possible diverse theoretical bases of differential threshold founded on quantum theory.

In conclusion, there is no experimental proof that quanta of light play a part at levels well above threshold, although this is not impossible. Actually, in all fields in which, only a few years ago, photochemical theories were predominant, quantum theory now seems to outweigh them, probably equally excessive in exaggeration as the concepts that it claims to replace.

CHAPTER FIVE

Vision of Details

Following the simpler problem of vision of a point of light, more complicated problems of vision of details will be considered. The scope of this chapter will be limited to white light and (in general) to natural conditions of vision, that is, without an artificial pupil or a fixation point.

Black Point on a Light Background

The problem of a black circular test of apparent diameter u seen on a background of uniform luminance L has been investigated for a long time, and the usual results are of the order of 20 to 30″ for the threshold. The International Commission for Optics has adopted 36″ for the normal observer between 10 and 100 cd/m². Under the best conditions, Hecht et al. (1947) obtained $u = 18''$ for 95% probability of seeing (four observers, blue sky of 3000 cd/m² luminance) and $u = 14''$ for 75%.

The deduction of these results could be attempted by the following reasoning. If the absorption of light in the transparent media of the eye is neglected, the retinal illumination produced by a uniform background of luminance L is

$$E_0 = LS\frac{n'^2}{p'^2} \tag{36}$$

where S is the area of the entrance pupil, n' the refractive index of the vitreous body, and p' the distance between the second principal point of the eye and the retina, i.e., the second focal length if the eye is emmetropic and unaccommodated (22.3 mm in the schematic eye). If a small area subtending a solid angle Ω is isolated on this background, the illumination in the plane of the entrance pupil produced by this area is $L\Omega$ and the corresponding luminous flux which enters the eye is $L\Omega S$. In equation (18) representing the maximum illumination in the center of the Airy disk, this same flux is equal to $\pi d'^2 E'/4$ and the equation can therefore be expressed

$$E_m = \frac{\pi d'^2 L \Omega S}{4x'^2 \lambda'^2}$$

Taking equation (15) into consideration, it follows that

$$\frac{E_m}{E_0} = \frac{\pi d'^2 \Omega p'^2}{4x'^2 \lambda^2}$$

Assuming an entrance pupil 4 mm in diameter, $d' = 3.68$ mm and the distance x' between the exit pupil and the retina is 20.5 mm. Finally, letting $\lambda = 589$ mμ, the above expression will be equal to

$$\frac{E_m}{E_0} = 3.627 \times 10^7 \Omega \tag{37}$$

If the given area is circular and the apparent diameter is u radians,

$$\Omega = \pi u^2/4 \tag{38}$$

A radian is equal to 2.06×10^5 seconds of arc, so that equation (37) becomes

$$\frac{E_m}{E_0} = 6.69 \times 10^{-4} u^2 \tag{39}$$

when u is expressed in seconds of arc.

If this small area that we have theoretically isolated is a perfectly black spot, E_m can be subtracted from E_0. This means that in the uniform retinal illumination E_0 there is a central area of minimum illumination equal to $E_0 - E_m$ on the image of the spot. The ratio E_m/E_0 is nothing but the *differential threshold* (contrast threshold). The best experimental value ($u = 14''$) corresponds to 0.13 for this threshold, according to equation (39). This value is important, particularly if it is compared to the usual values of the threshold for extended fields, which is of the order of 0.01. But as the intensity discrimination of the eye decreases with the test size, the value of 0.13 is not surprising a priori for a point.

In the photopic domain, the vision of a point of light of apparent brightness E, on a background of luminance L considered in Chapter 4, is somewhat complementary to the problem of the black point, since the superimposition of the two gives a uniform background. When L is sufficiently high, equation (27) becomes

$$E = abL$$

where ab plays the same part as the angle Ω of the area in the above theory. With the experimental values ($a = 7 \times 10^{-8}$, $b = 0.4$), equation (37) gives approximately $E_m/E_0 = 1$, which means that in the center of the image pattern the illumination would be nearly twice that of the background. This same intensity discrimination would correspond to $E_m/E_0 = 0.5$ for a black point on a white background and furthermore corresponds to $u = 27''$ according to equation (39): This is a very common value. It is therefore possible that the same mechanism applies in both cases.

In the preceding theory, we have assumed that the threshold was obtained for a given relative value of the diminution of illumination in the center of the image pattern. But the numerical coefficient of the second part of equation (37) is proportional to d'^2, that is, the square of the diameter of the pupil, and it follows that for constant retinal illumination E_0 (which is easily achieved by varying the luminance inversely to the pupillary area), the product ud should remain constant at the threshold. But this does not happen. According to measurements by Arnulf et al. (1954), u remains constant when d varies between 0.5 and 4 mm; this means that the threshold is obtained when the total *flux* removed from the image pattern has a certain ratio with the uniform retinal illumination. Of course, the measurements of Arnulf and his collaborators were done in an extrafoveal area where summation is more marked than in the fovea. If one assumes for high luminances a value of $u = 20''$ as the normal threshold of a black point and a differential threshold of 0.01 for flux integrated in the area of total summation, this results in a value of $10u$ for the diameter of this area, that is, about 3′. This is certainly a maximum. With a differential threshold of 0.02, the result would be about 2′.

For small grey disks on a white background (or white on grey), contrast will be introduced. It is defined by the formula

$$C = \frac{L - L'}{L} \tag{40}$$

with L the luminance of the background and L' the luminance of the object. If the object is darker than the background, this contrast is positive and less than 1; if the object is lighter than the background, C is negative and can have any value. Equation (37) remains valid by replacing Ω by $C\Omega$. It follows that the limiting angular diameter u must vary inversely to the square root of C. This relation was verified by Byram (1944). The results with a grey patch on a white background are slightly better than with white on grey. Byram also noted that the dimension of the background played a part when its diameter fell below 20′. For 10′ and 5′, the size of the black point must be increased by factors of 1.3 and 2, respectively, to make it visible.

At low luminances, the intensity discrimination decreases considerably and the black or grey point must be larger. Arnulf proposed the following formula, which represents reasonably well his experimental results:

$$u = a\left(1 + \frac{1}{\sqrt{bE_0}}\right)^2$$

where E_0 is the retinal illumination produced by the background.

Linear Source of Light

This case is of little practical interest. It corresponds in the laboratory to the vision of a luminous slit, and some research has been devoted to it.

Fry and Cobb (1935) determined the maximum width of a linear source (on a dark background) which produces complete summation, that is, such that the luminance is inversely proportional to the visual angle u of this width. They obtained $u = 30''$. Above that value summation is only partial and ceases when u is greater than $4'$. The luminance at the threshold then remains constant.

Niven and Brown (1944) studied the threshold of vision of a slit of $29'$ height, as a function of the luminance, the width of the slit, and the time t of exposure. The product of the luminance multiplied by the exposure time t remains more or less constant up to $t_0 = 0.19$ second.

A similar study was done by Bouman (1953), but instead of the eye being left free to move, the center of the test was situated at 7° eccentricity from the fixation point. There was complete temporal summation for $t < 0.1$ second. The slit was $2'$ wide, but its height varied between 2 and $250'$; complete spatial summation was found up to $8'$ for red and $32'$ for green, after which the summation is partial, the intensity of each flash varying with the square root of the height (Piper's law). Bouman also carried out some measurements with a light background. Finally, a theoretical study of distribution of light in the absence of aberrations was done by Shade (1954) and by Fry [18a].

Vision of a Line

A straight line on a light background, such as a telegraph wire against the sky, is a reasonably important object in practice. The calculation of the distribution of retinal illumination in the image would be very easy if the pupil were a slit parallel to the line (see, for example, Reese, 1939), but with a circular pupil, numerical integrations have to be used. Regarding this subject, the tables by André (1876) are often referred to, but they have many errors, and those of Danjon and Couder [13] or of Selwyn (1943) will serve the purpose better. The method of serial computation proposed by Byram (1944) can also be used. Using the same symbols as in the preceding section, it is found that

$$\frac{E_m}{E_0} = \frac{0.89ud'p'}{x'\lambda}$$

where u represents the apparent width of the line in radians. With the same numerical values as above, but expressing u in seconds of arc, we obtain

$$\frac{E_m}{E_0} = 2.94 \times 10^{-2}u \tag{41}$$

For a good many years (Barnard, 1897), it has been observed that the visibility of lines of angular width u is of the order of 1 second of arc and sometimes even less. Hecht et al. (1947), with 10 trained observers looking at lines on a very bright background (14,000 cd/m^2), obtained 95% probability of seeing for $u = 0.48''$ and 75% for $u = 0.43''$. But the length of the line must subtend at least 1°. The smallest value corresponds to approximately $E_m/E_0 = 0.01$ according to equation (41). This result may be surprising since the intensity discrimination for black points was not so good, but it must be kept in mind that a line is easier to perceive than a point. For instance, a row of points is sometimes easily seen, whereas each of these isolated points might not be. This can be accounted for by an effect of retinal cooperation similar to summation at the absolute threshold. The threshold of a collection of points (nonlinear) was studied by Jainski (1955). The most favorable conditions are with horizontal and vertical lines. This fact actually applies for many tests; the reason for this preference is not known, at any rate it does not seem to be related to eye movements (Nachmias, 1960).

The case of grey lines on a white background or white lines on a grey background was studied by Byram (1944). The results are slightly better in the former case. If C represents the contrast [see equation (40)], theory predicts that u must vary as $1/C$, and experiment confirms it. Of course, beyond a critical value of u, the minimum contrast becomes independent of u; this value is 2′ according to Hartridge (1947) and 4′ according to Fry (1947).

Finally, Hecht and Mintz (1939) investigated the variation of u as a function of the luminance of the background. Their results can be represented satisfactorily by the formula below, as was shown by Moon and

Spencer (1944):

$$u = u_\infty L^{-1}(a + L^{1/3})^3$$

If L is expressed in cd/m^2, $a = 0.75$.

Vision of a Rectangle

The case, already a little more complicated, of a dark rectangle on a bright background was studied by Jones and Higgins (1947) on a background of 10° diameter and 89 cd/m^2 luminance. The mean for three subjects for threshold visibility is given in Table 9; it is for the square that the area is least.

Lamar et al. (1947), then Hendley (1948), studied the case of a rectangle brighter than the background at 60 and 10^4 cd/m^2 luminance, with a length/width ratio varying between 2 and 200. The results tend to show that the efficient luminous flux that makes the test just visible would be limited to a band of 1 to 1.5′ width, and would therefore depend upon the perimeter of the test more than on its surface. The judgment of recognition therefore involves the concept of *contour* between the object and its background. This notion of contour is discussed here for the first time, applied to a simple figure, but it will play a much more important part for complicated figures.

Let us also mention a statistical study by Otero and Aguilar (1950) on the vision of a black rectangle of which the length/width ratio is 9, on a background of 30° diameter and 3×10^{-4} cd/m^2; at threshold the length of the rectangle is of the order of 80 to 100 minutes of arc.

Vision of Two Points

The discrimination of two stars is a common problem in astronomy; Hooke (1705) found that it was very unusual to discriminate between two stars when their angular separation was 1′ and relatively easy around 3′. A criterion of sharp vision sometimes considered—the vision

TABLE 9

Minimum Values of the Width of a Black Rectangle on a White Background as a Function of Its Length

Length, minutes of arc	0.36	0.81	1.33	1.75	3.6	5.2	7.5	15	32	39	46
Width, seconds of arc	20	9.4	7.0	5.3	4.3	3.1	2.1	0.83	0.74	0.70	0.68
Relative area	1	1.05	1.3	1.3	2.1	2.2	2.2	1.7	3.3	3.8	4.3

of the four larger satellites of Jupiter (of which the separation from the planet is between 2 and 8′)—is too complex to be utilized.

Formerly, the smallest angle between the center of the point sources that can be seen as separated through an instrument was called the *resolving power.* This terminology is defective because the numerical value of this "power" decreases as the instrument improves; therefore, the term *limit of separation* (in Latin, *minimum separabile* according to Giraud-Teulon, 1879) is now favored. An equivalent term is the *limit of resolution.*

While the absolute threshold *(minimum visible)* of a source is obtained more easily when the apparent brightness increases, Volkmann [69] and Aubert [2] noted that the limit of resolution s of two point sources of light of equal apparent brightness increases slightly with the luminance, at least in the photopic range in which foveal vision is possible. The reason is that each source seems to enlarge as it becomes more intense. More recent measurements, in particular those of Berger (1936), van Heuven (1937), and Fiorentini (1950), have confirmed this variation, the minimum values of s varying between 1.7 and 3.5 minutes of arc, depending upon the subjects.

This criterion of "separation" of two points is of course subjective. Another method consists of presenting randomly either both points or one point of doubled intensity. The subject identifies the duality then by the elongation of the image in the direction of the two points, so that it is not necessary to see them separated. The threshold is of course smaller with this method. Martin et al. (1950) thus obtained $s = 1.3'$ between 10^{-7} and 10^{-6} lux, with 0.7 second exposure time. The limit of separation s increases below and above (in the first case because the dark-adapted parafovea is probably utilized, and in the second case for the reason indicated above). The same authors have also studied the threshold with very short exposures as small as 0.005 second.

A theoretical criterion proposed by Airy (1834) is usually adopted for optical instruments: Two points are separated when the maximum of the diffraction image of one point falls in the first dark ring of the other point. In a perfect eye the separation s would be theoretically equal to some 40″ (for a pupil of 4 mm and a radiation of 589 mμ), which is much too small. Yet under these conditions the contrast of the image (i.e., the difference between the maximum and minimum illumination divided by the maximum illumination) is on the order of 0.25. It is therefore very unlikely to be only a matter of intensity discrimination in this case. The theories proposed to explain the limit of separation of the eye will be dealt with in Chapter 6.

Tonner (1943) and, under better conditions, Ogle (1951) studied the case of two identical point sources, each producing on the subject's pupil E lux of illumination and seen on a uniform background of luminance L cd/m^2 (between 0.2 and 200 cd/m^2). As long as E is less than a limiting value E_0 (which is approximately $2 \times 10^{-7} L$), the limit of resolution remains constant ($s_0 = 105''$ with the particular observer of the experiment) and the following expression applies:

$$s/s_0 = (E/E_0)^{0.135}$$

Here again the limit of resolution increases with the illumination, and probably for the same reason as above. (The reasoning of Ogle to explain this result by differential threshold as compared to the background does not seem convincing; see also Ogle, 1962 and [5a].)

The case of small bright squares on a dark background was studied by Berger (1939); the separation between the two edges (not between the centers as before) is 3′ for the sides of the square of up to 40″, and diminishes to 1′ when the sides of the square are 5′. Certainly this decrease must continue, because when the sides reach 1°, it comes back to the case of a line on a white background.

The case of two black dots on a white background studied by Hofmann [23] gives separations on the order of 3′ (between the centers of the dots), up to a diameter of 2′ for the dot. As a function of the luminance of the background, a variation is noted similar to that of intensity discrimination.

Vision of Two Lines

To study experimentally a test made up of two bright lines, parallel and identical on a dark background, the simplest way is to observe a luminous slit through a birefringent system which produces two lines; the separation can be adjusted by rotation of the system. Between two infinitesimal lines, the limit of separation is of the order of 40″ to 1′ under the best conditions (Roelofs, 1918; Wilcox, 1932; Fry and Cobb, 1935). There is a certain optimum level and beyond that level s increases with intensity for the same reason as in the case of points. With regard to the diminution of intensity, intensity discrimination probably plays a more important role in this case than in the case of points. According to Coleman and Coleman (1947), this test arrives more or less at the theoretical values (Airy's criterion).

A point should be mentioned regarding this subject: as Ditchburn (1930) rightly pointed out, trained spectroscopists may disclose an inhomogeneity by the superimposition of two lines very close in the spectrum, even when there is no diminution of intensity in the center of the image, simply because the observer is well aware of the apparent width of a line. Similarly, it would be possible theoretically to recognize a double star by the elongation of the image in a certain direction, although there may not be a relative minimum intensity in the center. When observing two bright parallel lines through a telescope, Selwyn (1943) noted that some subjects pretended to see a lower intensity in the center of the image although it presented a plateau. As Sparrow (1916) pointed out, there is here an effect of contrast and it may imply that even for such simple patterns some psychophysiological phenomena complicate the physical interpretation of the distribution of illumination on the retina.

The same principle applies in the case of two parallel black rectangles on a light background. Fry and Cobb (1935) experimented with rectangles 33′ long and 2.8 or 17′ wide. With the larger rectangles the limit of separation (between the parallel lengths) diminishes regularly as the background luminance increases. Conversely, for thin rectangles there is a minimum separation at about 30 cd/m^2, then a slow increase which is interpreted by most authors as a retinal interaction of one edge upon the other. Similar phenomena occur in Figure 23 (Wilcox and Purdy, 1933), where the concept of separation gives way to one of visibility when the width of the black rectangles increases.

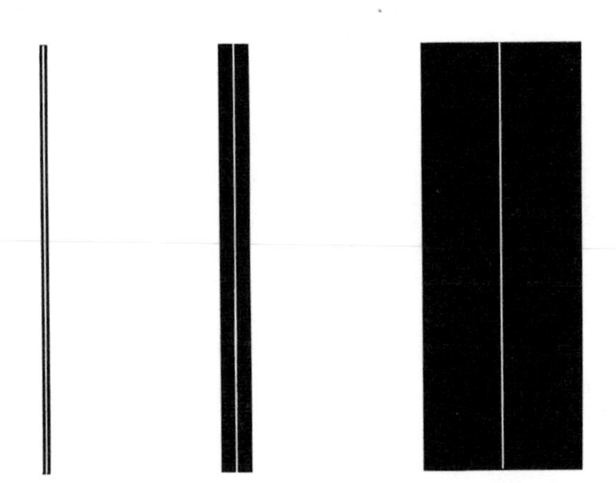

FIG. 23. The separation between the lines and the two pairs of rectangles is the same (0.2 mm). However, if the distance from the observer is increased, the separation on the left disappears rapidly, for the rectangles in the middle more slowly; for the rectangles on the right the criterion is no longer the separation but the visibility of a bright linear source on a dark background and it is much more difficult to see it disappear (Wilcox and Purdy).

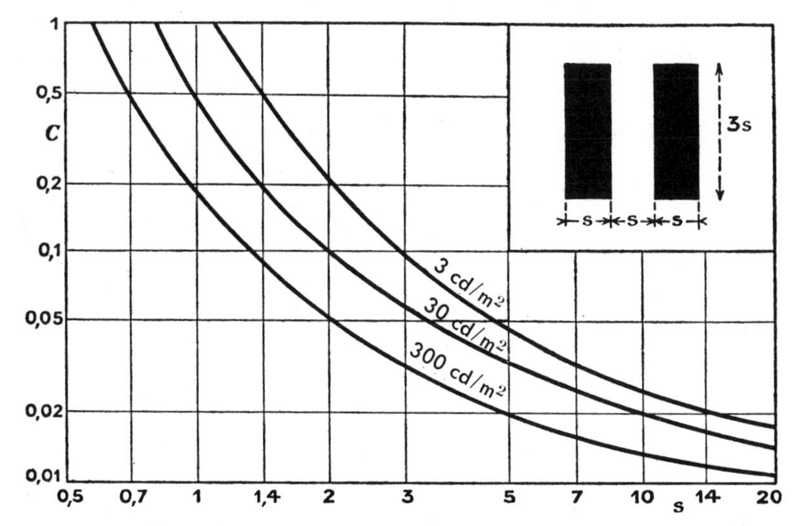

FIG. 24. Relationship between the contrast *C* and the limit of separation *s* (in minutes of arc). The contrast *C* represents the value necessary to perceive the separation between the two rectangles (shown in the upper corner), for three luminances. Both axes represent the log of the values.

Two black or grey rectangles on a white background (Fig. 24) constitute a visual acuity test often utilized. The experimental relationship between the contrast and the limit of separation for three luminances of the background was established with nine trained observers by Cobb and Moss (1928). The curves of Figure 24 were determined with 0.3-second exposure. Reducing the exposure time to 0.075 second has more or less the same effect as dividing the luminance by 10. Luckiesh and Moss (1935) devised a "visibility meter" using this test of two rectangles.

Foucault Grating

The idea of utilizing a grating of black and white bars of equal width as a visual acuity test is very old (Mayer, 1754). The celebrated optician Foucault (1859) applied this test to study the limit of resolution of optical instruments. It is not a very satisfactory visual acuity test because the results will vary a great deal according to the orientation of the bars in cases of astigmatism. Even in the absence of astigmatism, the limit of resolution seems greater (7 to 20%) if the bars are oriented at 45° than if they are vertical or horizontal (Higgins and Stultz, 1948–1950; Leibowitz, 1953). This strange phenomenon persists with momentary illumination (flashes of 10^{-3} second), which eliminates eye movements as a possible cause. It increases slightly with the pupillary diameter, which suggests that part of it at least may be accounted for by some dioptric effect. The luminance hardly alters this phenomenon.

Another drawback of this test is the well-known "false resolution" of microscopists when they look at diatoms. This is due to the fact that a periodic structure may produce a periodic distribution of retinal illumination even when the bars have ceased to be seen individually (refer to Fry, 1961, on this subject).

In France it has long been customary to measure the limit of separation by the angle subtended by the *grating interval* (that is, the distance between the centers of two consecutive black bars). Anywhere abroad, the angle considered is that subtended by the width of a bar, black or white, and we shall adopt this convention henceforth. The only advantage of the French custom is that it applies to gratings in which the ratio of the width of white bars to black bars differs from 1, for Shlaer (1937) has shown that the limit of separation of the grating interval remains approximately constant, varying by only 6% when the ratio varies from 1 to 7. But its serious disadvantage is that it leads to limiting angular values nearly twice as great as those given by other visual acuity tests, such as Landolt rings.

Generally, the grating is made by drawing black bars on white paper and the angle is varied by moving the chart farther away or by rotating it around an axis parallel to the bars of the grating or else by means of an optical system of variable magnification. A test can also be drawn so that the separation of the bars decreases according to a given rule, and only a portion of the test is uncovered (Washer and Rosberry, 1951).

It has been known for a long time that the level of illumination of a chart (Jurin, 1738) and its contrast (Aubert, 1865) are important factors in the perception of details. Recall that the contrast C is defined as the difference between the luminances of the white and black bars divided by the luminance of the white bars; this is actually represented by equation (40) if white is considered to be the background. To produce a variation of contrast, either the gratings can be drawn as grey and white bars, or rotating disks can be used.

A paper by Danjon (1928) can be cited as an example of a study of the limit of resolution s of a Foucault grating as a function of the luminance L of white bars and the contrast. When the angle inceases, the subjective experiences of the subject are: uniform aspect, occasional slight marbling effect in any direction, transient striations parallel to the bars, more precise striations appearing about every second, striations still uncertain but always visible, then black and white bars perfectly steady. Adopting as a criterion the average of the third and fourth aspects, Danjon obtained results that can be represented by the empirical formula

$$2s = 71 + [8 \times 10^{-4} + (C + 0.07)(0.033 \log_{10} L + 0.04)]^{-1}$$

where s is expressed in seconds of arc and L in cd/m² ($10^{-3} < L < 2 \times 10^3$).

In Figure 25 (curve a) are represented the results found by Shlaer (1937) with two trained observers, with a grating of contrast equal to 1. The asymptotic value of s was of the order of 38″ at high luminances (the original data were obtained with an artificial pupil 2 mm in diameter, the scale of the axis of L has been altered in Fig. 25 to apply to a natural pupil). At about 10^{-2} cd/m², there is a kink which corresponds to about $s = 7'$ and indicates the transition between cone and rod vision. At very low luminances, the limit of resolution s becomes of the order of 1°. At about the absolute threshold (10^{-6} cd/m²) $s = \sim 2°$, according to Arnulf [1a].

The effect of the exposure time t has been the subject of some research. Niven and Brown (1944) found that in the interval between 0.19 and 0.004 second variation of t shifts the curve parallel to the axis of the

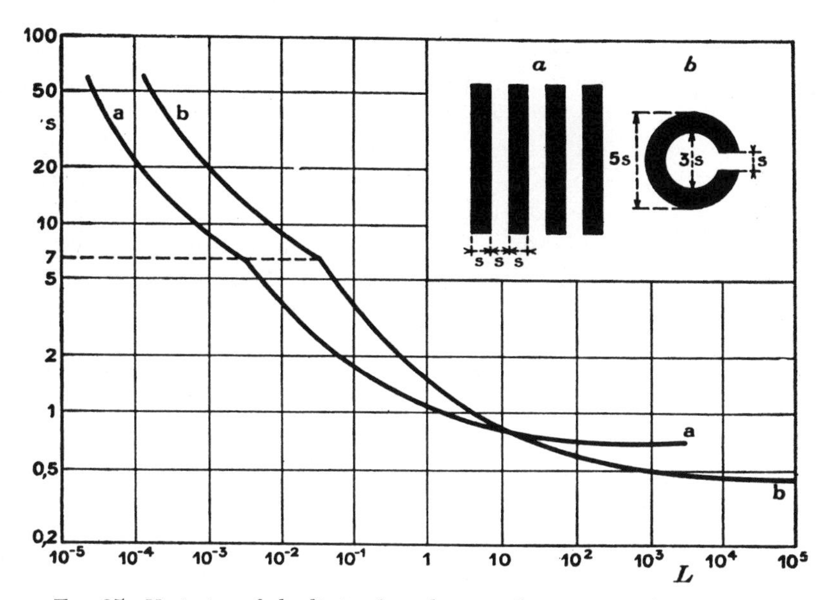

Fig. 25. Variation of the limit of resolution s (in minutes of arc) with the luminance L (in cd/m²) for a Foucault grating (a) and a Landolt ring (b) according to Shlaer ($C = 1$). The scales of both axes are logarithmic.

luminance—that is, diminution of t can be compensated for by an increase of L, more or less proportional if t is small, and then it varies with a slackening similar to Blondel and Rey's law. The results of Martin et al. (1950) arrive at the same conclusion: For $t < 0.1$ second, the shortness of the exposure does not affect s directly; it mainly modifies the apparent brightness. With regard to the critical value of the exposure time beyond which there is no more variation, it would be of the order of 0.18 second according to Graham and Cook (1937), but it may vary slightly inversely to the luminance.

Landolt Rings

Visual acuity tests, which to a certain extent require *form vision* and are usually referred to as *optotypes*, will now be considered. The simplest and most utilized is the Landolt ring (1874), which is represented in Figure 25*b*. It is a complex figure and perception of the gap (which can be presented in four or eight different positions) does not depend only upon this gap. Guillery (1899) and Pergens (1906) have shown that the thickness of the ring diametrically opposite to the gap plays an important part, and Schober (1954) showed that the position of the gap influences the results to a great extent. A test object similar to the Landolt ring is the Snellen hook, in the form of a U, the opening of which can be placed in four different positions (Fig. 26). For these optotypes it is common to specify the limit of resolution not by the angle s but by the *visual acuity* V defined as the reciprocal of the angle s in minutes; the acuity is therefore a greater number the smaller the test. An acuity of 1 corresponds by convention to $s = 1'$, but no physiological or statistical significance should be attached to this value. Another visual acuity scale, proposed by the U.S. Army [78] and defined by the quantity $10 - \log_2 s$, is not very often used.

The first classical study of the variation of V with the luminance L of a white background on which there is a black test is the one of Uhthoff (1886–1890), completed by König (1897), with the Snellen hooks. The luminances have again been calculated by Hecht (1928), who showed that when V is plotted against log L a simple graph is obtained (Fig. 26) —two lines, corresponding to photopic and scotopic vision, intersecting each other at about $V = 0.16$. König and Uhthoff had actually confirmed that rod monochromats, that is, those who have rod vision only, show a curve of visual acuity which prolongs the scotopic line toward high luminances with an asymptote at about $V = 0.20$.

If the luminance increases greatly, V tends toward a plateau, or at least it increases only very slightly, as was shown by the measurements

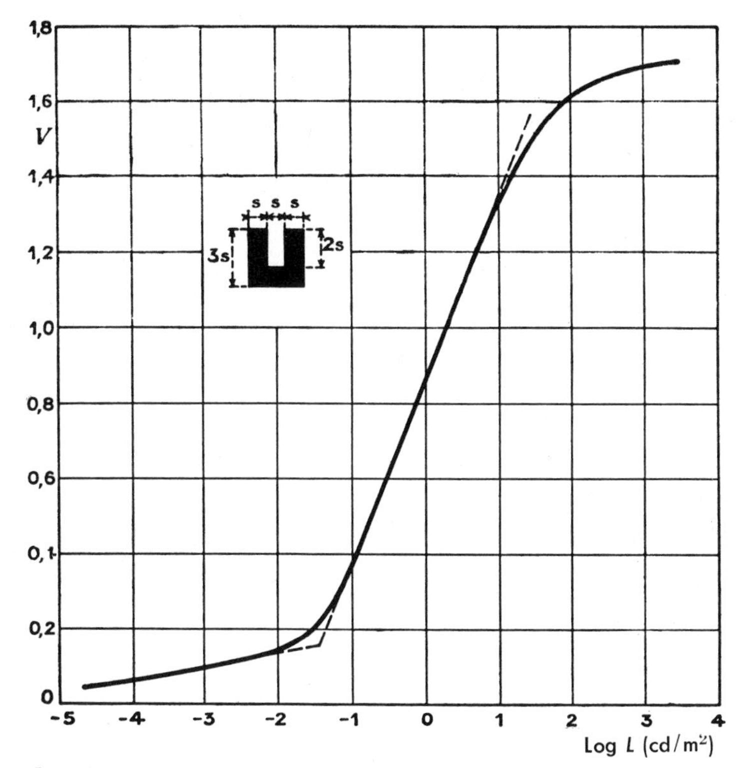

FIG. 26. Variation of the visual acuity V with the luminance, according to König-Hecht.

by Lythgoe (1929–1932), Shlaer (1937; see Fig. 25*b*), and Siedentopf et al. (1941). Moon and Spencer (1944) have condensed all results into an empirical formula which is a satisfactory representation of the variation of acuity in the photopic region ($L > 0.1$ cd/m^2) assuming contrast equal to 1 and natural vision:

$$V = V_\infty L\,(0.28 + L^{1/3})^{-3} \tag{42}$$

L is in candelas per square meter. The asymptotic acuity V_∞ depends upon the subject, as will be seen in Chapter 6, but Crouch (1945) found for the best eyes $s = 24''$, that is, $V_\infty = 2.5$.

In the range 0.06 to 40 cd/m^2, the following law, proposed by Lythgoe, is much simpler:

$$V = a + 0.49 \log_{10} L \tag{43}$$

where a is a constant which depends upon the subject.

TABLE 10

Visual Acuity Measured with Landolt Rings as a Function of Luminance L and Contrast C

	L, cd/m^2							
C	4.5×10^{-4}	3.4×10^{-3}	3.4×10^{-2}	6.9×10^{-2}	0.15	0.34	1.1	3.4
0.929	0.055	0.114	0.335	0.459	0.638	0.741	0.850	0.965
0.762	0.044	0.089	0.268	0.404	0.508	0.657	0.715	0.863
0.394	0.030	0.055	0.194	0.267	0.370	0.443	0.538	0.632
0.284	0.023	0.041	0.131	0.198	0.291	0.359	0.414	0.584
0.155		0.025	0.071	0.105	0.160	0.198	0.258	0.335
0.096			0.040	0.064	0.114	0.125	0.158	0.203
0.063			0.034	0.054	0.087	0.119	0.141	0.168
0.040				0.036	0.039	0.048	0.060	0.085
0.018					0.028	0.033	0.046	0.071

In the scotopic range, the variation of acuity is much slower, according to Roelofs and Zeemann (1919), and proceeds regularly until the absolute threshold is reached. According to measurements by Hamburger (1949) on 150 young subjects, the value $s = 1°$ ($V = 1/60$) is reached between 5 and 8×10^{-6} cd/m^2. Pirenne et al. [53b] have also measured the acuity of 20 young subjects down to the absolute threshold.

Measurements of visual acuity at low luminances may differ greatly according to whether night myopia is corrected or not, as was shown in experiments by Jiménez-Landi and Cabello (1943), Ronchi (1943), Schupfer (1944), Katz (1945), and others. Generally, improvement by the correction is maximum at about 10^{-3} cd/m^2; the improvement is least as the threshold is approached, whereas the night myopia increases (on the average, -1 diopter at 10^{-3} cd/m^2 and -2 at 3×10^{-5}, according to Morris, 1953).

The effect of contrast on visual acuity has been studied particularly by Houstoun and Shearer (1930) and by Conner and Ganoung (1935), using Landolt rings. Table 10 shows the results by Conner and Ganoung, who carried out their experiments on seven subjects under the following conditions: white light (2760°K) used as the standard source in the calculation of luminances; total adaptation at each level; 3 seconds exposure time, experimental threshold corresponding to 66% correct responses (the gap placed in eight different positions).

The findings of Fortuin [18] complete those above for high luminances ($1.3 \leqslant L \leqslant 1300$ cd/m^2). These results, which were taken on 228 emmetropic subjects, can be well represented by the following empirical formula:

$$\log V = g + 2.17\frac{\log C - 1.57}{\log L + 3.96} \tag{44}$$

The coefficient g varies between 0.5 and 0.9, depending upon the subject.

The effect of exposure time t has given rise to some important studies, but as the goal was to determine the best conditions of illumination, they will be discussed mainly in Chapter 17, and only the curves by Ferree and Rand (1927–1928) will be mentioned now. They were established with a Landolt ring of 0.94 contrast and a trained observer; the gap was presented in eight different positions and the threshold corresponds to 60% correct responses (Fig. 27). When t is small, Durup and Fessard (1929) found that visual acuity remains constant if the apparent brightness is kept fixed and is about two thirds the acuity obtained for a prolonged exposure at the same apparent brightness. Other investigations in this field were also carried out by Graham and Kemp (1938), Keller (1941), Nachmias (1958), and Gibbins (1961).

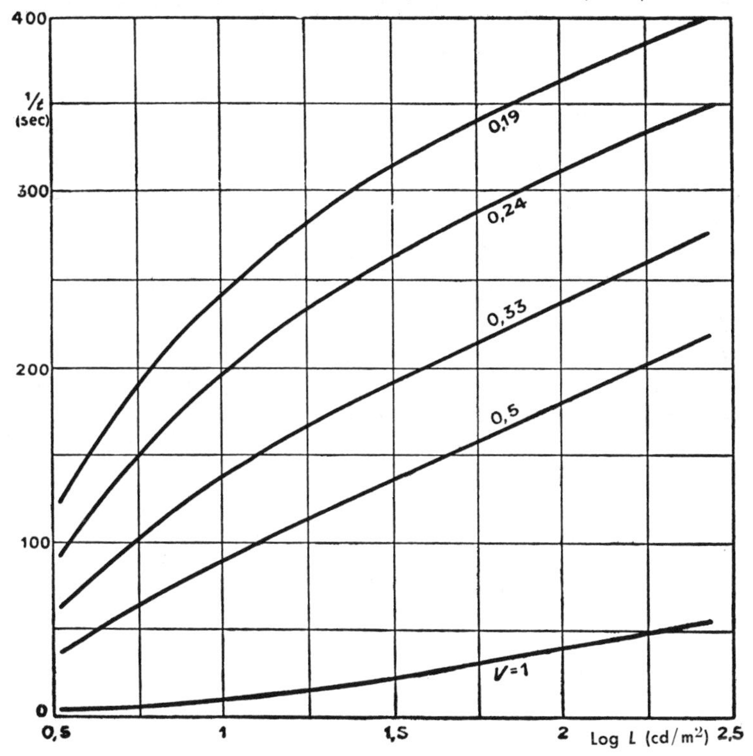

FIG. 27. Minimum exposure time t of a Landolt ring as a function of luminance L (Ferree and Rand).

Other Optotypes

A good number of optotypes have been designed and utilized to determine visual acuity. For instance, Daza de Valdès (1623) had his clients select spectacles by counting grains of mustard in a line. A similar type of test was designed by Burchard (1882), but the discrimination of the number of points in a figure involves complex processes of eye movements and perceptive integration (Ancona, 1952); it is therefore not a very good retinal criterion. Goldmann (1943), to determine visual acuity in some pathological cases, devised a test consisting of black and white dots regularly distributed. This test was improved by Bouman et al. (1951).

The principle of the checkerboard pattern proposed by Mayer (1754) and Goldmann (1943) has been utilized by Giese (1946) and Morris et al. (1955). Very careful measurements at low luminances were made by Morris and Dimmick (1950) using a checkerboard of 0.98 contrast subtending a total angle of 9° (the center was at 10° eccentricity from the fixation point); the squares of the checkerboard varied between 8 and 175′ and the test could occupy one of four different positions (top, bottom, right or left), the other three being occupied by checkerboards having much smaller squares and consequently seen as a uniform greyish area (Fig. 28*a*). By the method of constant stimuli, six subjects measured the threshold for 50% correct answers and the visual acuity could be represented by the following law:

$$V = 0.033 \log_{10} L + 0.19 \tag{45}$$

where the luminance L is expressed in cd/m² (white light, 2750°K). The measurements were made in the interval $1.6 \times 10^{-5} \leqslant L \leqslant 1.6 \times 10^{-2}$, but at 5×10^{-3} cd/m² some signs indicated the beginning of cone vision.

The use of printed letters is ancient and still very prevalent in optometry. The first charts, in gothic letters of decreasing size, were published by Kuchler in 1843 [18]. Snellen (1862) adopted the convention of a visual acuity equal to 1 for a letter of 5′ with details of 1′ (Fig. 28*b*). He simply mentioned that such letters are "easily seen by most normal eyes." All optometrists know that not all capital letters are recognized equally. The subject's experience is also an important factor: Kreiker (1928) obtained an acuity 2.3 times better with the same observer using letters rather than unknown figures of an equal limit of resolution. Therefore the notion of the *minimum legibile* (Giraud-Teulon) is not very meaningful in spite of its practical importance. The conventions underlying the drawing of the letters play a great part in their legibility,

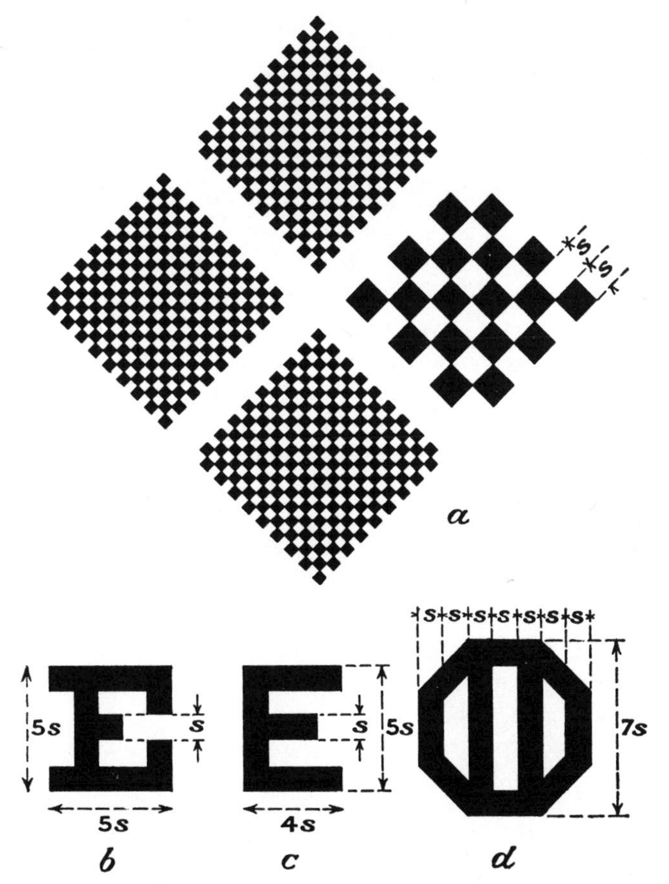

FIG. 28. Examples of various visual acuity tests: *a*, checkerboard; *b* and *c*, letters; *d*, ISO character. Visual acuity is defined in each case by the reciprocal of the limit of resolution *s*, in minutes of arc.

and it is obvious, for instance, that the letter E is easier to recognize in *c* than in *b* (Fig. 28). Berger (1948–1950) showed how the legibility of letters and numbers varied with their height and width (as a function of the thickness of the line). Lucia Ronchi (1950–1951) established that legibility was maximum when the line was equal to one fifth of the small side of the rectangle in which the letter is contained. She also classified several letters according to the variation of their legibility with the distance. Finally, she established (1952) that white characters on a black background were (depending upon the subjects) equally visible or less visible than black characters on a white background but never more visible.

The International Organization for Standardization (ISO) proposed a new test in 1954 designed to check the resolution of instruments used to read microfilms. This "ISO character" is an octagon crossed by two black bars which can be placed in four different directions (vertical, horizontal, 45° to the left or right; Fig. 28*d*). This test was compared to the Landolt ring by Aguilar (1955) with 20 observers. The variation with luminance and contrast is more or less the same for the two test objects, and, for the best observers, values of *s* are comparable, but if there is astigmatism the ISO character is more difficult to discriminate.

The comparison of several visual acuity tests has given rise to many studies. Monjé and Schober (1950) have experimented with several types of letters, numbers, and Landolt rings on 100 subjects. The most complete study comparing these tests [78] concluded that the checkerboard test is superior, as it seems to be a sufficiently objective criterion and less sensitive to astigmatism than other tests. A photographic study by Perrin and Altman (1953) (although based on a criterion which is not only visual) came to the following conclusion: if a visual acuity of 1 is obtained for the Foucault grating, the acuity for the Landolt ring is on the average 0.91, for the Snellen letter *E*, 0.87, and for Cobb's double rectangle, 0.76, but these comparisons apply for a given level of luminance. When the luminance varies, these ratios may be reversed, as in Figure 25. It is therefore very difficult to formulate a general comparison of visual acuity tests; each has its own laws. At very low luminances near the absolute threshold, according to Dratz (1945), the acuity with a Snellen letter is 0.62 times that obtained with a Foucault grating.

CHAPTER SIX

The Basis of Visual Acuity

We studied in Chapter 5 several visual tests of the vision of details under natural conditions of observation and in white light. In order to search for the theoretical basis of visual acuity we must now increase the number of variables (so far we have considered only luminance, contrast, and exposure time).

Pupillary Diameter

The variation of pupillary diameter plays an important role, which was suggested by Lister (1842), unraveled by Hummelsheim (1898) and Cobb (1915), and studied systematically by Arnulf [1a]. Arnulf used gratings of contrast C and variable luminance L with four subjects. If $L > 1$ cd/m^2, the limit of resolution s remains constant if the pupillary diameter d is greater than 2 mm and if $C > 0.1$. For low contrasts there is an optimum around $d = 2$ mm at high luminances and it shifts up to $d = 6$ mm when $L = 1$ and $C = 0.02$. If, on the contrary, $L < 0.1$ cd/m^2, there is an improvement of the acuity (diminution of s) as d increases.

For a reason that will become clear later, Arnulf presents his results by giving the *specific limit of resolution* defined by the expression

$$\sigma = sd \tag{46}$$

With luminance and contrast fixed, σ passes through a minimum at about $d = 0.5$ mm for large L and about 1.5 mm for low L and C (Fig. 29).

It seems that these results did not get across the Atlantic, for several American authors repeated the same experiments. Byram (1944) also obtained a minimum of σ around $d = 0.5$ mm, with $C = 1$. Coleman et al. (1949) found for $L = 1700$ cd/m^2 and $C = 0.94$ that the specific limit of resolution is independent of the observer (on 32 subjects of different ages) up to $d = 0.75$ mm.

It could be objected that all these experiments use a given luminance instead of a constant retinal illumination (since the pupil size varies). To avoid this criticism, Leibowitz (1952) used a different method. By binocular photometry, the subject matched the apparent brightness of the test seen through an artificial pupil of variable diameter d with that of a field of luminance L seen by the other eye through a fixed pupil 2 mm in diameter. Table 11 contains the mean results of two subjects with a grating of contrast 1. (The visual acuity 1 corresponds to a grating with equal white and black bars, each subtending an angle of 1′.)

In Maxwellian view, that is, when only a small part of the pupil is utilized, acuity varies with the point of incidence in the eye because of the heterogeneous nature of the crystalline lens. This phenomenon has been known for a long time (Le Grand, 1936) and was christened the "Campbell effect" by Dunnewold (1964).

Adaptation

It is a common experience that when one enters a dim room, the adaptation of the eye to darkness renders the eye susceptible first to

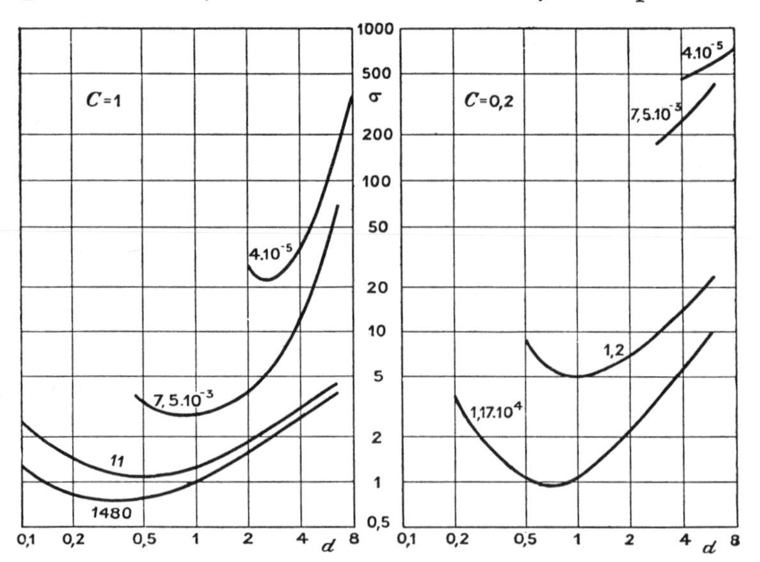

FIG. 29. Relationship between the specific limit of resolution σ (in mm $\times$ minutes) and the pupillary diameter d (in mm) for one of Arnulf's subjects for several luminances (indicated in cd/m² along the curves) and two different contrasts C. The scales along both axes are logarithmic.

TABLE 11

Variation of the Visual Acuity with the Pupillary Diameter d *at a Constant Brightness (Matched with the Luminance L Seen through a Pupil 2 mm in Diameter)*

	d, mm						
L, cd/m²	1.0	1.4	1.6	2.0	2.77	3.86	4.75
318	0.94	1.30	1.50	1.75	1.93	1.99	2.00
31.8	0.91	1.22	1.41	1.69	1.88	1.92	1.83
3.18	0.81	1.06	1.23	1.46	1.52	1.49	1.39
0.318	0.72	0.95	1.03	1.15	1.18	1.12	1.07
0.0318	0.53	0.59	0.64	0.67	0.70	0.63	0.60

faint impressions of light, then to forms. In the same way as there exists a photochromatic interval between the absolute threshold of light and that of the perception of color, there also exists a similar interval between the absolute threshold and the attainment of a given acuity. In both instances the laws are very similar.

This phenomenon was studied by Hecht et al. (1937) using a black cross as a test; by Craik and Vernon (1942) with an arrow on a dial; by Miles (1943) with the silhouette of an airplane; and by Wolf and Zigler (1950) and Marshall and Day (1951) with a grating. The most complete study is that of Brown et al. (1953). Their subjects were previously light adapted for 5 minutes to a square 10 cm in width, at a distance of 30 cm from the subject and a luminance of 4800 cd/m². Then the subjects were placed in front of gratings of contrast 1, seen at a distance of 57 cm in order to compensate for night myopia. The grating was seen in a total field of 7° diameter with a red fixation point in its center. The minimum luminance L of the white bars of the gratings necessary to obtain a given acuity was determined as a function of the time t of presence in the dark, counting from the end of the light adaptation. This luminance L appeared in flashes of 0.016-second duration, through an artificial pupil 3 mm in diameter. Figure 30 shows the results for one of the subjects. For high visual acuities ($V > 0.15$, only cones function and the adaptation curve is of the photopic type. For low acuities the rods commence functioning after some 10 minutes, and the curve takes its classical slope with the transition between the photopic and scotopic regions. One can note (for $V = 0.62$) an irregularity which often occurs after 4 to 8 minutes adaptation and which has already been pointed out by Ivanoff (1948). It is possible that the course of adaptation of the rods, even if they are not used effectively, inhibits cone vision

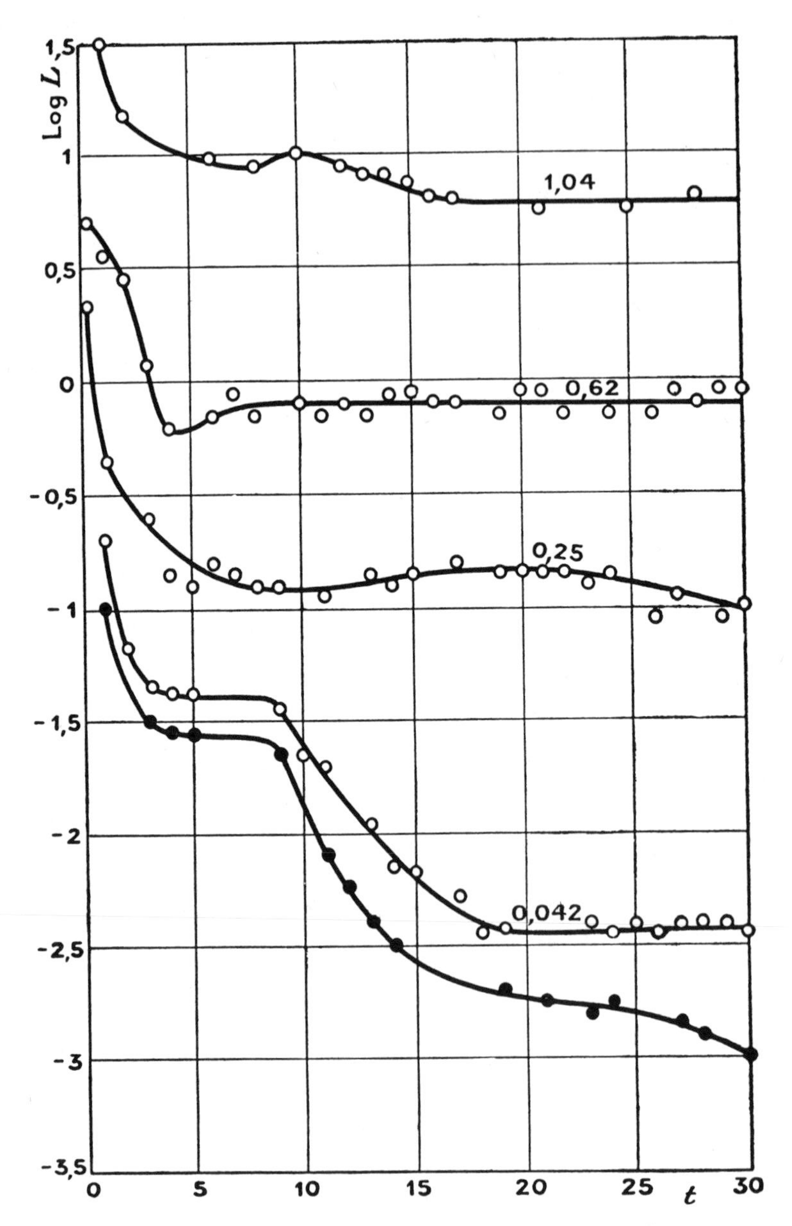

FIG. 30. Threshold of luminance L (in cd/m²) necessary to obtain a given acuity (indicated along each curve) as a function of the dark adaptation time t (in seconds). The lower curve (black circles) represents the absolute threshold, without the grating.

to a small extent; the subject feels at that moment a rather peculiar impression of discomfort. Mote (1955) noted, at about the seventh minute of dark adaptation, a large dispersion of the absolute threshold on three subjects.

Brown (1954) completed this study by varying the intensity of the light adaptation between 3.1 and 35,600 cd/m², during 5 minutes. For cone vision, with each acuity the curve simply moves in a direction parallel to the axis of luminance, but it is less obvious for rod vision (see Brown, 1962).

If the retinal adaptation is imperfect the acuity is reduced. This explains why the first observers believed in an optimum level of luminance beyond which the acuity diminishes. It only diminishes if the test is too bright as compared to the surrounding visual field. According to Lythgoe (1929), if the foveal test is surrounded by darkness the maximum acuity (1.75) is obtained around 50 cd/m². If the background surrounding the test has a luminance of 0.04 cd/m², the maximum (1.95) is obtained around 200 cd/m². Similarly, Fisher (1938) noted that for a test of 2° the acuity increased or decreased with the dimensions of the surrounding field, whether the latter was of the same order of luminance or much greater than the luminance of the test. In this instance there is glare (see Chapter 17). The influence of the surrounding field has also been studied by Ronchi (1949) and Foxell and Stevens (1955). The latter employed a Landolt ring of which the diameter subtended 0.5° and was surrounded by a background 6 to 120° in diameter. For the low luminances of the test, the background has no effect except when its luminance is much larger than that of the test. But when the luminance of the test is greater than a few candelas per square meter, the best acuity is obtained for a background of about equal luminance; the enlargement of the field beyond 6° has very little effect. The best acuity would be obtained around 3000 cd/m² and would decrease slightly beyond that value. This result may be due to an imperfect adaptation; Eguchi (1931) observed a continuous improvement of acuity up to the highest luminance used (250,000 cd/m²) if the whole field is of equal luminance and the subject is well adapted progressively.

Color

The first measurements made by Macé de Lépinay and Nicati (1881–1883), confirmed later by König (1897), showed that at equal luminance, lights of different colors produced approximately the same acuity, so that the criterion of equal acuity has often been suggested for heterochromatic photometry. Of course, in scotopic vision, the Purkinje phenomenon must be taken into account or the comparison becomes meaningless. However, König, then Roaf (1930), proposed an exception for blue light, which in their opinion gave a lower acuity than white and other colors. This idea, which is still often expressed, is founded on two basic concepts: first, the usual weakness of the luminance in blue light (artificial sources generally radiate little energy in the blue, filters which isolate the short wavelengths have a large absorption, and relative luminous efficiencies are low in the short wavelengths), and second, the myopia produced by chromatic aberration in the blue and violet (Bell, 1911). But if all precautions are taken to assure a constant level of luminance and to be exactly in focus at each trial, the measurements of Arnulf et al. (1950) prove that between 2 and 4 mm pupil diameter all wavelengths (including 405 mμ) give approximately the same acuity, which is also equal to that obtained in white light.

It is only with thorough statistics that an effect of color on acuity can be demonstrated and those that we know of are not very conclusive. The most prevalent opinion (Ferree and Rand, 1931; Conner and Ganoung, 1935; Shlaer et al, 1942; and others) is that white light gives better acuity than all monochromatic lights, except perhaps yellow light, whereas red is least satisfactory. However, Martin and Pearse (1947) found on the contrary an acuity 4% better with red than with white at a luminance of about 5 cd/m^2 in near vision.

The assertion, often advanced, that to cut out the short wavelengths by means of a yellow glass will improve acuity by reducing the effects of chromatic aberration does not seem proved either. In spite of a subjective impression favorable to yellow light, the statistics (Richards, 1953; Blackwell, 1953) give the same acuity if the luminances are equalized (in the interval 0.03 to 30 cd/m^2) and there is a slight deterioration of the acuity with yellow if the filter is simply interposed in front of the source or the eye. With some subjects, however, the measurements of Pagès (1954) indicate a slight superiority of yellow light.

In short, the problem of acuity in colored light is extremely confused. The only point on which authors seem almost agreed is a slight superiority, at equal luminance, of yellow monochromatic light, as, for example, the light emitted by sodium vapor lamps, over other monochromatic radiations and even over white light (Luckiesh and Moss, 1933; Arndt, 1934; Klein, 1934; Weigel, 1935; Bouma, 1936; Netusil, 1941). The speed of discrimination of details would also seem in favor of yellow light (Ruffer, 1928).

An interesting problem on which there are still few data is the determination of visual acuity with two colors on a chart, instead of white and black. One could, for example, use optotypes drawn in blue on yellow paper, or a grating in which the bars are red and green. A report from the Eastman Kodak Company on the effect of color contrast in camouflage deals with this question [76]. The limiting case of pure chromatic contrast (object and background of equal luminance) has given rise to some investigations, using Landolt rings, by Fazakas (1928), Koffka and Harrower (1931), MacAdam (1947), and Foxell and Stevens (1955). Acuity seemed slightly less than with white and black figures.

This subject has also attracted color television technicians. Bedford (1950) measured acuity on four subjects with a test similar to a grating in white and black, red and green, red and blue, and green and blue. He obtained the following relative values: 1, 0.40, 0.23, 0.19. But it has been pointed out (Le Grand, 1951) that whereas the luminance of white was 70 cd/m², that of red and green was 20 cd/m² and that of blue only 2 cd/m². Moreover, the test was surrounded by a background of 34 cd/m² luminance, which decreased the acuity in the latter case.

By projecting gratings on a screen, Boutry et al. (1954) obtained on four observers the acuities shown in Table 12.

With a grating illuminated by two monochromatic lights, Cavonius and Schumacher (1966) found the same acuity as with black and white, providing the two wavelengths were far enough apart in the spectrum.

With pure radiations there is some difficulty due to chromatic aberration of the eye, which by shifting the image of the test on the background may replace the chromaticity contrast with a luminance contrast. To remedy this drawback, O'Brien and Miller (1952) proposed making the test of two overlapping systems of interference fringes on the retina itself, with two monochromatic radiations having wavelengths in a simple ratio such as 5:4. This problem of chromatic acuity has also interested the theorists of color vision because it is capable of providing information on the retinal distribution of various types of receptors, for example, the three types of cones sensitive to red, green, and blue that are postulated by Young's theory. But under normal conditions even with monochromatic tests, at least two types are stimulated simultaneously and the practical equality of acuity for all wavelengths does not

TABLE 12

Visual Acuity for Colored Gratings of Luminance L Seen at a Distance Δ

L, cd/m²	Red–black	Green–black	Blue–black	Red–white	Green–white	Blue–white	Red–green	Green–blue	Blue–red
1	1.2	1.2	1.0	0.9	0.9	1.0	1.2	1.2	1.3
9	1.5	1.5	1.3	1.3	1.2	1.3	1.5	1.5	1.5
30	1.8	1.7	1.5	1.4	1.3	1.5	1.7	1.7	1.7
Δ, m	1.7	1.7	0.6	3	3	3	3	3	3

prove that the three types possess identical distributions. It is known, for example, that fundamental blue has an intensity discrimination some five times less than fundamental red or green (Stiles, 1946) and this does not prevent the differential threshold being practically independent of the wavelength. The same reasoning could apply to acuity.

There exist, however, particular conditions which permit the isolation, almost complete, of one of the types of receptors, for example, the artificial monochromatism induced by an adaptation to very intense colored sources (Brindley, 1953). In the case of violet monochromatism produced by observation for 10 seconds of a yellow source (578 mμ) of luminance over 40,000 cd/m^2, the limit of resolution of a grating is as high as 12′. Or one can consider the case in which the threshold of one color is projected onto a background of a different color (Stiles, 1949; Brindley, 1954) and here again the isolation of the "blue" receptors leads to poor acuity (7.5′). This does not necessarily mean that the "blue" cones are more rare than others. It seems more likely that there is a different convergence (on the bipolars and ganglion cells).

Other Factors

By measuring acuity with optotypes illuminated by the spark of a Leyden jar, Aubert and Förster (1857) noted that in peripheral vision small characters were more easily discernible at near than large ones at distance, with characters of the same angular size. In foveal vision Biedermann (1927) and Freeman (1932) observed the opposite effect; that is, acuity is slightly better at distance than at near. It is essential in these measurements that the background on which the optotype appears subtends a constant visual angle (Musylev, 1937). With seven subjects, Luckiesh and Moss (1941) obtained on the average an acuity of 1.33 at 10 m and 1.13 at 50 cm. With checkerboards, Giese (1946) obtained with a luminance of 250 cd/m^2 on 500 young people the following mean values of acuity: 1.32 at 10 m; 1.61 at 5 m; 1.63 at 1 m; 1.40 at 50 cm; 0.94 at 20 cm. At very low luminances, Rouleau (1934) and Dratz (1945) also noted a decrease of acuity at very small distances although because of night myopia the opposite could have been expected.

This peculiar phenomenon depends much upon the observer, according to the measurements of Blackwell (1948), and can even disappear or reverse from one subject to the next. The physical factors (variation of pupillary diameter and accommodation) do not seem very important in central vision. In peripheral vision the dioptrics of the eye may be important. The influence of ametropia and its correction was studied by Sloan (1951) and by Ronchi and Zoli (1955). Many psychologists have, of course, taken much pleasure in tackling this problem. Jaensch (1922) suggested a "shrinking" effect of the attention directed toward near objects and MacFadden (1940) noted that even a change in perspective of the surroundings modifies acuity by varying the apparent distance of the test.

It has long been known that binocular vision gives a slightly better acuity than monocular vision. According to the very recent determination by Campbell and Green (1965), the ratio of binocular to monocular acuity is $\sqrt{2}$. Barany (1946) showed that this is due to a statistical summation of two independent receptors. Improvement of acuity when the dimensions of the test vary can, at least beyond a certain value, be due to the same cause and so may the interaction noted by Travis and Martin (1934) between two objects projected on the same retina, nasally and temporally. Let us note that there is always a reduction in acuity of an isolated target when it is surrounded by black and white contours (see, for example, Flom et al., 1963). This difficulty of separation becomes very important with amblyopic subjects (Avetisov, 1965).

Poor physiological conditions are, of course, detrimental to acuity. For example, lack of oxygen (MacFarland and Halpern, 1940); prevention of blinking of the eyelids, which alters the cleanliness of the cornea (according to Newhall, 1935, there is a rapid decrease in acuity of about 10%, then a progressive and slow decline which after 20 minutes produces an additional reduction of acuity of 5%); fatigue due to a flickering light (Feilchenfeld, 1904) and, to a slight degree, visual fatigue [according to Peckham and Arner (1952) the decrease in visual acuity of 24 truck drivers who drove for 6 hours at a time in the sun-drenched deserts of Arizona without protective lenses was hardly noticeable].

On the other hand, acuity can be improved by respiratory exercises, by increase of oxygen in the air, by ingestion of strychnine or helenin (Hamburger, 1949); it is mainly the scotopic acuity which thus improves. The influence of mydriatics was studied by Ronchi (1955). Experience is an obvious factor, of which Walls (1943) gives an amusing example. On a printed list of names placed at a distance, one reads the familiar first names easily, whereas not one last name can be distinguished.

Finally, there are very marked individual variations among subjects. Age is an important factor, as was established by de Haan (1862) and Galton (1885). The visual acuity of a newborn baby seems very poor as far as can be assessed. Acuity is maximum in the child about 10 to 15 years of age, often reaching $V = 2$ or even $V = 2.5$. Around 20 years of age, with Landolt rings, the best emmetropic subjects still obtain $V = 1.5$. It is around 50 to 60 years of age, according to Boerma and Walther (1893), the $V = 1$ is obtained on the average; that is the so-called "normal vision," the 20/20 of optometrists! Then acuity still decreases, owing to sclerosis of the crystalline lens and reduction in pupil size. According to Bordier (1893) a man possesses in his youth sight slightly better than a woman but the effect of age is less pronounced in a woman, who thus overtakes her companion in old age.

From the statistical study by Fortuin [18] coefficient g of formula (44) is seen to vary with age, as is shown in Table 13. Fortuin proposed the term *visual power*, G, defined by log $G = g$, and he called the unit "snellen." In short, it comes to (A being the age)

$$G = 9 - 0.1A$$

From another statistical study by Chapanis (1950) it has been found that maximum acuity is found at about 20 years of age and then reduces first slowly up to 50 years of age, then more rapidly. Life in a submarine also slightly affects acuity, according to Schwartz and Sandberg (1954). On the other hand, it is often noted that ametropia, even perfectly corrected, reduces acuity. According to Gilbert and Hopkinson (1949) variations of luminance of the test affect the subjects with poor acuity more than the normal ones. This justifies the use of low levels of illumination to reveal low ametropias and particularly astigmatism. Conversely, to detect acuities lower than normal, high luminances of the order of 1000 cd/m^2 [1b] must be used. With young subjects, statistics give the standard value of acuity of 1.4 for two thirds of the subjects, whereas three quarters stand in the region between 1 and 1.4 and only 6% below unity. This applies to well-corrected subjects. Optimum correction shifts acuities below 1 to values above 1 in about 17% of corrected ametropes without pathology and below 1.4 to above 1.4 for the same percentage of subjects. Young people therefore must always be advised to have a thorough visual examination if the acuity is below 1.4.

Rowland [76] found a high correlation of acuity among subjects at luminances of 32 and 0.12 cd/m^2 but lower at 10^{-2} cd/m^2. In scotopic

TABLE 13

Variation of Visual Power with Age (Fortuin)

Age group	Mean age	Number of subjects	g	G
5–9	8.2	27	0.88	7.6
10–14	11.7	48	0.91	8.1
15–19	16.8	37	0.86	7.3
20–24	22.2	34	0.79	6.2
25–34	28.8	29	0.81	6.5
35–44	38.8	26	0.75	5.6
45–69	53.8	27	0.53	3.4

vision, Beyne and Worms (1931) found hardly any correlation with photopic acuity. There is, however, a high correlation between scotopic acuity and the absolute threshold (Ogilvie et al., 1955).

The effect of age is more marked at night than in daytime. Around 4×10^{-4} cd/m^2 after 20 minutes dark adaptation, the acuity of a young subject is between 0.06 and 0.09, whereas after 60 years of age it is around 0.04. According to Morris and Dimmick (1950), at low levels of luminance pupillary variations constitute the principal factor of individual differences in acuity and it is known that the pupil reduces with age. The maximum diameter beyond 50 years of age is less than 6 mm, whereas it is more than 8 mm in youth; this reduces the retinal illumination by half.

There are very few data on the variation of acuity as a function of race. Colored people are often believed to have an advantage over whites, perhaps because of their heavier choroidal pigmentation (Helson and Guilford, 1933), but an official American statistical study of 7000 subjects arrives at the opposite conclusion [79]. With regard to traveler's reports describing extraordinarily keen vision in certain natives, they should be treated with caution (Cohn, 1895, pretended to have found $V = 8$ in an Egyptian!). The highest visual acuity scientifically observed is of the order of 3 (Banister, Hartridge, and Lythgoe, 1926) and the "records" of natives indicate more a training to identify things at distance than an acuity as such. The legend according to which our ancestors would have had better sight than we have is contradicted by the description of the stars in the sky seen with the naked eye, which has not changed since antiquity.

To end this section it is perhaps not superfluous to recall an important cause of the variation of acuity—the technique of measurement. Different tests and optotypes give different results, as we have seen previously. And for a given test—say, Landolt rings—one can still obtain different acuities even at constant contrast and luminance. Thus the "clinical" acuity defined by the correct reading of a line of 12 letters is for a young emmetrope less by 0.5, on the average, than the "statistical" acuity defined by 50% correct answers regarding orientation of the ring (Barany, 1946). Therefore it is safe only to compare acuities which have been obtained with the same technique. Even in this case there would be diurnal variations of acuity in any given subject, according to Schupfer (1943).

Theories of Acuity

The description of facts in this chapter and the previous one suggests that acuity is a complex function involving dioptric, retinal, and perceptive processes. Depending upon the tests utilized, these factors take on a different relative importance, and there is very little hope that any scheme may include, even grossly, the bulk of experimental facts. Many attempts, however, have been made and we would like to summarize succinctly the fundamental concepts that have been advanced on this matter.

Continuous Theories

By this term we refer to the theories which voluntarily neglect the discontinuous receptive structure of the retina. The retina is therefore assumed to be a sensitive continuous surface, gifted with a certain intensity discrimination—that is, it can distinguish two contiguous fields when their retinal illuminations differ, in relative value, by a certain amount (differential threshold), this amount varying with the region of the retina which is stimulated. This hypothesis is in fact nothing but an extrapolation at the limit of the properties of contrast perception as they are provided by experiments with extended surfaces.

An obvious difficulty of this theory is that the spread of illumination on the retina is also continuous. There are no uniform areas but a more or less rapid variation of the illumination. The differential threshold should give room to the notion of *gradient* of retinal illumination. It is true that this notion can also demand a large experimental study of its own (contours between wide fields), as we shall see in Chapter 7.

We have previously seen that a continuous theory explained satisfactorily the vision of points or linear sources. Helmholtz, then Hartridge (1922), attempted to extend it to more complex tests, as for example a grating. But here the continuous theories stumble over insurmountable difficulties.

Let us consider a grating observed by a perfect eye (free of aberration). The retinal image is then a reduced grating but with the image contrast lower than the object contrast and with a continuous variation of illumination instead of the abrupt variation of luminance on the grating. Calculation [1] shows that the image contrast is nil when the visual angle u of one bar (black or white) is 0.41 times the angle corresponding to the radius of the Airy disk. With an object contrast $C = 1$, the image contrast will become distinguishable as soon as u is greater than the preceding value. The image contrast is 0.02 for 0.42 times the radius of the Airy disk. As we have already seen in the study of vision of a line, a retinal contrast of less than 0.02 permits the recognition of a line, so we will take this value.

If we use equations (11) and (20), we arrive at a value of the visual angle u which corresponds to the threshold and is nothing but the limit of resolution:

$$s = \frac{0.42 \times 20.3}{4.85} \frac{\lambda}{d} = 1.76 \frac{\lambda}{d}$$

where s is expressed in minutes of arc, λ in microns, and d in millimeters (mathematicians will cover their faces before this orgy of units . . .). Hence, finally, we obtain the specific limit of resolution:

$$\sigma = 1.76\lambda \tag{47}$$

For example, with $\lambda = 0.58\ \mu$, we obtain $\sigma = 1.02$ (mm $\times$ minute). Referring to Figure 29 we see that for high luminance, σ is actually of the order of 1, and this result was true with all of Arnulf's subjects but only within an interval of pupillary diameter

$$0.4 < d < 0.7 \text{ mm} \tag{48}$$

In this region (very limited) the eye behaves as if it were perfect and for high luminances (no matter what the object contrast C is) the threshold is obtained when the image contrast is around 0.02. The limit of luminances for which this property is verified is approximately 500 cd/m^2 (through a reduced pupil size of about 0.6 mm diameter; that is, about 15 cd/m^2 through a pupil of 3.7 mm, which is usual for this luminance). In this same region (48), the variation expected from equation (47) if the wavelength is changed is also verified. Another consequence of this theory which is also verified with these same conditions is that, if the equal black and white bars are replaced by infinitely narrow luminous lines on a dark background, the value of σ is diminished by half. All these consequences are, on the other hand, in contradiction to the experiment using normal pupils 4 mm in diameter, for which we know that the variation of λ and of the ratio of white to black bars does not have any large effect.

This theory can still account for some known facts. For example, the variation of retinal contrast is much more rapid when increasing the distance of a grating than when moving a single line. In fact, the determination of threshold is much more precise in the former instance than in the latter; a slight ametropia also has a greater effect. Finally, the false resolution of the gratings can be easily explained by the reversal of the image (the white bars becoming darker), as would be expected from the calculations, between 0.41 and 0.27 times the radius of the Airy disk.

The large discrepancies between theory (σ = constant) and the experimental curves of Figure 29 have several causes:

1. For $d < 0.4$ mm, the important values of s which follow (since $s = \sigma/d$) indicate that the fovea contains very few bars of the grating and that these conditions are less favorable. Moreover, the luminance decreases with d. If the retinal illumination is kept constant, the minimum of σ occurs around 0.2 mm.

2. For $d > 0.7$ mm one might wonder if the alteration of the image due to aberrations of the eye were not the cause of the increase of σ. It is certainly not due to chromatic aberration, since acuity with monochromatic light is not appreciably better than with white light, and since optical systems calculated to correct chromatic aberration of the eye do not improve vision (Helmholtz; Lapicque; Hartridge; Thomson and Wright, 1947). Their only effect is to increase, very slightly, the apparent contrast of the test. Similarly, attempts at correcting regular spherical aberration (van Heel, 1935–1946) have only given the test an unreal appearance. The subjects had to wait for a moment before they could see the test clearly when they put on or removed the corrective system, but they did not see more clearly in one case than in the other.

By forming interference fringes directly on the retina, as, for example, by Young's method (two small apertures in a screen), the limit of resolution of the retina can be measured regardless of the ocular dioptrics (Le Grand, 1935; Byram, 1944; O'Brien, 1952; Rosenbrück, 1959; Westheimer, 1959; Arnulf and Dupuy, 1960; Campbell and Gregory, 1960). This method seems to have returned to fashion since the development of the Fourier method in optics, first enunciated by Duffieux (1946), which attributes a special importance to sinusoidal distributions of illumination on the retina; we shall return to that point at the end of the chapter. With these interference fringes, at high luminances, practically the same limit is obtained as in natural vision of an identical objective test.

The humorous remark attributed to Helmholtz—that an instrument as defective as the eye (as far as the optics is concerned) would shame its constructor—is unfounded, because the visual acuity is the same as if the eye were free of aberration, at least at high luminances. It is even possible to increase the aberration of the eye, for example, by using poorly corrected instruments that increase the defects of the resulting image, without altering the limit of resolution of a grating of contrast 1, at high luminances.

Arnulf et al. (1951) thought that it was possible to rescue the continuous theory by involving scattering occurring within the retina as well as microfluctuations of accommodation (1955). To do so they utilized the values that they had determined of the retinal distribution of illumination around a point source; by double integration (for the rectangular source of a white bar of the grating) they obtained results in reasonable agreement with experiment. But a more rigid study of the problem has shown (Flamant, 1955) that the retinal image contrast only retained its constant value (about 0.02) for relatively poor acuities and that it increased progressively with acuity. Is this not a proof a posteriori that the continuous theory becomes inadequate when the details diminish in size so that they are comparable to the size of the retinal receptors?

3. Finally, this theory does not at all account for the variation with luminance. As a matter of fact, one would have to assume an increase of the image contrast of an order of magnitude considerably greater than variations obtained experimentally with extended fields. When luminance decreases, aberrations and defects of the image begin to display their effects. The method of interference then gives values of thresholds s which are smaller than in natural vision, and the results in Table 12 show also that the spread of the image due to dioptric effects diminishes acuity with large pupils at levels of a few candelas per square meter and below.

Obviously, in the scotopic region the receptor changes its nature and it is not surprising that acuity follows different laws. A continuous theory in this case would actually stumble against the same difficulty as above in accounting for the variation as a function of luminance.

From a purely logical point of view the continuous theories of visual acuity seem to me to depend on a question of principle. By experiment it has been proved that the differential threshold $\Delta L/L$ between two contiguous surfaces of luminance L and $L + \Delta L$, assuming that these surfaces are presented (for example) as rectangles of fixed height and variable width u, is a function of u which varies slowly when u is large, and more and more rapidly as u diminishes toward a limit of the order of 1 minute of arc. At the limit when u is confounded with the limit of resolution s, the differential threshold $\Delta E/E$ of the retinal illumination can be deduced from the experimental values of $\Delta L/L$ only with hypotheses (somewhat uncertain) regarding the actual distribution of retinal illuminations. The numerical coincidence of the calculated $\Delta E/E$ with the experimental $\Delta L/L$ relative to much larger values of u seems to me artificial and without the weight of a proof. Actually, the problem of differential threshold for very small values of u and the problem of visual acuity cannot be deduced one from the other, as they constitute two different aspects of a single concept, according to whether there is more emphasis on contrast or on angle.

Anatomical Theories

Theories of acuity which place emphasis only upon the discontinuous structure of the retina are also insufficient, and for reasons which are contrary to those which condemn continuous theories. Thus in the first publication of his *Physiological Optics*, Helmholtz justified the limit of resolution, which he assumed to be of the order of 1 minute of arc, by the diameter of foveal cones, which according to Kölliker (1854) was 5 μ. The measurements of Schultze (1866) reduced this value by half and established that the distance between the centers of two neighboring foveal cones is 2 to 2.5 μ near the fixation point. Helmholtz, however, kept his anatomical explanation (after having modified it, as was suggested by Weber, 1846) and assumed now that for two points to be perceived as such, their images must fall on two cones with a third cone between them. This reasoning appears clear and obvious and consequently has acquired over the last century the solidity of classical errors, but it cannot withstand critical examination. Helmholtz also attributed to the retinal mosaic the wavy appearance of the grating which is sometimes perceived at the limit of vision, but Bourdon [6] explained this phenomenon much more simply by slight heterogeneities of refraction. Helmholtz was so obsessed by this anatomical basis of 1′ that he even introduced it for stereoscopic acuity, which is utterly wrong. But the authority which–rightly so–is attached to the name of Helmholtz is such that this dogma of the resolution limit of 1 minute of arc for the human eye is still referred to in many texts and even sometimes in the case of gratings to indicate either the width of a white bar or the grating interval (white bar + black bar), depending upon the usage in the writer's country. . . . In fact, the optimum values that have been found with simple tests are approximately the following: black point, 14″; black line, 0.5″; 2 points, 1.3′; 2 lines, 40″; grating, 35″; Landolt ring, 24″. There is not one which is close to 1 minute of arc, but old legends die hard.

In the central bundle of particularly thin foveal cones Rochon-Duvigneaud [59] determined on seven human eyes that 100 cones extended over a distance varying between 150 and 200 μ, that is, 1.5 to 2 μ for the separation between the centers of two contiguous cones. Similarly, Polyak [55] found 1 μ for the minimum diameter of cones and 0.2 to 0.3 μ for the separation between cones. The measurements of O'Brien (1951) arrive at a value between 2.0 and 2.3 μ for the distance between the axes of two neighboring cones. The common doctrine now attributes to each foveal cone its own nervous conduction to the brain.

However, we have seen that the Airy disk has a radius of the order of 3 μ. The fineness of the foveal cones is therefore sufficient to utilize the optics of the eye, but it is probable that it imposes a limitation on the keenness of resolved details and that it comes into play in the asymptotic acuity observed at high levels of luminance and contrast. This was already the conclusion of de la Hire (1660), who calculated that the limiting fineness of vision was 1/8000 inch on the retina (3 μ) and he remarked: "... such is the smallness of the net of which it is made up. . . ."

If it appears certain that the structure of the retina must be one component of theories of acuity, a purely anatomical theory stumbles over the fact that acuity reduces regularly with the luminance. Indeed, the cones cease to see at a sufficiently low level and the utilization of rods, elements which only function in groups, as we will see later, reduces acuity, but when acuity has already decreased to 0.2 and even less, colors are still seen, which shows that cones are still functioning.

Hecht (1928) attempted to resolve this difficulty by an ingenious theory of *recruitment*. He assumed that individual foveal cones all possess different absolute thresholds distributed normally over a wide range. With increasing field luminance the number of active cones increases and the limit of resolution, which is proportional to the mean distance between neighboring functioning cones, also decreases. There is a simultaneous variation of intensity discrimination since each differential step corresponds to the activation of one more cone. This theory runs into great difficulties, as analyzed by Wright and Granit (1938). First, the differential threshold should pass through a maximum at the point of inflexion of the visual acuity curve and then decline as the luminance increases, but this does not occur. Moreover, to explain the variation of photopic acuity, the absolute thresholds of the cones must be assumed to vary in a ratio of 1000 or even more, and this result is contradicted by experiment. O'Brien and O'Brien (1951) have measured the threshold of two equal point sources separated by 4 to 8′ on a black background of 15′, which is also surrounded by a uniform background of a luminance of 0.03 to 300 cd/m^2. Two levels of the brightness of the sources were determined, one at which neither of the sources was perceived 8 times out of 10 (each flash was of 2×10^{-5} second duration, and an artificial pupil was utilized, while at the other the two sources were seen at the same time 8 times out of 10. The ratio of these levels was 1.2 for high levels of the adapting luminance and 1.4 for low levels. This slight variation of sensitivity from one foveal cone to another eliminates Hecht's hypothesis.

Another difficulty of anatomical theories is in explaining why acuity is practically independent of wavelength. In Young's hypothesis, where the retina contains three types of cones sensitive to red, green, and blue, respectively, acuity should be much better with white light, which stimulates all types of cones, than with monochromatic light, which only stimulates one type of cone. It is true that the existence of many "dominators" (receptors sensitive throughout the spectrum) could account for this result. Another hypothesis maintained by Hartridge [19d] is of the existence of foveal sensory units, each consisting of a group of red, green, and blue receptors. Some of these units are nearly circular, but others may be elongated and consequently could introduce a *retinal astigmatism* that has been described by Shlaer (1937). But this phenomenon changes with the luminance level and it is more likely that it could be explained by the microfluctuations of accommodation which produce, as Arnulf noted, changing images often presenting astigmatism.

Another type of asymmetry is the superiority often noted (particularly by Wertheim, 1894) of the superior and inferior parts of the visual field, even near the fovea. Acuity is better in the upper part. According to Vilter (1954) this asymmetry can also be found in the density of the cones and ganglion cells which are more numerous in the lower part of the retina.

Physiological Theories

The discontinuous structure of the retina can be considered a fundamental basis of acuity, but some mechanisms modifying the effect of this structure, especially as a function of the luminance, must be conceived. For example, Broca (1901) thought that migration of the epithelial pigment situated between the retina and the choroid, more or less isolates the cones. These migrations, influenced by the luminance level, exist in some animals but have never been demonstrated in man. Lythgoe (1932) assumed that light adaptation may inhibit some lateral connections among cones and therefore isolate them more. Buddenbrock (1937) provided good arguments in favor of this notion of receptive units, constituted by the groupings of cones varying with the intensity level. In peripheral vision such units certainly exist (we will return to this point later). In the fovea, where the effects of summation are much less marked, this phenomenon is less certain.

A theory by Marshall and Talbot [39] places in the brain the variations of connections among elements, which is, of course, possible but cannot be verified. In any case, a limitation applying at the level of the retina cannot then be taken away in the brain in spite of the multiplication of the size of the "cortical" image as compared to the retinal image.

Instead of placing the emphasis on spatial interaction, other physiological theories call upon temporal phenomena. For example, it has often been thought that rapid and involuntary movements of the eyes, which we will study later, play the essential role in acuity by converting the spatial variations of illumination of the retinal image into temporal variations of the flux received by each cone. Already suggested by Hering (1899), this *dynamic theory* of acuity has been developed principally by Jones and Higgins (1948). The fact that acuity with brief illumination hardly differs from acuity with steady exposure does not favor this hypothesis very much; experiments with stabilized retinal images that we will discuss later provide evidence against this theory (Keesey, 1960), but a more recent study by Millodot (1967) provides evidence supporting it. It is likely that in certain cases inhibition phenomena (off responses) linked to eye movements do, to a certain extent, play a role in visual acuity (see, for example, Pirenne [53c]).

Another temporal hypothesis was proposed by Granit and Therman (1935). They stipulated, on the basis of electrophysiological experiments, that the retinal receptors present activity that is not continuous but alternating and separated by refractory periods of insensitivity. Berger and Buchtal (1938) thought that the increase in luminance diminishes the refractory periods and therefore ensures an increased number of active receptors per unit of time and per unit of retinal surface.

Other physiological theories call upon the photochemical concepts of Hecht's school; there is, for example, a study by Hecht and Mintz (1939) on the vision of a line, where it was hardly necessary, as it is in fact a case where the continuous theory suffices. Let us also cite the theory of Byram (1944), which supposes that retinal receptors are sensitive to variations in amounts of photochemical substance, and that of Jahn (1946), which also assumes a variation of concentration of photosensitive pigment with the level of illumination. In all these theories many constants have to be introduced, which makes the relationship between calculation and experiment somewhat illusory.

Quantum Theories

The absolute threshold of rod vision involves, as we have seen previously, the concept of quantum fluctuations of light. It is possible that acuity, at least scotopic, is also related to the concept of the quantum nature of light, as was suggested by de Vries (1943). Experiments by Pirenne (1945) and Bouman and van der Velden (1947) seem in fact to confirm the role of these quanta, but as these studies were relative to the peripheral retina they will be referred to in a later chapter.

This problem has been discussed from a very general and theoretical point of view by Rose (1948); in any receptor the absorption of N quanta (as a mean) is accompanied by statistical fluctuations of which the quadratic mean is theoretically $\sqrt{N}$. The threshold of the *signal*, that is, the smallest perceptible change, is expressed by

$$\Delta N = k\sqrt{N}$$

where k is the minimum value of the ratio of signal to "background noise." For a photographic film and for television, k varies between 3 and 7. If for the eye, one assumes, by analogy, $k = 5$ and an accumulation time of 0.2 second, Rose showed that the experimental values of acuity correspond, in white light, to a quantum efficiency of the order of 0.05 at low levels and decreasing to 0.005 at high levels (up to 100 cd/m^2). These orders of magnitude are quite reasonable (see also Gregory and Cane, 1955).

The great superiority of the eye over artificial receptors at low luminances could be due to the fact that the eye constitutes an "ideal" receptor over a considerable range of luminances; i.e., if the luminance L decreases, the signal received by the retina decreases linearly with L, whereas the background noise decreases as $\sqrt{L}$. In photography or in television this property is only verified over a small range of luminances with a quantum efficiency of the same order as that of the eye. But there remains a constant background noise (for example, the veil in photography) which does not exist in the eye. The slight variation, in the ratio of about 10, of the extreme quantum efficiencies of light upon the eye would prove that the mechanism of adaptation is only partially photochemical (concentration of visual purple in the rods), and that the main part would be provided by a system of variable amplification between the visual cells and the optic nerve. The following fact could be explained by this concept: at high luminances (> 10 cd/m^2) the apparent brightness of the radiations, the differential threshold, and the acuity all follow more or less the photopic relative luminous efficiency curve. It is the same at low luminances ($< 10^{-3}$ cd/m^2) where these functions follow the scotopic curve. In the *mesopic* region the brightness of red begins to fall below that of blue, whereas the two other functions remain the same for these colors; this is because the gain from amplification becomes less for red than for blue at that level.

Lamar et al. (1948) also analyzed from the point of view of quantum theory their measurements of the thresholds of rectangles and arrived at the conclusion that vision of a contour would necessitate some integration of excitation of *all* the cones situated on the edge of the given figure. Various other authors [5a] have also recently presented fluctuation theories of visual acuity.

Conclusion

At the end of this long account on the problems and theories of the vision of details, the reader will probably be disappointed and will reproach us for not taking any particular point of view and so leaving him guessing. We believe that it is more honest not to attempt a synthesis, which could be only approximate and which risks concealing the actual complexity. Obviously, many factors contribute to limiting the perception of details, and it is illusory to attempt too simple schemes which are of ingenious construction, but artificial.

Information theory, which has enabled us to demonstrate some very general theorems regarding the limits of the quantity of information transmittable by a given mechanism, independently of the practical application of this mechanism (as thermodynamics limits the efficiency of machines, whatever their structure), could not have failed to be concerned with the problem of visual acuity. In optical instruments, in order to express the limit of resolution as a function of aberration and diffraction, it is becoming more common to utilize the method called "frequency response technique" or "modulation transfer function," which, owing to Fourier transform, enables the properties of the system to be represented by a spatial spectrum. Descriptions of this method can be found in, for example, Maréchal and Françon [38b], Hopkins [23a], and Westheimer (1963). Applications in the case of the eye are based on the perception of gratings whose intensity variations are sinuosoidal, obtained either by interference fringes or by other methods (de Palma and Lowry, 1962; Bryngdahl, 1964). But the essential role of contrast in retinal perception complicates the phenomena for the eye and according to Kelly (1965) it is the bandwidth rather than the spatial frequency which becomes the essential element.

With regard to information theory, these methods have, up to the present, been used only to represent known concepts in a new form, as, for example, in the numerical evaluation of the information capacity of the human eye (Toraldo di Francia, 1954; Benari, 1961). But further research will surely revive the ancient question of visual acuity. For example, the reader may find mathematical applications of the modulation transfer function to the spatial resolution of the human visual system in a paper by Patel (1966), or in another by Campbell and Green [5a].

CHAPTER SEVEN

Vision of Forms

The distinction between the problems of the vision of details, normal visual acuity, and those of the perception of forms is artificial. The recognition of the position of the gap in a Landolt ring or the reading of a Snellen letter is a complex process in which psychological concepts play an important role. In this chapter several examples of form vision will be discussed briefly, beginning with the simplest and most symmetrical.

Aligning Acuity

The alignment of two lines (Fig. 31*a*) is accomplished by the eye with extraordinary precision. This is referred to as *vernier* acuity, as this ability is utilized in reading a vernier scale. It is not, in fact, so much a problem of separation as of spatial perception. At a sufficient distance the segments *b* (Fig. 31) are not seen separated, but the upper segment is localized in space more to the left than the lower segment.

Wülfing (1892) was the first to note the remarkably small visual angle corresponding to the threshold of alignment, which he found of the order of 10″ for bright slits as well as for black bars on a white background. Stratton (1900) lowered this to 5″ and Best (1900) to 2.5″ for 80% correct responses. The precision depends very little upon the orientation of the test but seems, however, slightly better for vertical lines. Laurens (1914) found a variation with luminance (less than for the minimum separation of two parallel bars). He found a foveal threshold of about 100″ at the lowest luminance. French (1920) studied systematically the most favorable conditions for vernier acuity. The thinner the segments the quicker the alignment but it is not better in Figure 31*b* than in *c*. On the other hand, the wider the separation of the segments

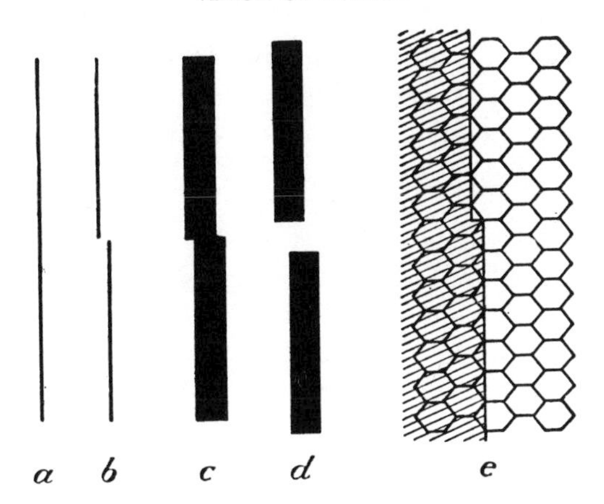

FIG. 31. Vernier acuity. (*e*) Relates to the explanation of Hering based on the retinal mosaic.

(as in *d*), the lower the precision. French found an average error of 0.5″ for thin lines of 12′ of arc length with the two segments touching. This value is 1.1″ for a separation of 4′ between the two segments and 5″ for a separation of 19′. A similar study was carried out by Berry (1948). Furthermore, systematic error occurs, which depends upon the direction of the lines and may be as high as several seconds of arc. There are individual differences that vary with the subject and even with the eye he uses. According to Emsley (1944), the left eye usually places the upper segment too much toward the left and vice versa for the right eye if the lines are vertical. This alignment plays an essential role in the use of coincidence rangefinders, and according to Mazuir [40] the standard deviations for trained observers are 2 to 3″, but with individual constant errors which may reach 10″.

Baker (1949) studied the effect of luminance and time of exposure. Black lines of 7′ width and 2° height appeared in the center of a 12° field of variable luminance. With an artificial pupil of 2.4 mm, the results remained constant between 200 and 6000 cd/m^2 with steady exposure. With an exposure of 0.02 second the maximum was reached for 800 cd/m^2 and the accuracy was reduced by half. With colored lights the red (690 mμ) and the yellow (575 mμ) were 10% better than white and the blue (490 mμ) 10% worse than white. But if the chromatic aberration of the eye is corrected, the best results are obtained with blue.

The astonishing accuracy of vernier acuity presents a difficult problem, at least as far as anatomical theories are concerned. Hering (1899) suggested that the cone mosaic was so regular that a line could be situated between a series of cones whereas another line would just overlap it. But the physical nature of the retinal image makes this hypothesis and that of Polyak [55] very unlikely. Polyak suggested that the thin separation of neuroglia (0.3 to 0.5 μ) which isolates the cones from one another would be responsible. Weymouth et al. (1923–1928) proposed a dynamic theory which, unlike Hering's, places the emphasis on the irregularity, of the retinal mosaic, which is scanned by the image because of the constant movements of the eye. Motokawa (1949) proposed a physiological explanation based on the phenomenon of *retinal spatial induction*, which he describes as follows. It is known that a luminous sensation (phosphene) can be evoked by electrical stimulation of the eye. After a short illumination of the retina, Motokawa determined the threshold of electrical excitability, which is decreased by this preillumination (whereas the photosensitivity is reduced). Suppose the subject fixates steadily a given figure and that at the same time the sensitivity of the retina at various points is analyzed by the above method (small spot of intense illumination, followed by the measurement of the threshold of the phosphene). It will then be possible to draw curves of equal spatial induction around the figure. For the case of aligned and nonaligned bars the curves are as represented in Figure 32. Motokawa interprets these deformations of the field of spatial induction as an amplification of the nonalignment of the bars.

Actually, if knowledge of the retinal anatomy is omitted, vernier acuity becomes much less extraordinary. The threshold of nonalignment of two lines (0.5″) is the same as the threshold of seeing a black line on a white background, and it is probable that the continuous theory, which places the emphasis on the differential threshold, may be valid in both instances, but along with it is a remarkable spatial discrimination, of which more examples will be encountered later.

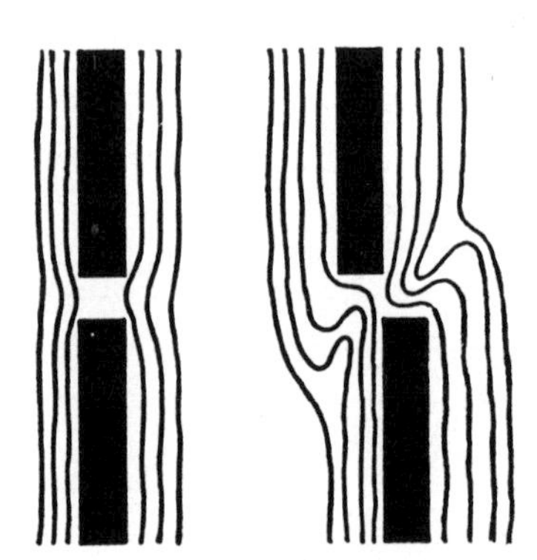

FIG. 32. Explanation of vernier acuity by the deformation of the lines of equal spatial induction (Motokawa).

Accuracy of Sightings

An important practical problem is the one of sighting, by means of a line in the reticule of a telescope, a problem which is analogous with vernier acuity. The best results are obtained at high luminances with lines 1 to 2′ in diameter. Sighting with only one line, as, for instance, by bisection of the image of the star, is affected by a systematic error which depends upon the observer and even upon the eye that he utilizes. It would be advisable for the observer to bracket the object with two parallel lines whose separation is slightly more than the width of the object. The standard deviation is then of the order of 15″, according to Mazuir [40].

Placing a vertical line on the crossing of two other lines gives the best results (according to Guild, 1930) when the angle of the two lines is between 40 and 50°, the vertical line bisecting this angle. The standard deviation is about 2″ and individual variations are rarely beyond 5″. Luminance seems to have very little influence over a large range.

Measuring Ability of the Eye

Many experiments have been conducted in an effort to study the appreciation of the equality of two lengths. For the moment we will only consider figures placed in a vertical plane perpendicular to the observer's direction of gaze.

Volkmann [69] used three vertical parallel lines and adjusted the position of the central one until it appeared to lie at equal distances from the other two. According to Weber's law the *relative* precision of this procedure is a constant. It was found that it held reasonably well, the average error being of the order of 1% except when the lines were very close (5% error for less than 2′ between the central line and the other two). To compare very small dimensions, the photometric criterion overtakes the criterion of separation, as Bourdon [6] has demonstrated.

Chodin (1877) noted that the comparison of the length of two horizontal parallel lines was more precise than that of vertical lines (1 to 2% error in the former and 2 to 3% in the latter). It did not follow Weber's law and the optimum precision was obtained, in relative values, for segments subtending an angle of 3°. Of course, precision becomes greater as the lines become closer and when they are adjacent it is equal to 0.6% (Merkel, 1894; Reese, 1953). The equalization of two arms of a cross is obtained with 1 to 2% error, according to Fischer (1891).

The bisecting of a horizontal line is a very precise procedure. According to Kundt (1863) the difference between the two halves may be of the order of 0.3 to 0.7%, which means that on a total segment of 10° the central point is determined with 1′ error. If the experiment is monocular there is a systematic error. The point which appears in the center is in fact toward the left if the right eye is fixating and toward the right if the left eye is fixating. This *retinal asymmetry* plays a certain role in binocular vision which will be considered later. Subsequent research has shown that the retinal asymmetry does not present the constancy that Kundt had believed. Münsterberg (1889) found subjects who showed an overestimation of the nasal part using the right eye. Meyer (1936) and Kleint (1940) pointed out the fact that the subject being right-handed or left-handed may influence the bisection even in binocular vision. The level of luminance also has some influence (Pfeiffer, 1937). A study by K. T. Brown (1953) on six subjects confirmed the asymmetry described by Kundt but with a more or less regular variation from day to day. The variations occur at the same time for both eyes, if the observation is monocular. This suggests a central origin (K. T. Brown, 1955).

According to Takala [63], in the bisection of vertical lines there seems to be a tendency to place the central point slightly too high. The precision is of the same order as for the horizontal line. In all these experiments the eye was free to move at will.

Geometric Figures

When a very small possible number of forms is presented to an observer, such as a circle, a square, and a triangle, recognition is effective for very small angular values of the order of 3′ for the side of the triangle and 4′ for the side of the square and the diameter of the circle, all at high luminances (Martens, 1919). Variation is, in fact, small in photopic vision, and thresholds are nearly twice as high at 10^{-2} cd/m^2 (Berger and Buchthal, 1938). According to Dratz (1945), at very low luminances recognition of a circle or a triangle occurs for dimensions which are in the ratio 6:1 with the limit of resolution of a grating (width of a white or black bar).

If the possible number of forms increases, visual discrimination becomes a difficult psychological problem which is beyond our scope. The reader may refer to a study by Casperson (1950), who shows that there does not exist a simple criterion to predict, a priori, the relative ease of recognizing geometric forms.

Recognition of the distortion of a geometric figure constitutes an interesting practical problem. Under optimum conditions an ellipse is recognized as such, and differentiated from a circle, when the long axis is greater than the small axis by 1%, and a rectangle is differentiated from a square when the length is 1.4% greater (Veniar, 1948). But, under normal circumstances, it is prudent to predict values at least 10 times greater, according to Sleight and Mowbray (1951).

It has been assumed since Fick's work that there is generally an overestimation of the heights of figures. For instance, to see a square, one must look at a rectangle which is longer horizontally than vertically. The experiments of Sleight et al. (1952) tend to bring this rule under suspicion.

Perception of Forms

It is only with caution that we will venture into the difficult field of psychology, but it is essential to convince the reader that it is not only the retinal image which governs vision. Opin [47] stated ". . . Since antiquity one has been struck by the discrepancy between external reality and the representation that our visual sense gives us. The main portion of the fourth volume of *De Rerum Natura* deals with the amazing errors to which our sense of vision subjects us. But nothing is more improper than the term 'illusion' of senses as if, to use Piéron's expression, the senses had to give not biological responses adapted to the stimuli, but exact representation of external reality."

It was believed for a long time that perception of the world of forms was a simple association of visual sensations put together by experience and memory. This idea dominates the entire third volume of Helmholtz's *Physiological Optics* [20]. Bartlett (1916) was one of the first to bring the attention of the psychologists to the importance of *original construction* which occurs in the most common perception. This construction is utilized to preserve stability and constancy in the chaotic world of visual sensations while keeping the very high sensitivity necessary for a response adapted to variations of the external world.

The "Gestalt" school under the influence of Wertheimer, Koffka, and Köhler has codified this rebellion against associationism. The visual field is not made up of a juxtaposition of luminances and colors but by *objects* separated and perceived as such. The parts of these objects are not only contiguous but ultimately linked. The object is a superimposed structure, a perceptive *schema* which enables the subject to recognize, according to his past experience, what is familiar and permanent and to isolate eventually what is new, inconstant, interesting, or dangerous. These schema do not exist in the small infant, who has no true perception. Piaget (1937) has nicely analyzed the gradual transition in the infant from a chaotic world of sensations without significance, to a universe of stable objects keeping their individuality in spite of the changes in perspective, movements, etc. Furthermore, the data gathered by Senden [61] on people blind from birth, due to congenital cataract, to whom an operation restored their sight, prove that the visual identification of objects remains slow and inaccurate for several weeks while the schema of the new universe are being elaborated.

These schema or "Gestalten" are configurations limited by *contours*. The first characteristic of this configuration is the opposition of the *figure* and the *ground* from which it becomes separated, and according to Senden it is the first distinction that the congenitally blind acquire after their operation, long before they perceive objects as such. The figure has a form, color, solidity, and shape, whereas the ground remains undifferentiated and its color, in particular, is poorly localized. This solidity of the figure is due essentially to the contour, and Liebmann (1927) showed that a figure which separates itself from the ground only by its color, without a difference of luminance or a marked contour, tends to reunite itself with the ground.

It is probable that the individual variation of perceptive schema is considerable, but this classification of *perceptive types* is somewhat deceiving. The most classical distinction is that of the "synthetic" and "analytical" attitudes. The former integrates the whole visual field, whereas the latter examines more the details. Ninety percent of all children would be of the synthetic type, but this diminishes with age. An intermediate type has also been described as "plastic," of which the structure would be more differentiated than that of the synthetic type. Other classifications have been proposed by the German typological school but are now less in vogue. It is actually not clear whether or not the differences correspond to specific modes of perception.

The Gestalt school has now lost a great deal of its dogmatism and the specialists have only retained some of their useful concepts. The reader who would like to become more familiar with the divergences on this subject could compare two books, one by Vernon [68], who maintains that the perceptive universe differs greatly from that of the stimuli, and the other by Gibson [19], to whom the perceived universe faithfully reflects that of the stimuli and who further explains the inconsistencies, illusions, and contradictions exploited by the Gestalt school, as the result of the artificial conditions of laboratory research.

Nowadays psychologists tend toward three schools of thought on this important problem of form. Some still search for rigorous general theories (Dodwell and Sutherland, 1961; Sutherland, 1963). Others adhere to information theory and present the perception of visual patterns as operational schema. This trend has developed considerably since the advent of supersonic airplanes, as the piloting of these requires rapid discrimination of forms. Some excellent monographs exist on this subject [2a, 19a′]. Still others try to link the discrimination of forms with somewhat complex mathematical formulations (see, for example, Singer, 1961) in view of its application to the recognition by machines of forms such as printed letters [17a, 60a].

Optical Illusions

There exists considerable literature about optical illusions [63a] and two examples among the classical ones will be cited. In the figure of Zöllner (1860), parallel lines seem to diverge when they are crossed by oblique lines. This is interpreted as an overestimation of acute angles as compared to obtuse angles. In the Hering illusion (Fig. 33) the parallel lines are bent by the same phenomenon. The illusion is more marked if the subject scans the figure, and it disappears in monocular vision if the page is slanted and one looks in the direction of the parallel lines. If the

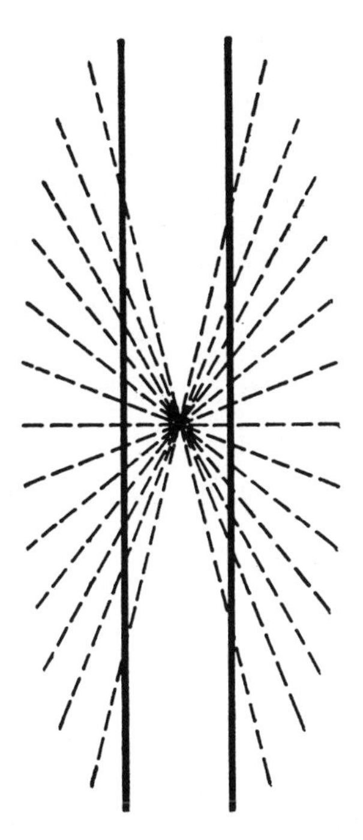

FIG. 33. Hering illusion.

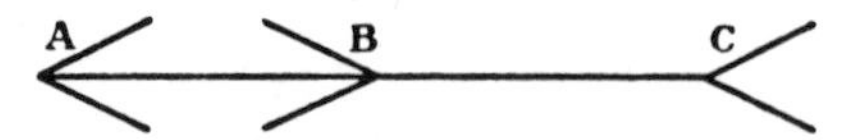

FIG. 34. Müller-Lyer illusion.

two parallel lines and the dotted lines are drawn separately and their induced afterimages are superimposed, the illusion persists, which proves the origin central rather than retinal. It is linked to "field-deforming forces," which arise from lines drawn on the ground and which produce distortion of the figure (Orbison, 1939).

The well-known illusion of Müller-Lyer (1896), where the arrows give the impression that *BC* is longer than *AB* (Fig. 34) is interpreted by the Gestalt school as proof that the line prolonged by the outgoing angles constitutes a whole form which is longer than the other one. According to Benussi (1914), the time necessary for the illusion to be perceived depends mainly upon the subject. It is less than 0.1 second for the synthetic type and sometimes more than 1 second for the analytical type. According to Gemelli (1928), the fact that the figure of Müller-Lyer could be split into pieces and reshaped binocularly, retaining the illusion, proved a central origin. Motokawa (1950) tried to explain the illusion by his

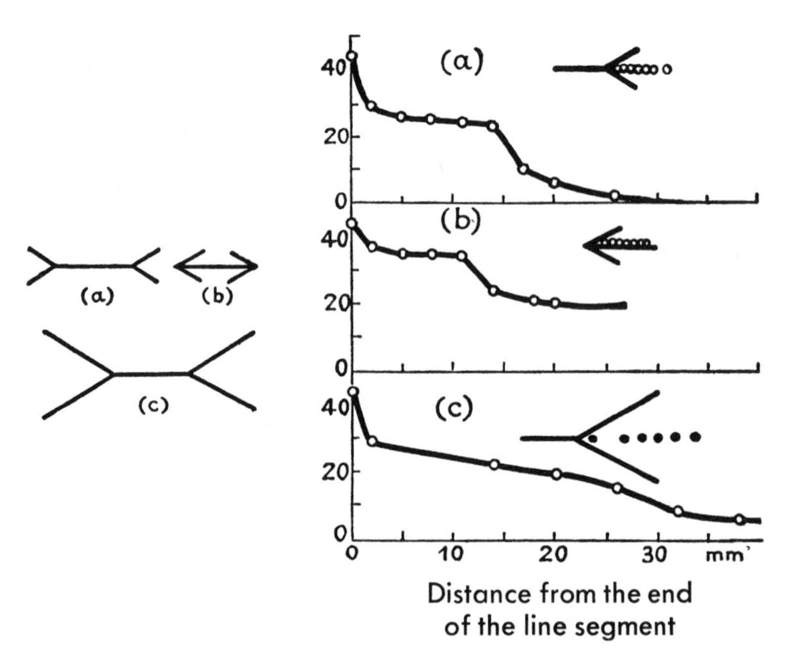

FIG. 35. Interpretation of the Müller-Lyer illusion by Motokawa. The small circles indicate the points where measurements were made. Ordinates: Contrast effects.

Fig. 36. Hering's grid. Grey spots are seen at the intersection between the black squares, especially in peripheral vision.

experiments on spatial retinal induction. The subject looks at the intensely illuminated figure at 30 cm for 2 seconds with a fixation point a few millimeters below. After 3 seconds in darkness, a small circular test spot of 2 mm diameter appears outside the preceding figure. The electrical excitability is then measured. Comparison of the results obtained with and without the figure enables one to ascertain the contrast and to draw the curves of spatial induction. The Müller-Lyer illusion (Fig. 35) would be due to a discontinuity in the slope of the curve of spatial induction at a distance which is greater in the case of the outgoing angle. (Fig. 35*a*, *c*). In any case, eye movements are not necessary because the techniques of image stabilization which we will discuss in Chapter 10 do not alter optical illusions (Pritchard, 1958). According to Gregory (1965) the natural tendency to see a plane figure in depth may play a role in the Müller-Lyer illusion.

Contours

We have mentioned that sharp gradients of luminance in the visual field play an important role in the materialization of an object. It has been known for a long time that they modify the apparent contrasts by introducing a *border contrast*. The word contrast is used here as Chevreul (1849) defined it—a subjective exaggeration of the real contrast C as defined in equation (40). This border contrast can be observed easily in Hering's grid* (Fig. 36). Grey spots appear at the white inter-

* This figure was first published by Hering in his *Grundzüge der Lehre vom Lichtsinn*, 1905. Hering gives credit to L. Hermann, who first called attention to this phenomenon.

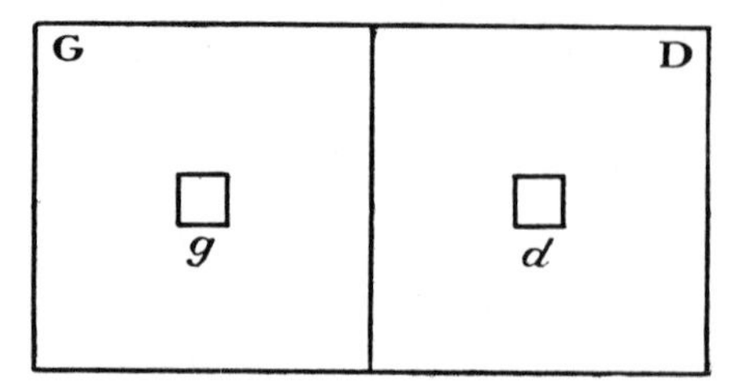

FIG. 37. Contrast by spatial induction (Hess and Pretori).

sections, more so when the eyes move. Fröhlich (1921) believed this to be caused primarily by some afterimage phenomena. A detailed quantitative study of border contrast was undertaken by Harms and Aulhorn (1955).

Border contrast, which enhances white next to black and vice versa, can extend over the entire area if it is not too large, as was shown in the experiments of Ebbinghaus (1887) and of Hess and Pretori (1894). The latter presented squares of 1° on a background of 10° (Fig. 37); the luminances G, g, D, and d are variables. For example with $G = 0.03$ cd/m² and $g = 0.6$ cd/m², the apparent equality of the luminosity of g and d gave the results shown in Table 14.

The influence of the respective sizes of the test and the background, which is important, was studied by Kirschmann (1890).

The interpretation of this phenomenon is contested. At any rate, the afterimage effects due to movements of the fixing eye are not essential, as Wirth (1930) showed that instantaneous illumination gave the same results. According to Bartley [3] the contour itself has an inductive action and even the shape of the contour may play a part. With sinuous or jagged contours, Fry (1947) discovered a direct effect of the contour on contrast threshold.

Contrast was formerly interpreted as a perceptive effect of central origin. Nowadays it is believed to be caused mainly by retinal interaction. Electrophysiology of the retina of animals has in fact brought more evidence in favor of this point of view. Thus Kuffler (1953) recorded with microelectrodes the local responses of a cat retina on which

TABLE 14

Measurements of Contrast by Hess and Pretori

D, cd/m²	0.52	1.1	1.7	2.1	4.0	5.2	9.6
d, cd/m²	1.2	1.4	1.8	2.0	3.2	4.1	6.1

he produced a point image. He noted that in general the retinal receptors which were illuminated gave an "on" response (a series of spikes at the beginning of the stimulus followed by a reduced response as long as the light is kept on), whereas the surrounding receptors gave an "off" response (inhibition of the receptors by the illumination and a response when the light is switched off). This antagonism, which may constitute the physiological basis of contrast, was demonstrated on the frog by Barlow (1953). The projection on the retina of an annular stimulus of small diameter (about the size of the sensory units) inhibits the central "on" response.

Perception of a contour is not necessarily linked to a discontinuity of luminance; a rapid change is sufficient. Helmholtz [20] rotated a white disk with a black circle near its edge. Instead of an annulus of graded intensity, one sees a more or less uniform grey annulus with its sharply defined edges tangent to the circle, because at this point the gradient of luminance is infinite. The perception of a contour as a function of the gradient of retinal illumination was studied in particular by Ludvigh (1953). When a disk with black and white sectors was rotated, with the white dominating in the center and the black in the periphery, with an intermediate annulus of continuous variation between black and white, Mach (1886) observed a dark band at the edge of the central area and a light band at the edge of the peripheral area, at the points of change of the gradient of luminance. Kühl (1928) studied this effect quantitatively.

Finally, border contrast can, in some cases, give the illusion of a contour in a field of uniform brightness. This is very obvious when looking at the sky through a circular diaphragm a few millimeters in diameter held 2 or 3 cm in front of the eye. The edge of the diaphragm is seen blurred and a bright circle separates it from the dark surrounding. Described by Mach (1865), this effect was studied by many authors (see Fiorentini, 1955); some attribute it to the aberrations of the eye (Kincaid and Blackwell, 1952), others to the movements of the eye (Toraldo di Francia, 1952), others to retinal interaction (MacCollough, 1955). Fiorentini et al. (1955) studied the differential threshold in the presence of a gradient of luminance and obtained results which explain, on the whole, Mach's bands. The effect on Mach bands of the stimulus duration (down to 0.002 second) has been investigated by Thomas (1965) and Matthews (1966), while measurements by Watrasiewicz (1966) show the influence of retinal illuminance, wavelength, object contrast, and polarization. A full discussion of the theory of the production of Mach bands by retinal interactions is given by Ratliff (1965).

The effect of contrast certainly plays an important role in the sharpness of the image, particularly of the retinal image itself. Tschermak

[65a] even infers that contrast is a physiological mechanism to correct the aberration of the image. We believe this hypothesis superfluous, as we have seen that aberrations play only a small part in retinal imagery. It is correct, however, that by increasing the physical differences of illumination of the image, contrast may eliminate part of the veil of diffusion.

If the contrast of a large object with its background is near the differential threshold, the contour tends to disappear and one sees an illusion of diffused edges, which explains the blurred appearance of objects when seen through fog (Langsroth et al., 1947).

This concept of sharpness is also most important in the observation of photographs. Obviously, the resolution of the objective by which the picture has been taken is fundamental, but Higgins and Jones (1952) have proved that the best photographic focus is not necessarily the one that provides the best resolution. The magnification with which the slide is observed must also be taken into account. The investigations of MacDonald (1953) and of Wolfe and Eisen (1953) have established the complexity of the concept of sharpness, which involves, besides the resolution and the gradient of luminance, some psychological factors.

To conclude this discussion of the prominent role of contours we will recall an amusing experiment of Creed and Granit (1928) which reveals the action of contour in the evolution of afterimages. Imagine two small identical luminous disks very close to each other. One is fixated. After they are switched off at the same time, one notes that the afterimage of the fixated disk appears approximately 2 seconds before the other and these images change independently of each other. If the two disks are united by a luminous line, the afterimages evolve simultaneously, or at least show a very marked tendency to synchronize. According to Bartley [3] the variation of the latency of the afterimages with the size would not be an effect of summation but a direct action of the contour uniting the two disks.

Irradiation

The fact that a very bright surface appears larger than a dark surface of the same size has been well known since antiquity. Galileo observed that a planet is smaller in front of the sun than in front of a black background and Kepler (1604) explained this phenomenon as it is still done, merely by the optical magnification of the retinal image due to the defects of the eye, including scattering. In fact, the inaccuracy of accommodation increases this effect greatly, especially when the pupil is dilated, which enlarges the circle of diffusion. Even if only diffraction were involved, it could produce part of the phenomenon. If the retinal illumination in the uniform image of an extended source is unity, the illumination at the edge of of the geometric image is 0.5 and it is still 0.01 at a distance from this edge equal to 5 times the radius of the Airy disk [13]. Consequently, as luminance increases, this edge becomes visible and diffraction alone could account for an enlargement of several minutes of arc.

The first measurements of irradiation were made by Volkmann [69] and Aubert [2], who adjusted the separation of two wires until it seemed equal to the thickness of the wires. They also observed a negative irradiation; that is, a black band on a white background may seem to enlarge when the luminance increases but much less than with white on black. Plateau (1838), following Descartes, thought that a physiological irradiation could occur by the interaction of the retinal receptors. Since then, some cortical processes have been held responsible. Van Heuven (1926) has even attempted to separate the optical and psychological components of irradiation.

The variation of irradiation with luminance was studied by Wilcox (1936) by means of two vertical rectangles of 2.4′ width. The subject adjusted their separation until it seemed equal to the width of the rectangles. With bright rectangles on a dark background there was a minimum of irradiation for a certain luminance. With black rectangles on a light background, negative irradiation occurred at very low intensities, whereas at high intensities the usual phenomenon occurred. Byram (1944) criticized these results. He observed that if the two halves of a line (where one half is more illuminated than the other) are adjusted in width, one observes a positive irradiation (if the mire is light) which increases regularly with the luminance. If the line is black, one notes a maximum in the width at a background luminance of approximately 10 cd/m^2, then a regular decrease as the luminance increases.

Fry and Bartley (1935) brought attention to the role of contours in the effects of irradiation and Kravkov et al [3] confirmed it. For instance, whereas a luminous rectangle 7′ wide seems to enlarge constantly as the luminance increases, a rectangle of 3.4′ presents, on the contrary, a minimum at a certain level because one of the edges affects the other. Fry (1963) studied this physiological irradiation in the case of a disk-shaped test object. We see that these phenomena are much more complex than simple optical theory would predict and that one must be very cautious when utilizing the irradiation of a source in the calculation of the distribution of retinal illumination.

CHAPTER EIGHT

The Visual Field

The vision of details and forms as we have studied it so far did not impose on the observer any obligation with regard to fixation. He looked where he wished to, which means that he utilized his fovea in photopic vision and his parafovea in scotopic vision. We will now extend the scope of our study to the whole *visual field*, that is, all that is seen by the subject for a given position of his head and eyes.

Even the simplest experiment shows the extreme heterogeneity of the visual field. According to the comparison by Hirschberg (1878) the visual field resembles a painting of which a very small central area would be carefully done whereas the rest is barely sketched and becomes rougher and rougher with increasing distance from the fixation point. But this coarseness does not mean uselessness, for this peripheral field is very important. The least change in this field initiates a fixation reflex, a shift of the gaze to bring the fovea onto the object of attraction.

In this chapter a point of the visual field will be defined by its *eccentricity*, that is, the value of the angle η which separates the given point from the fixation point. Moreover, the terms specifying the localization (upper, lower, temporal, nasal) will indicate a direction in the field where it is seen and not on the retina. For example, for the right eye, the temporal field is on the right side of the subject, whereas it is projected on the nasal (left) half of his retina because of the reversal of the image.

Peripheral Dioptrics

The first question to treat is that of the quality of the retinal image as the eccentricity η varies. We have seen that chromatic and spherical aberration practically do not alter the sharpness of the image. This remains true in indirect vision but the other aberrations which are added tend to bring a different conclusion.

Let us, first of all, consider *astigmatism*, which affects the pencil of light because of the oblique incidence of the rays on the corneal and lenticular optical surfaces, as Young (1801) had already considered. The calculation using the schematic eye is easy using the classical formulas. The path of a ray is shown in Figure 38, and the trigonometric calculations are boring but not difficult. It is found that the pencil coming from a point object at infinity passes, after it has reached the vitreous body, through two curved image surfaces; the sagittal surface Σ'_3 which is behind the retina and the tangential T'_3 in front of the retina. Assume 12 mm as the radius of the sphere which is supposed to represent the retina, the eye being emmetropic for $\eta = 0$. To evaluate in diopters the separation of these focal surfaces from the retina, the calculations must be conducted in the opposite direction, that is, to determine in air the conjugates of M_4, one in the sagittal plane, the other in a plane perpendicular to it. On the other hand, it must be kept in mind that the fixation point does not correspond to $\eta = 0$ but to the value of the observer's angle α. Taking $\alpha = 5°$, the amounts of astigmatism calculated are given

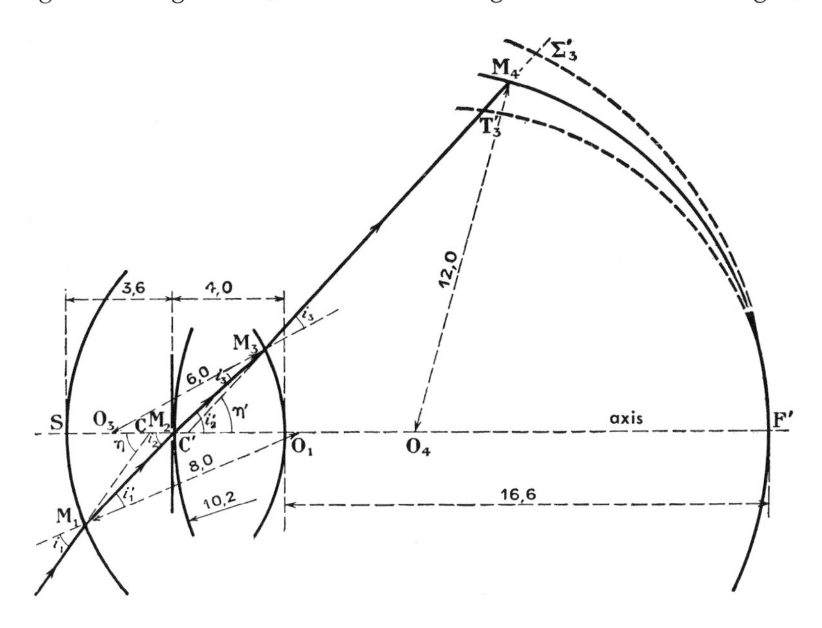

FIG. 38. Calculation of the astigmatism of the image in the periphery (schematic eye). The distances and radii of curvature are indicated in millimeters. The following values of the refractive indices have been utilized: aqueous humor, 1.3374; crystalline lens, 1.42; vitreous body, 1.336. As a simplification, the cornea is considered as a single optical surface. The O's are the centers of the spheres (ocular surfaces and retina). C and C' are the entrance and exit pupils. The second focal point F' is assumed to be on the retina.

TABLE 15

Astigmatism of the Peripheral Image (in diopters)

Eccen-tricity η, deg	SAGITTAL SURFACE				TANGENTIAL SURFACE			
	Nasal side		Temporal side		Nasal side		Temporal side	
	Calc.	Exptl.	Calc.	Exptl.	Calc.	Exptl.	Calc.	Exptl.
0	0	0	0	0	0	0	0	0
10	0.28	−0.05	0	0.04	−0.47	−0.44	0	−0.15
20	0.85	−0.02	0.28	0.43	−1.40	−1.33	−0.47	−0.56
30	1.71	0.27	0.85	0.50	−2.81	−2.19	−1.40	−1.52
40	2.84	1.02	1.71	0.75	−4.68	−3.48	−2.81	−2.60
50	4.27	2.39	2.84	1.43	−7.02	−4.35	−4.68	−3.39

in Table 15, assuming a slight corneal astigmatism (0.05 diopter) which cancels out exactly the astigmatism for $\eta = 0$.

It will be noted that these values are considerable, but beyond $\eta = 30°$ the point of incidence on the cornea is more than 2 mm from the corneal pole, that is, out of the optical zone, and the cornea then flattens, which improves the imagery. The same probably occurs for the posterior surface of the lens, which also becomes flattened away from its pole. The experimental values should then be less than the theoretical values. Let us also note that the aberration called *curvature of field* has no effect in the eye, as the curvature of the retina compensates for this aberration because the retina is situated between the two focal surfaces, which is the best theoretical position.

What do experiments tell us? The first observations made by Weber [71], then by Landolt and Nuel (1872), Druault (1898), and Groenouw (1899), consisted of observing the retinal image on enucleated eyes of albino rabbits. This image is visible through the sclera, owing to the lack of pigments. They observed a satisfactory quality of the image, even for large values of η. Hidano has photographed the retinal image through small apertures made in the sclera of enucleated eyes. He found it to be of good quality even in the periphery. Parent (1881) carried out some measurements by retinoscopy on man and found 0.5 diopter of astigmatism for $\eta = 15°$ and 2.75 at 45°. Druault (1900) noted that the retina was usually situated between the sagittal and tangential surfaces and closer to the former.

The most complete study is that of Ferree et al (1931), who examined 21 subjects with the Henker refractometer. Twelve of the subjects had their retinas between the sagittal and tangential surface and closer to the former. It is the mean of the results obtained with these subjects that is given in Table 16. On the whole, the calculations and the experiments

TABLE 16

Area of the Entrance Pupil in Relative Values for Several Diameters d and Eccentricities η

η, deg	*d*, mm			
	8	6	4	2
0	1	1	1	1
25	0.94	0.94	0.94	0.94
50	0.76	0.76	0.76	0.76
75	0.44	0.44	0.44	0.44
85	0.25	0.27	0.29	0.30
95	0.094	0.114	0.132	0.15
100	0.030	0.042	0.053	0.064

are in agreement, but the experimental values are lower than would be expected theoretically. This is either due to the peripheral flattening of the cornea and of the posterior surface of the lens or to the layered type of structure of the crystalline lens, which may provide a favorable action, as Hermann (1882) believed. For these subjects, who seem to be the most frequent type, if the eye is assumed to be emmetropic for the fixation point it becomes slightly hyperopic when the test is radial, i.e., a line passing through the fixation point and the considered point. It becomes appreciably myopic for a test consisting of a circle with its center at the fixation point. For six other subjects the retina was inside the two surfaces, and the eye was then hyperopic for any peripheral tests. Finally, the last three subjects had a large asymmetry between the temporal and nasal parts. In all cases when the eye presents some astigmatism for $\eta = 0$, this astigmatism becomes nil for a certain eccentricity, which means that the surfaces always intersect.

It is possible to observe on oneself the astigmatism of a pencil corresponding to a large eccentricity. On myself, a myope, a grating invisible in central vision sometimes becomes perceptible peripherally when its orientation is radial. But it would be difficult to make exact measurements by this method because of the difficulty of perception in peripheral vision.

Besides astigmatism, peripheral imagery involves, of course, the same aberrations (chromatic and spherical) as the foveal image. There may be, in addition, *coma*, which is a pear-shaped distribution of light. Tscherning [67] mentioned having noted on himself a pear-shaped coma with its narrow end toward the fixation point, but there are few data on this question. An outline of the calculations has been attempted by van der Pijl (1929). Finally, there is *distortion*, which will be treated in the next section.

In conclusion, it is probable that the peripheral image retains more or less the same structure as the foveal image up to eccentricities of the order of 30°. Beyond this value a high astigmatism is manifest.

Distortion

When the shape of an image on a plane is not identical to that of its plane object, which is perpendicular to the axis, there is *distortion*. Thus, depending upon the sign of the distortion, a grid is deformed into a "barrel" or "pincushion" shape. Tscherning [67] ascribed a certain importance to the barrel distortion by the cornea, but it must be kept in mind that, on the one hand, the retina is not plane and therefore identical shape with the object is not very meaningful and, on the other hand, retinal images are not made to be looked at.

In fact, the notion of a straight line is very precise when it passes through the fixation point. According to Guillery (1899) and Bourdon [6], if three points are illuminated in the dark at the same distance from a subject, he can align the central point (which is fixated) with a precision of 1% of the angle which separates it from the two other points when these three points are along a horizontal or vertical meridian. If the points are along an oblique the errors are much larger.

When a vertical or horizontal line is placed laterally in the field of view and the gaze follows this line, it appears straight when it is in reality convex toward the center of the field of view. This systematic error is associated with Listing's law (see Chapter 10). The same phenomenon occurs in indirect vision with the eye immobile. According to Bourdon, a square of 75 cm width seen 1 m away must, when it is fixated in its center, possess in the middle of each side a sag of the order of 5 cm to appear square. Helmholtz thought that this phenomenon was a consequence of the preceding. R. Fischer (1891) assumed that the lines which appear straight are those which are projected on the retina along the small circles whose planes are parallel to great circles passing through the fovea. In any case distortion does not seem to play a role in these experiments.

Vignetting

If the retina were plane, retinal illumination would decrease as $\cos^4 \eta$ when the subject looks at a field of constant luminance filling the whole visual field.

Because of the curvature of the retina, the retinal image reduces in size as η increases, and this effect compensates almost completely the factor $\cos^4 \eta$ which is due to the oblique incidence of the light on the entrance pupil. Taking the theoretical diagram (Fig. 38) and calling the foveal retinal illumination unity, this illumination reduces to 0.97 for $\eta = 30°$, and to 0.91 for $\eta = 50°$. This variation is negligible and it is even possible that it is masked by the diminution of light absorbed in the transparent media of the eye, the oblique course being shorter. Moreover, Weale (1955) has shown that variation of the reflection factor on the cornea is negligible up to $\eta = 50°$.

When the light incident on the pupil of the eye is very oblique, a *vignetting* occurs which certainly reduces the retinal illumination markedly. Spring and Stiles (1948) have measured the relative area of the pupil seen from different angles and the values they obtained are given in Table 16. They have been verified by Sloan (1950). Some direct measurements of the illumination have been carried out by Gross (1931) through the sclera of an albino rabbit but are to be considered with caution.

Absolute Threshold of Small Sources

We have already mentioned that at photopic levels it is the fovea which is most sensitive; this simply means the source must be fixated to be seen. According to Tousey and Hulbert (1948), if E is the illumination produced in the plane of the subject's pupil by a point source which is just visible to the fovea on a uniform background of 3900 cd/m^2 luminance, it must be increased to $2.5E$ for $\eta = 2°$, $7E$ at 4°, and over $50E$ at 10°.

This is, of course, no longer correct in scotopic vision and the region of maximum sensitivity is around $\eta = 15°$, according to Livingstone (1944), but there are very marked individual variations and frequent retinal asymmetries. Some subjects present a crescent-shaped region where the sensitivity is best (Pickard, 1935). Weinstein and Arnulf (1946) found practically a plateau between 5 and 35° and Cabello and Stiles (1950) found it between 7 and 20°. Figure 39 represents the measurements of Sloan (1947) on 101 subjects between the age of 14 and 70 years, in the horizontal meridian after complete dark adaptation. The experiment was performed in white light with a test of 1° in diameter. Mention should also be made of measurements by Meur (1965) on 50 young soldiers.

The classical method of *isopters*, which consists of moving a small white test on a black background until it is seen (coming from the periphery) or no longer seen (coming from the center), is not a measurement of the absolute threshold but a complex function involving the

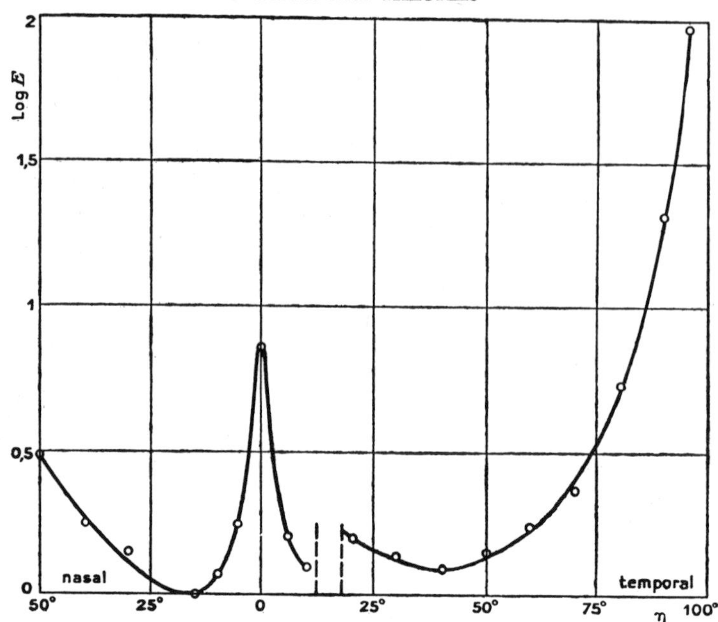

FIG. 39. Variation with eccentricity η of the absolute threshold E of a point source in the dark (after Sloan). By convention, $E = 1$ at 15° nasally. The temporal region where the curve is interrupted is the blind spot.

differential threshold and the acuity (see Osterberg, 1933). The reflection factor of the "black" background intervenes as well as the illumination (see Goldmann in [15]). The adaptation of the subject is not usually well specified either. However, to give the reader a better insight into this question, we show in Table 17 the maximum eccentricities at which a normal subject still sees a circular test of diameter u of white paper of 0.8 reflection factor on a black background of 0.04 reflection factor. The

TABLE 17

Maximum Eccentricity η at Which a White Test of Diameter u on a Black Background Is Seen

u, min	η, deg			
	Temporal	Inferior	Nasal	Superior
1	10	8	6	4
2	21	20	20	18
4.8	36	32	31	29
7.8	60	44	44	40
10.6	62	48	48	42
19	85	65	60	56
275	104	76	62	69

TABLE 18

Effect of the Reflection Factor ρ of Background on the Peripheral Perception of a Circular Test of Diameter u and Reflection Factor 0.78

	η, deg	
ρ	$u = 1°$	$u = 10'$
0.04	71.3°	57.3°
0.19	68.3	43.7
0.26	66.9	37.2
0.41	63.0	24.4
0.49	60.3	16.8
0.60	55.8	
0.64	44.0	
0.69	28.3	

illumination is of the order of 75 lux. The values in the table are the findings of many authors, compiled by Traquair [64].

Table 18 shows the effect of contrast between the test and the background. These are the results of Ferree and Rand (1931). The circular test of diameter u and reflection factor 0.78 was seen on a background of variable reflection factor ρ. The numbers represent the mean of the maximum eccentricity in 8 meridians. The illumination was 75 lux.

Vision of a Black Point

The peripheral perception of a black point on a light background was studied by Hueck (1840), Groenouw (1893), Rönne (1915), and by many who wrote on perimetry, but unfortunately with a lack of thorough specification of conditions of the experiment, which makes most of the results almost useless. On a background of 3800 cd/m^2 luminance, Byram (1944) found the diameter of a just perceptible black point to increase from 28″ at the fixation point to 1′ at $\eta = 1°$, 3′ at 10°, and 30′ at 30°. The effect of exposure time was studied by P. W. Cobb (1923) for $\eta = 2°$ and by Bouman and Blokhuis (1952) for $\eta = 7°$.

In scotopic vision, Pirenne (1945) determined on two subjects the smallest area of a just-detectable black disk seen in the center of a screen of luminance L, 46° in diameter. The center of the black disk was placed at a constant eccentricity $\eta = 20°$. The subject looked through an artificial pupil 2 mm in diameter. Flashes of 0.03 second duration illuminated the field, and the threshold was determined for a 50% probability of detection of the disk. The results (Fig. 40) are in good agreement

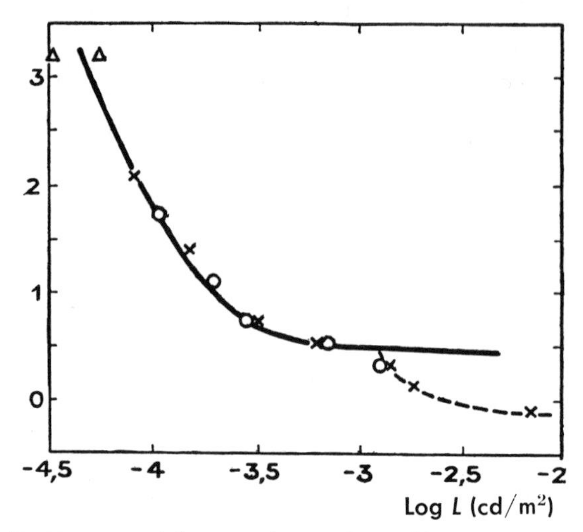

FIG. 40. Variation of the area of a just-detectable black disk presented on a background of luminance L. The ordinates represent the log of the area of the disk in square degrees. The full curve was drawn using the hypothesis that each functional unit receives at least five quanta. The circles and crosses are the experimental values, the triangles the absolute threshold of the background (after Pirenne).

with quantum theory. Assume that a flash of light leaves, in the shadow cast by the disk, an area containing N retinal receptoral units. If in each flash the probability of exciting one of these units is p, the number of stimulated units (or, on the contrary, remaining unstimulated) is pN. This product varies with p, therefore with the luminance, which means that the apparent fineness of the retinal mosaic increases with the level of luminance. By subjecting this hypothesis to calculation, the theoretical curve agrees well with the experimental points if it is assumed that five quanta are necessary to stimulate a functional unit. This curve tends toward a plateau for a disk diameter of about 2°, which, at low luminances, is the dimension of a functional unit. But when L is above 10^{-3} cd/m^2 (which would correspond to about 10^{-5} cd/m^2, with steady exposure and no artificial pupil), another mechanism intervenes and the experimental curve (dashed line) deviates from the theoretical curve.

Very near the absolute threshold of the background, the black disk becomes large. The International Commission for Optics (1950) has defined a "normal" observer for night vision as one who is capable, after complete dark adaptation, of perceiving a black disk 10° in diameter on a uniform background of 2×10^{-6} cd/m^2 when the center of the disk is 15 to 35° from the fixation point. At very low luminance, longer exposure time is recommended, of the order of 5 seconds. Furthermore, the edge of the disk looks blurred and progressively blends into the luminance of the background. Pirenne (1948) showed that with a disk of 20° of which the center is at $\eta = 20°$, a subject is unable to distinguish between the disk and each of its component halves (horizontal diameter) presented separately to him in flashes.

Different Tests

All the tests which have been studied in central vision in the previous chapters, without imposed fixation, have also been studied as a function of eccentricity. We will review them briefly and all the results will be expressed in *relative acuity*, unity being ascribed by convention to the central acuity.

1. Two points of light on a dark background. This test was studied by Wertheim (1887), then in greater detail by Clemmesen [12] on five emmetropic subjects with an artificial pupil of 3 mm and a diurnal adaptation maintained constant by a peripheral field ($\eta > 30°$, $L = 60$ cd/m^2). Individual variations are considerable. When $\eta > 5°$ the limit of resolution becomes much larger if it is measured by moving the points apart until they are just seen separated than by moving them toward each other until they just appear to be in contact (Fig. 41). The best results are obtained with just visible points. At a level 100 times higher, the acuities reduce by about 25% for all eccentricities. Finally, the fact that the test is exposed for only 0.1 second instead of a steady exposure changes practically nothing in foveal vision but improves the peripheral results greatly. Clemmesen explained this curious result by a theory of rhythmic activity of the receptors. Flamant (1950) used two points of 7″ which produced an illumination of 10^{-4} lux in the plane of the pupil. Oliva and Aguilar (1956) used two points producing an illumination of about 10^{-7} lux and presented for 0.3 second.

2. Two black points on a bright background. This case was studied by Aubert and Förster (1857). They found that acuity is slightly better along the horizontal meridian passing through the fixation point than in the vertical meridian. This result is most commonly found.

3. The case of two black lines on a white background was studied by Wright [74], but under scotopic conditions (dark-adapted eye, luminance 2×10^{-3} cd/m^2). In this case the best results are obtained for $10° < \eta < 15°$ with a separation of 30′ between the lines. At the fovea Wright found 80′, which was also found at $\eta = 40°$.

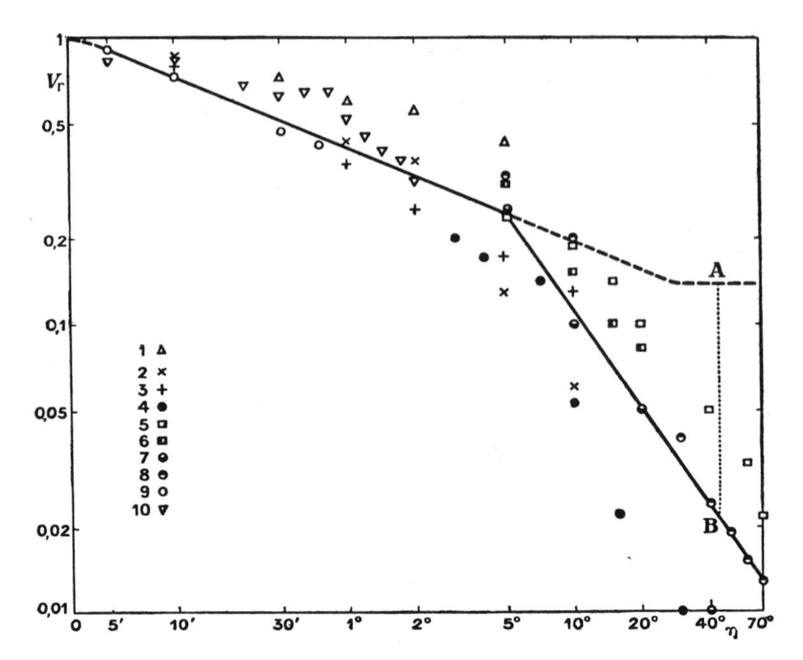

FIG. 41. Variation of relative acuity V_r with eccentricity in photopic vision. The points are borrowed from the following authors. (1) Wertheim (two points of light); (2 and 3) Clemmesen (two points of light moved toward and away from each other); (4) Aubert and Förster (two black points); (5) Wertheim (grating); (6) Fick (grating); (7) Hirschberg (Landolt ring); (8) Ruppert (Landolt ring); (9) Jones and Higgins (Landolt ring); (10) Flamant (two points of light).

The solid curve represents equation (49); the dashed curve represents equation (50). The two scales are logarithmic, but by convention we have represented $\eta = 0$ (which should be at infinity to the left).

4. Dor (1873), Wertheim (1894), and Fick (1898) utilized a grating. The acuities along the horizontal meridian are represented in Figure 41. Wertheim also determined curves of equal acuity in the visual field by the classical representation of *isopters* of Hirschberg (1878). The curves obtained have roughly an elliptical shape with the long axis horizontal. The ratio between the long and short axes is about 10:7. Weymouth and his collaborators (1923–1928) have studied the case of a small grating (square of 10′) at very slight eccentricities, i.e., in foveal vision. The limit of resolution increases more or less linearly up to $\eta = 1.5°$, and for this eccentricity the limit of resolution is about twice what it is at the fixation point. Weymouth thought that this very regular variation, observable even at $\eta = 10'$, was the fundamental basis of fixation. This idea had been already put forward by Javal [26]. There would not be one special cone at the fixation point but a gradient of discrimination which is so sharp that it would be practically always the same point on the fovea ($\pm 1'$) which is used to fixate. Opposed to this thesis, Hartridge [19d] maintains that there is an anatomical center of fixation, or rather several centers, depending upon the wavelengths, the two most extreme bands of the visible spectrum (corresponding to blue and red) being separated by a distance of several minutes. But the experiments that he describes on this matter can be explained by the chromatic variation of the angle α, which we discussed in Chapter 1 (Fender, 1955).

5. Landolt rings were employed in peripheral vision by Hirschberg (1878) and Ruppert (1908) and followed by many other authors (see Fig. 41). Training has a considerable effect, as for most tests of peripheral vision (MacQuarrie, 1955). Conner and Ganoung (1935) have shown that the variation of luminance and contrast had less effect at 5° eccentricity than in foveal vision, as is shown in Table 19. At scotopic luminances peripheral vision is better because the apparent brightness of the test is greater. At equal apparent brightness the acuity of a given region of the retina remains nearly constant; this can be verified by having one eye light adapted and the other dark adapted (von Kries, 1895). Bloom and Garten (1898), however, found that the visual acuity of the dark-adapted eye was invariably less than that of the light-adapted eye.

Mandelbaum and Sloan (1947) studied the effect of luminance with complete adaptation to each level using a Landolt ring of contrast 1. The ring was flashed for 0.2 second duration. The curves of Figure 42 are

TABLE 19

Visual Acuity Using Landolt Rings (Reciprocal of the Size of the Gap That Can Be Resolved, in Minutes of Arc) as a Function of the Luminance L of the Background and the Contrast C between the Ring and the Background in Foveal Vision and for $\eta = 5°$. *Two Subjects Were Utilized*

	L, cd/m^2							
C	4.5×10^{-4}	3.4×10^{-3}	3.4×10^{-2}	6.9×10^{-2}	0.15	0.34	1.1	3.4
$\eta = 0°$								
0.93	0.040	0.090	0.396	0.515	0.609	0.710	0.960	0.930
0.76	0.034	0.073	0.350	0.427	0.501	0.663	0.849	0.865
0.39	0.028	0.043	0.188	0.309	0.340	0.417	0.628	0.689
0.28	0.014	0.032	0.139	0.220	0.268	0.348	0.529	0.633
$\eta = 5°$								
0.93	0.058	0.100	0.146	0.162	0.189	0.227	0.252	0.289
0.76	0.046	0.089	0.119	0.131	0.162	0.187	0.225	0.234
0.39	0.028	0.053	0.101	0.105	0.113	0.113	0.138	0.154
0.28		0.040	0.070	0.093	0.083	0.112	0.105	0.121

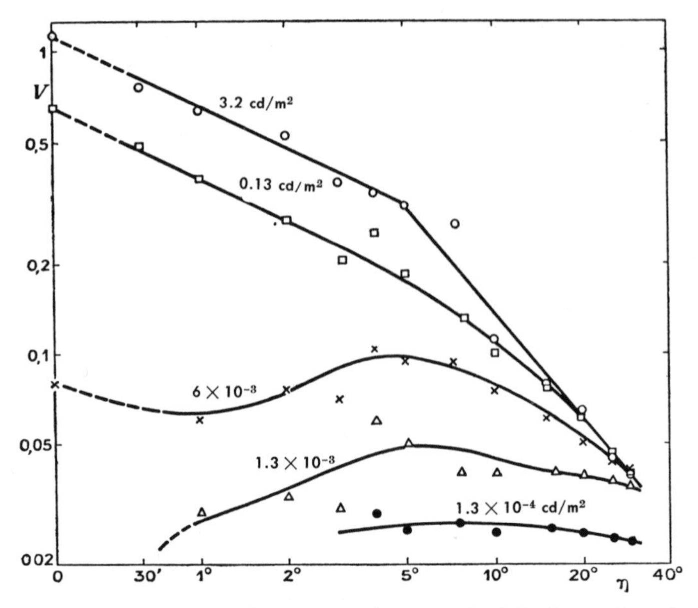

FIG. 42. Variation of the acuity V (reciprocal of the limit of resolution of a Landolt ring in minutes of arc), with eccentricity η for several luminances (from Mandelbaum and Sloan). The conventions along the axes are the same as in Figure 41.

the findings of one trained observer, but these results have been verified on 18 subjects and in spite of large individual variations, they always resemble the curves of Figure 42. The fovea becomes blind at very low luminances (a few thousandths of a candela per square meter).

Jones and Higgins (1947) studied the variation of acuity around the fixation point with a Landolt ring of contrast 1 on a background of 10° and luminance 90 cd/m² (Fig. 41).

Finally, measurements have been made at very low luminances by Pirenne and Denton (1952). At a luminance five times the absolute threshold, acuity was 0.003 and became 0.1 around 10^{-3} cd/m². In this interval calculations show that the number of quanta which reached the retina coming from the gap of the ring was almost constant.

6. Snellen letters have been tried but they quickly become impossible to recognize. According to Königshöfer (1876), the acuity at $\eta = 1°$ falls to between 0.3 and 0.2. Generally speaking, the recognition of forms in peripheral vision is very difficult and depends considerably upon experience and also on the complexity of the figure. A triangle, for example, is recognized as such at 5° eccentricity with a dimension no more than 2.5 times the size necessary in foveal vision [12] at high lumi-

TABLE 20

Threshold of Nonalignment as a Function of Eccentricity

	η, deg					
	0	1	2	5	10	20
According to Bourdon	7″	23″		4′	7′	14′
According to Hofmann	9″		54″	2′	4′	

nances. Psychologists and medical men have taken a great deal of interest in this problem of form perception in peripheral vision (Zigler, Galli, Collier, Renfrew, and others) but it is beyond the scope of our subject. The measuring ability of the eye is very poor in peripheral vision. According to Münsterberg (1889), the comparison of two parallel segments 5° apart is seven times more erroneous if a fixation point is imposed in the middle of the interval between the two segments than if fixation is free to move. Finally, aligning acuity has been measured in peripheral vision by Bourdon [6] and by Hofmann [23]. The thresholds of nonalignment are given in Table 20 for subjects not trained to use their peripheral vision. Le Grand (1963) showed that prolonged training could improve the results by a factor of 2 at an eccentricity of 10°.

Of course, the inferiority of the lateral retina for the recognition of forms is, at low luminances, amply compensated by an increase in apparent brightness. For example, in the recognition of the silhouettes of ships, Sperling and Farnsworth (1950) showed that for all luminances below 10^{-2} cd/m² and all contrasts between 0.97 and 0.10, the most favorable region of the retina is 6° away from the fixation point. It is therefore necessary for night watchers to practice using this retinal region. According to Sperling (1954), the orientation of a grey rectangle of $1° \times 0.33°$ on a white backgound is best perceived between 2 and 8° eccentricity, for any contrast.

In conclusion, examination of Figures 41 and 42 shows that, in spite of the large differences among authors and the tests employed, the variations of visual acuity can be described as follows, provided the perception of the form has little effect: Using logarithmic coordinates for both acuities and eccentricities, one obtains two straight lines in photopic vision, intersecting around $\eta = 5°$. The relative acuity can therefore be expressed by the formula

$$\log V_r = a \log \eta + b \tag{49}$$

By convention $V_r = 1$ for $\eta = 0$, but this formula is then no longer applicable. The constants have the following values, if η is expressed in degrees:

$$0.1° \leqslant \eta \leqslant 5° \qquad a = -0.30 \qquad b = -0.39$$
$$\eta \geqslant 5° \qquad a = -1.12 \qquad b = +0.18$$

In scotopic vision the kink in the curve disappears and the acuity becomes nearly constant all over the retina (except for the fovea which is blind) with a slight maximum around $\eta = 6°$. Obviously, equation (49) must only be considered to give an order of magnitude. From one subject to another there are, in fact, enormous differences, as is shown, for example, by the statistical study of Low (1946) on 100 subjects and seven different retinal regions. In peripheral vision, in fact, practice plays a considerable role, and according to Dobrowolsky and Gaine (1876), 6 weeks of repeated exercises are needed to obtain the best acuity. Strabismic patients who develop a lateral false fovea—that is, a region which poorly replaces the true fovea—do so only with much training. Careful observation of a central test enables one to improve perception of it in the periphery (Hartridge). Without moving the eyes one can actually shift attention in the visual field. Purkinje [57] noted that if two tests were presented, one to the fovea and the other to the peripheral retina, the acuity of each changes with attention, although the gaze has not moved. Holm (1923) and Clemmesen [12] have verified this; for $\eta = 2.5°$, central acuity decreases by half if attention is placed upon the eccentric point and peripheral acuity decreases by 25% if the attention is placed upon the central point. Guratzsch (1929) and Grindley (1931) have also noted that placing attention upon a peripheral point of the visual field improves the acuity less at that point than it alters the vision in the remaining field. According to Bethe (1908), the minimum time necessary to change attention from one point of the visual field to another is slightly less than 0.2 second.

To end this section we will cite a study by Arnulf et al. (1954) on the effect of pupillary diameter and focusing at very low levels in peripheral vision. It is found that a tolerance of 6λ (that is, 24 times Lord Rayleigh's criterion) is permissible in an instrument which is to be used at night, but of course this device would be useless even in crepuscular vision, that is, when the fovea just begins to play a role ($L > 5 \times 10^{-3}$ cd/m^2).

Theories of Photopic Acuity in Peripheral Vision

It is obvious that the dioptrics of the eye, which are of fairly good quality up to 30° eccentricity, have no effect whatsoever on peripheral acuity. The continuous theories based on intensity discrimination seem to provide little explanation in this case. Flamant (1950) has found a parallel variation between visual acuity measured with two points of light on a black background and the relative intensity discrimination of these two points when they are 10′ apart and situated on a circle of which the center is the fixation point. But this parallelism only persists up to $\eta = 40'$. Beyond this eccentricity the intensity discrimination remains almost constant (up to 2°), whereas the acuity still decreases. More peripherally, though, the measurements of intensity discrimination become meaningless, or rather they represent acuity measurements, which is a vicious circle.

Therefore the solution has usually been sought in anatomy and physiology. We will briefly summarize several points of view which have been advanced.

The simplest concept is that in photopic vision the cones would be the elements providing form vision. In the fovea the cone distribution has been described, following Schultze (1866), as a regular mosaic paving, hexagonal or pentagonal. The rows of cones would resemble circles intercrossing one another, the angle of intersection being about 60° (when looking at it much more closely, the pattern seems less geometric). Outside the fovea, the rods appear and quickly become more numerous than the cones, which become farther apart from one another and increase in size at the same time. The mean distance x between two neighboring cones increases with eccentricity. In Figure 43 we have reproduced the classical results found on human retinas by Oster-

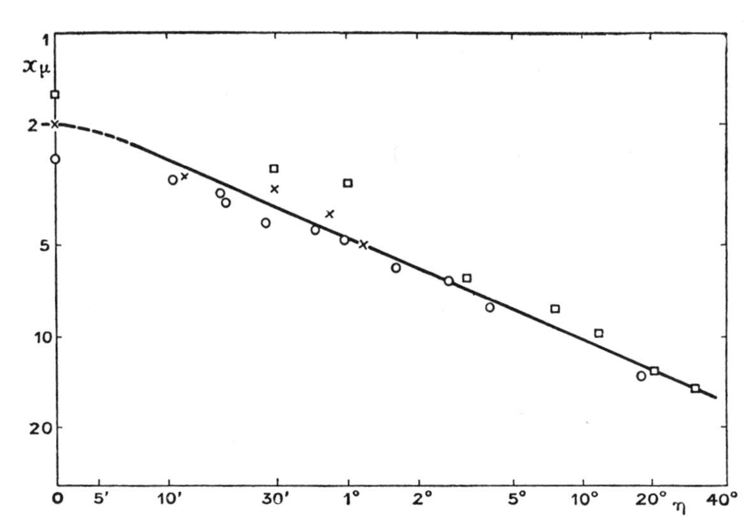

FIG. 43. Mean distance x between neighboring cones as a function of the eccentricity η, according to Osterberg (circles), Polyak (squares), and Rochon-Duvigneaud (crosses). Logarithmic scales.

berg [48], Polyak [55], and Rochon-Duvigneaud [59]. In logarithmic coordinates the points (for $\eta \geqslant 0.1°$) follow approximately the equation

$$\log x = 0.33 \log \eta + 0.68 \tag{50}$$

where x is expressed in microns and η in degrees. For $\eta > 30°$, x becomes more or less constant. Comparison with equation (49) shows that the curves have the same slope between 5′ and 5°. Hence, two conclusions present themselves. In photopic vision, the acuity is directly related to the density of cones up to 5° eccentricity. If it is agreed that in the fovea each cone possesses a direct nervous conduction to the brain, this prerogative must also be extended to parafoveal cones up to $\eta = 5°$ (the only difference is that some rods in the parafovea may use the same pathway).

It is rather remarkable that the limit of the foveal area does not appear at all from these findings of acuity. All those who have studied this problem have noted this peculiarity, especially Clemmesen [12], who showed a particular interest in it. The only authors who claim to have found, by measurements of visual acuity, a "physiological fovea" are Adler and Meyer (1935), who observed a constant value for the discrimination of the width of a line up to an eccentricity of 42′.

It must also be noted that our conclusion does not mean at all that acuity can be completely explained on purely anatomical bases. We mentioned in Chapter 6 that anatomy was only one element (necessary but not sufficient) of the theories of acuity. We only want to say that the "sensorial circle," to use Weber's terminology [71], or "receptive unit," as Kölliker (1854) called it, contains a constant number of cones up to $\eta = 5°$, whether this number is 1 as ten Doesschate (1946) assumed, 3 as Berger and Buchtal (1938) thought, or 4, as in Piéron's theory of the receptor tetrad.

With knowledge of the mean separation between cones, one can calculate the approximate number of cones in areas of the retina at given eccentricities. Table 21 contains the results of this calculation with the following assumptions; standard eye of Figure 38 (the retina is a sphere of 12 mm radius), the cones distributed in hexagonal paving; the separation is given by equation (50) up to $\eta = 30°$, then it is constant. The values in parentheses are hypothetical.

The calculated total number of cones is of the same order as the findings of Krause (1880), who found from 6 to 7 million.

Beyond $\eta = 5°$, the discrepancy between the curve representing the distance between cones (dashed curve in Fig. 41) and the acuity curve can be explained by assuming that a certain number of cones are connected with one optic nerve fiber. This hypothesis is necessary because

TABLE 21

Cones in Retinal Areas

Limits of the zones (η, deg)	Area on the retina, mm^2	No. of cones (calc.)	Mean no. of cones per optic nerve fiber	No. of optic nerve fibers
0–5	6.7	200,000	1	200,000
5–10	19.8	300,000	3	100,000
10–20	77.4	700,000	6	115,000
20–30	121	800,000	15	53,000
30–40	157	900,000	30	30,000
40–50	176	1,000,000	45	22,000
50–60	183	1,000,000	70	14,000
60–70	170	900,000	100	9,000
> 70		(700,000)	(100)	(7,000)
		6,500,000		550,000

the evaluation of the total number of optic nerve fibers in man ranges from 500,000 (Zwanenburg, 1915) to 1,000,000 (Krause, 1880; Bruesch and Arey, 1942), that is, considerably less than the number of cones. One can roughly evaluate, in each zone of Table 21, the mean numbers of cones connected to one optic nerve fiber in the following manner. Let us consider, for example, the zone 40 to 50°. At 45° a line *AB* is drawn between the two curves (Fig. 41) and it is measured on the logarithmic scale (that is, 0.82), it is doubled to obtain x^2, and the antilogarithm is then determined (that is, 44; we have rounded this number to 45). The values of Table 22 have been obtained in this manner. The total number of 550,000 optic nerve fibers is reasonable.

Similar calculations have often been carried out with very different results. Utilizing the same acuity results as Wertheim, Lythgoe (1938) found 250,000 sensorial units, Piéron [52] 200,000, and ten Doesschate (1946) 830,000. These discrepancies are caused by the initial hypotheses made on the relationship between acuity and receptive unit as well as on the number of cones assumed to be in that unit. The method presented above avoids uncertain hypotheses since it merely utilizes a parallelism, then a deviation between experimental curves.

It is known that the rods (110 to 130 million) are more numerous than the cones and consequently many more are connected to each optic nerve fiber. It is, in fact, totally unknown whether they have special optic nerve fibers or share a fiber with one or several cones. According to Chiewitz (1883), there would be about 11 visual cells (cones and rods) for each ganglion cell (that is, for each optic nerve fiber) at $\eta = 10°$, 42 at 16°, and 80 at 20° (see also Kolmer, 1936).

Theories of Scotopic Vision

The symmetric hypothesis to the one we used in the preceding sections consists of assuming that scotopic vision is provided by the rods. Figure 44 represents the distribution of rods in the horizontal meridian of the retina according to Osterberg (1935). It is obvious that the maximum density around $\eta = 15°$ must be related to the minimum absolute threshold that is observed in the same region when the retina is totally dark adapted. But to make the measurements comparable at high eccentricities, the diminution of illumination due to vignetting must be taken into account. Using the values of Table 16 for $d = 8$ mm, we have corrected the experimental curve. We know, on the other hand, that the effect of vignetting is partially compensated for by the closer distance between the retina and the exit pupil. Therefore it is to be expected that the absolute thresholds will be situated between the experimental curve and the corrected curve. This is noted, in fact, in Figure 44, where the values of the absolute thresholds of Sloan from Figure 39 have been included. The use of logarithmic ordinate scales facilitates these comparisons, since the contant factors (unknown) are reduced to simple translations of the curves parallel to the ordinate axis.

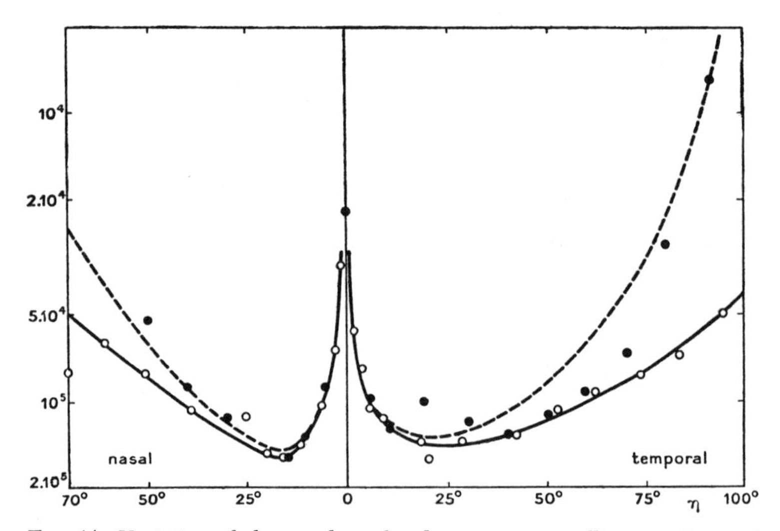

FIG. 44. Variation of the number of rods per square millimeter of retinal surface with the eccentricity η, according to Osterberg. The scale along the ordinates is logarithmic. The white circles represent the experimental points. The dashed curve is derived from the experimental one and corrected for the vignetting effect. The black circles represent the measurements of the absolute threshold (Sloan; see also Fig. 39).

The agreement, as satisfactory as we could possibly hope for, between the thresholds and the density of rods is one of the strongest arguments that we could possibly call upon in support of the famous *duplicity theory,* according to which the retina consists of two functional entities, one to be used during the day (composed of cones), the other at night (composed of rods).

Theories of scotopic acuity are less satisfactory. It is probable that, on the one hand, quantum considerations play a role of which we have already given a few examples, and that, on the other hand, the receptive unit varies with the level of luminance and becomes as large as about 2° in diameter according to the electrophysiological researches of Rushton (1949) and Kuffler (1953) on mammals. This unit would be made up of 40,000 rods, as an order of magnitude. At higher levels it would split into units of smaller dimensions (Pirenne and Denton, 1952; Pirenne, 1953). Several schemes have been proposed on this subject but are still very hypothetical.

CHAPTER NINE

Entoptic Phenomena

After the rather arduous preceding chapters we will take a little rest and describe some phenomena that anyone can observe without any apparatus, on himself, and which render perceptible certain peculiarities of the retina or the globe of the eye. They are referred to as *entoptic*, to recall their intraocular origin.

The Blind Spot

The region of the visual field which corresponds to the *papilla* of the retina, that is, to the entrance of the optic nerve, is called the *blind spot*, or *punctum caecum*. There are no visual cells in this region, hence an insensitivity to light and therefore a *scotoma* (which means a blind region of the field). This scotoma is very large and it may seem surprising that it was not observed until the seventeenth century. According to Evans (1938), it is not just a play on words to say that Hippocrates and Galen would have made dark allusions to it, had they known of it. Mariotte (1668) discovered the blind spot when investigating what was happening at the entrance of the optic nerve. This discovery became celebrated and Mariotte was called to the court of King Charles II of England to perform a demonstration. It has been recounted that the sovereign very much enjoyed seeing his courtiers without their heads. At a distance of 2 meters, a human head fits easily into the scotoma. The existence of one's blind spot is easily verified by looking monocularly at a drawing such as in Figure 45. It is not the eccentricity of the black or white circle which makes it invisible, because it is seen perfectly through the other eye.

Donders (1852) demonstrated that when an object is projected on the blind spot with an ophthalmoscope it is invisible to the subject. If the

FIG. 45. Demonstration of the blind spot. By looking at the upper figure (black cross) with the right eye at about 20 cm or the lower figure (white cross) with the left eye (the other eye being closed), one sees the black circle or the white circle, respectively, disappear from the visual field. By holding the book upside down, the role of the two eyes is reversed.

object is a very intense source—the full moon at night, for instance—scattered light is still seen because the papilla, being very white, scatters the light very much. By projecting a source on the blind spot, Asher (1951) recorded a noticeable potential in the electroretinogram, which demonstrates the important role played by scattered light in the electrical responses of the retina. If the source is small the scattered light is sometimes reddish when the image falls on one of the large retinal vessels which emerge at the papilla (Fick and du Bois-Reymond, 1853). It is probable that the importance of the scattering of light by the papilla is at the origin of all the legends regarding an alleged feeble sensitivity of the blind spot to light (Wolf and Morandi, 1962).

The blind spot is not circular. One can easily draw one's own blind spot by fixating monocularly a fixed point on white paper and moving a brightly colored pencil which has a very visible tip. The points where the tip of the pencil appears are marked on the paper moving from the center outward. Thus one obtains a form of an irregular ellipse longer in the vertical than in the horizontal with horns that are the beginnings of the large blood vessels.

According to Dubois-Poulsen [15], who compiled the results of several authors, including himself, the vertical diameter is of the order of 7 to 8° and the horizontal diameter 5 to 6° (that is, 11 full moons can be placed side by side, as Helmholtz stated). Anatomically the papilla is smaller, and it is rare that its diameter is larger than 5° (Nussbaum, 1920). The internal edge of the spot is between 12 and 13° from the fixation point. The center of the spot is not on the horizontal meridian of the fixation point but 2 or 3° below. Usually, the blind spot of the uncorrected myope is closer to the fixation point, and further away for the hyperope; the distance between the papilla and the fovea remains the same, but the eye is either elongated or shortened, depending upon the ametropia. The correction restores the customary position.

If one searches for the blind spot in the dark with a small source of low intensity it is noted that the spot is larger than in diurnal adaptation. The papilla is surrounded by an *amblyopic* region of about 1° width where the sensitivity is reduced but not nil. Therefore, the apparent dimensions are influenced by the luminance of the test, its contrast and size, in exactly the same way as the isopters of acuity in the periphery. Emotional or vascular factors also may influence the size of the blind spot; the spot would be larger in the morning than at night (from 2 to 5°). Binocular interaction also has an effect, as we will see in Chapter 12. Fatigue, fright, compression of the globe of the eye, stimulation of the nasal mucosa, and the effect of certain drugs can all slightly enlarge the blind spot (see Dubois-Poulsen).

The blind spot is a *negative* scotoma, as it is not perceived by normal subjects. Lagrange (1934), however, noted that some neuropathic subjects could perceive their blind spot as an annoying positive scotoma creating an apparent black hole in the visual field. With practice a normal subject can sometimes perceive a grey spot on a uniform background. According to Köllner (1923), if the eyes have been closed and then one eye is opened in front of a grey surface, one sometimes sees for a fraction of a second a darker disk corresponding to the blind spot of the open eye and a lighter disk corresponding to that of the closed eye. Rapid movements of the eyes, which slightly pull the optic nerve, produce phosphenes in the dark around the blind spot, which appears as a dark disk surrounded by light crescents.

Usually one is not even aware of the existence of blind spots. In binocular vision the blind area in one eye is compensated by vision in the other eye, but in monocular vision one has, not a perception of darkness in the region of the visual field which corresponds to the papilla, but no perception at all. This absence remains unnoticed, because the visual field surrounds the blind area. Thus when one looks at Figure 45, one sees either white or black, whether one looks at the pattern on top or at the bottom of the figures. This "filling-in" has given rise to many psychological studies (see Walls, 1954). The lines or forms which cross the blind spot are completed without the subject's awareness, exactly as the meteorologist who draws contour maps draws them by interpolation in the regions of the ocean from where he has not received information. Sometimes the blind spot produces deformations as if the spot inhibited the area around its edge (Bender and Teuber, 1946). Objects near the spot are slightly diminished; a square drawn by four points seems smaller when it surrounds the blind spot than elsewhere (Hofmann [23]).

Vascular Scotomata

The blind spot is not the only scotoma of the visual field. Aubert and Förster (1857) have described others in the peripheral retina which are either temporary or permanent. Bull (1895) has attributed them to the distribution of blood vessels which pass over the retina and cast a shadow upon it. With small test objects of low contrast, a moderate illumination, and lots of patience, Evans (1938) has been able to trace the vessels up to an eccentricity of 35° from the fixation point. In general it is difficult to reach 25°. Figure 46 represents a typical plotting (Weekers, 1945) of what ophthalmologists have named normal *angioscotomata.* The large trunks emerging from the blind spot are about 1° in diameter and then become thinner. No vessel is usually visible for $\eta < 10°$.

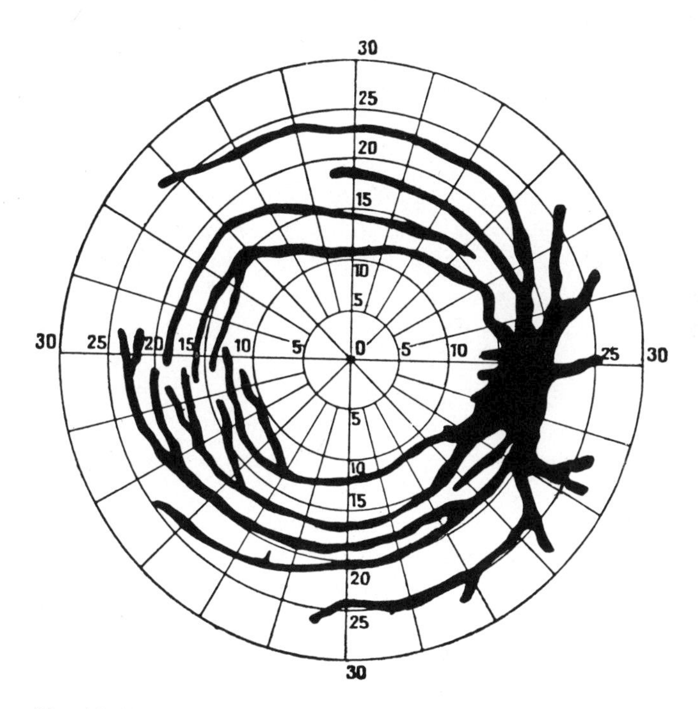

Fig. 46. Normal angioscotomata (Weekers). The fixation point is at 0 and the circles represent the loci of constant eccentricity (η in degrees). This is the visual field of the right eye.

As Goldmann (1947) rightly pointed out, the nature of the angioscotomata totally differs from that of the blind spot. The retina is not insensitive behind the blood vessels and these are not opaque; they only reduce the retinal illumination. Therefore the intensity must be nearly liminal to observe these scotomata, and the effects of inhibition are very marked.

When one looks at a background of uniform luminance, one does not see the shadow of the blood vessels (except sometimes for a short moment on waking) because it is a permanent feature of the visual field. But any modification of the usual shadow renders the vessels visible. Purkinje [57,58] has described three methods to produce this. The light may be thrown into the eye through the sclera by focusing the image of a small source onto it through a lens, or it may enter the eye obliquely through the cornea. The third method of observing the retinal blood vessels consists of looking through a small aperture at the sky, at the same time moving the aperture to and fro in front of the pupil; the shadow is smaller than with the full opening of the pupil and the movement makes the retinal tree more visible. O'Brien (1950) perfected the latter method by utilizing a hole which varies in size depending upon its position in front of the pupil in order to compensate for the Stiles-Crawford effect. Besides the vessels, one sees diffused spots which are not vascular but may be local variations of the Stiles-Crawford effect (groups of cones of abnormal inclination). Finally, Sewall (1934) showed that one could make the vessels visible by utilizing as a source a fluorescent light emitted by the crystalline lens when it is irradiated with ultraviolet light.

Recall that it was by measuring the parallax of the shadow of a blood vessel when the source is moving, with the light incident through the sclera, that H. Müller (1853) provided the proof that it is the layer of visual cells (rods and cones) which is the stratum of the retina sensitive to light. By parallax he assessed the distance between the vessels and the photosensitive layer to be of the order 0.2 to 0.3 mm, which corresponds to the average anatomical distance between the retinal vessels and the visual cells. At that time there were still some doubts even as to the photoreceptor role of the retina. Although Kepler had already suggested this hypothesis, Mariotte thought that it was the choroid which was sensitive to light, and Brewster (1835) still held Mariotte's idea on the basis that the retina is too transparent to be a receptor. Huschke (1835), on the other hand, restored to the retina its true function for reasons of embryonic development and structure. In spite of its poor precision, Müller's elegant experiment settled the matter. Later, Boll (1877) gave a more direct proof by looking into a microscope through an aperture of the sclera of an enucleated eye and noting that the retinal image was in focus on the rods (actually one must also assume that the eye was emmetropic and unaccommodated and that enucleation does not alter the focus).

Other Scotomata

At low luminance the cones become blind and there appears a central scotoma which covers and even extends beyond the fovea. Depending upon the size and luminance of the test, one finds diameters of the central scotoma up to 2°. The center is usually displaced with respect to the fixation point, but randomly, and according to Tibi (1951), this scotoma often has the appearance of a pear with its stem upward. Like the blind spot this is a negative scotoma, but the subject feels some unexplainable discomfort when he tries to fixate.

In many subjects (about 20%) there also exists a permanent scotoma around $\eta = 23°$ along the vertical meridian of the fixation point at the top of the visual field. According to Livingstone (1944), this would be a residue of the fetal retinal fissure. Wilbrand (1896) has described other small permanent scotomata, the position of which depends upon the subject.

Finally, Troxler (1804) has given his name to the following phenomenon. If several spots are drawn on a sheet of paper and one of them is fixated, the others disappear and reappear, now one, now the other, particularly with blinking of the eyelids or small eye movements (Holth, 1896). Dubois-Poulsen relates this effect to the variation of luminosity which occurs in peripheral vision after a prolonged fixation of large surfaces, alternately darkening and lightening, particularly in the lower part of the field. Clemmesen pointed out that the Troxler phenomenon is more pronounced at low luminances and he deduced from it that the explanation must be sought in the rhythmic activity of the visual receptors of the retina. On the other hand, Clarke (1961) showed the relationship between the Troxler effect and border inhibition. In collaboration with Belcher (1962), he provides several arguments which he feels prove that the origin of the phenomenon is not retinal but central. It must also be noted that attention plays a role in similar effects (B. B. Smith, 1961).

Muscae Volitantes

All the opaque bodies situated in the transparent media of the eye are easily perceived if one looks at a bright point source by interposing a lens which will produce a large circle of diffusion on the retina (for a myopic subject it suffices to remove his lenses). One can also place a small pinhole disk in front of the eye and look at a wide bright field. Many effects can be observed. The edge of the iris, slightly jagged, which limits the circle of diffusion and fluctuates constantly; the droplets of tears and mucus on the corneal surface which move as the eye blinks; the irregularities of the crystalline lens (radial lines, seams in the capsule, etc.) which take on a starlike appearance when the circle of diffusion becomes the image of the point source; we have discussed this phenomenon in Chapter 2, and from Hassenfratz (1809) to A. Monnier (1943) it has attracted interesting research. These radiating streaks can spread over more than 1° from the source and have an illumination between 1/100 and 1/1000 the illumination of the source itself. The contraction of these radiating streaks with a minus lens may perhaps play a role in night myopia (Monnier, 1945). Finally, one can also easily observe floating particles in the vitreous body which appear in various forms. When these particles are close to the retina they are visible on a uniform bright background even with a normal-sized pupil. The origin of these *muscae volitantes* has been explained by Dechales (1690). They are dead cells which are embryonic remnants and coagula of the proteins of the vitreous gel. They are lighter than the vitreous body and therefore tend to move upward, but appear entoptically to fall downward because of the reversal of the retinal image.

On a bright surface such as the sky, one sometimes sees small and mobile bright points that Purkinje ascribed to the refraction of light through the red cells of the retinal capillaries. As a matter of fact, they are seen more easily by looking through a dark blue lens which corresponds to the absorption band of hemoglobin, according to Rood (1860). Scheerer (1925) and Fortin (1927) have devoted extensive study to this phenomenon. The speed of circulation is about 0.5 mm/second, and with some imagination one can relate it to the heart rhythm for these movements. The fovea does not present this effect as there are no capillaries within the area of 1.5° around the fovea.

All these entoptic images are not merely simple amusing effects; they can impair vision in the case of very small pupils by superimposing entoptic images upon the real objects. Arnulf et al. (1949) have shown that vision could be much improved in a microscope (where the exit pupil is very small) by utilizing a rotating pupil which replaces the point of incidence on the observer's pupil by a circle and helps in eliminating the entoptic effects.

Haidinger's Brushes

So far the central retina has not provided us with any entoptic effect. With light incident through the sclera, Helmholtz [20] described the place of direct vision as a type of elliptical disk looking something like shagreen leather. He also sometimes noted a dark crescent-shaped shadow on the edge of the foveal depression, but these aspects are not constant.

The *yellow spot*, or *macula lutea*, a yellow area surrounding the fovea, sometimes gives a dark perception but of short duration when one opens his eyes in the morning in front of a uniform bright surface. The pigment of the macula lutea is similar to the xanthophyll of plants and probably helps to reduce the effect of chromatic aberration. Maxwell (1856) noted that if one utilized an extended source of uniform blue light, the macula appeared as a dark ellipse with its long horizontal axis of 2 or 3° centered on the fixation point. Duke-Elder [17] recommends looking at a bright white surface alternately through a purple and a yellow filter of equal luminous transmission factor. The phenomenon is thus very clear. The apparent colorations have been studied by Walls and Mathews (1952) with normal and color-defective subjects, and Judd (1953) explained these apparent colors by the absorption curves of the macular pigment. Loewe (1852) had also described a bright halo which surrounds the macula and has a diameter about three times that of the macula, but it is seen by few subjects.

The most remarkable entoptic phenomenon of the macula is that discovered by Haidinger (1844). When the eye looks at a field emitting white polarized light, Haidinger's brushes appear at the fixation point. The brushes are yellowish in color and relatively dark, with a light and bluish tint on each side (Fig. 47), and turn with the plane of polarization of light. With monochromatic light the pattern is only visible with blue

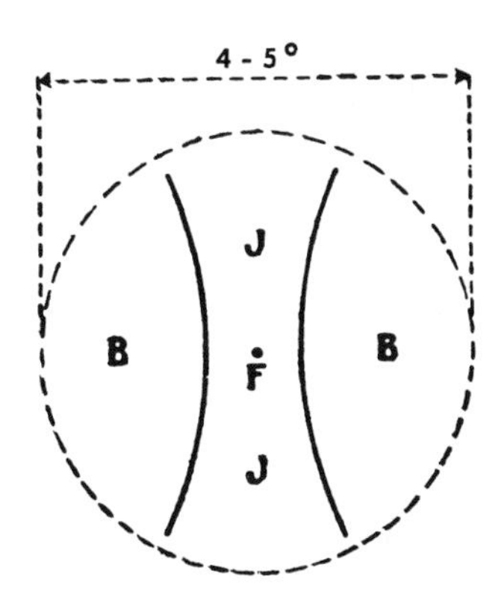

FIG. 47. Haidinger's brushes, with the plane of polarization of light vertical. *F*, fixation point. The area *J* is yellowish and dark and the areas *B* light and bluish.

light. Some young subjects cannot see them and when Haidinger published his description, Helmholtz tried in vain to verify it. Some twelve years later he saw them easily and explained the effect by a *dichroism* of the fibers which pass in front of the fovea and possess a regular inclination and disposition in this region (the word "dichroism" means an absorption which depends upon the direction of polarization of light with respect to the direction of the fibers). This is certainly not an effect of the crystalline lens since the phenomenon is centered around the fixation point which does not coincide with the axis of the lens and, furthermore, aphakic subjects are able to see Haidinger's brushes. Subjects to whom these brushes are very visible can recognize in each point of a blue sky the direction of polarization of light according to that of the brushes. Neuberger (1940) also noted that the patterns of a Savart polarimeter can be seen without an analyzer over a diameter which may reach 25°. Therefore the dichroism extends much further away than the macular region but it is only there that the patterns are easily visible and they are of complementary colors in the areas *B* and *J*. (Fig. 47) of the brushes, which proves a dichroism oriented in opposite directions in these two areas. Another theory was proposed by Stanworth and Naylor (1950), but that of Helmholtz still seems to be the best. Haidinger's brushes have been utilized in the study of squint by Tschermak (1899) and Malbran [38a]. Sloan (1955) studied the visibility of the brushes in normal and color-defective subjects.

Boehm (1940) described another entoptic phenomenon produced by polarized light. In scotopic vision a dimly illuminated surface seems to present a large cross in the periphery, of which one arm is dark and the other light and which turns with the plane of polarization. The rods are effectively dichroic (Schmidt, 1938; Denton, 1954), as can be observed when looking laterally at the retina through a microscope, because the molecules of visual purple are oriented regularly (this actually introduces, at equal concentration, a difference in absorption between the visual purple in vitro and in vivo). But as the light travels along the axis of the rods as long as the eccentricity is small, this cannot produce any visible effect. However, in the periphery the angle between the light and the normal to the retina (angle $C'M_4O_4$ of Fig. 38) is about $\eta/2$ and the dichroism may become visible. For a similar reason Haidinger's brushes are seen very slightly in circularly polarized light (de Vries et al., 1950; Shurcliff, 1955). It is also possible that corneal birefringence plays a role in this case (Seliger and MacElroy, 1965).

Finally, an entoptic effect has been pointed out (Le Grand, 1936) with polarized light incident on the blind spot. The scattering is spread unevenly and the maximum turns with the plane of polarization.

Besides these curiosities, polarized light does not present any visual properties which differentiate it from natural light. De Vries (1948) verified the photometric equality of two fields polarized at right angles. On the other hand, in bees, Von Frisch (1947–1949) found a special sensitivity which would enable these insects to utilize the polarization of the sky to orient themselves. The corneas of bees are very strongly birefringent (Berger and Segal, 1952). Many other animals with compound eyes seem also to possess this ability and Waterman (1950) has studied very carefully the polarizing properties of the *Limulus* eye.

Blue Arcs of Purkinje

When viewing monocularly a small weak source (preferably red) in the dark, 1 or 2° on the nasal side of the field, one observes two blue-violet bands arching out from above and below the light and converging toward the blind spot (Fig. 48). The above are the most favorable conditions, but, according to Dolecek and de Launay (1945), any point at eccentricity $\eta < 5°$ (up to 8° nasally) produces one similar arc and two in the particular case where the source is on the horizontal meridian of the nasal side. Stimulation of the fovea itself produces a bluish veil which fills the region between adjacent arcs. This phenomenon is very fleeting, it lasts a little less than a second but reappears with the slightest eye movement. If both eye are open, the arcs corresponding to the two retinas can be seen and the figure resembles the sign ∞.

These curious arcs were discovered by Purkinje in 1825 and gave rise to a lot of discussion. Most authors agree on the fact that they follow the course of the nerve fibers running from the image of the source toward the papilla, but the origin of these arcs is still uncertain. The photometric and colorimetric measurements of Newhall (1937) give as

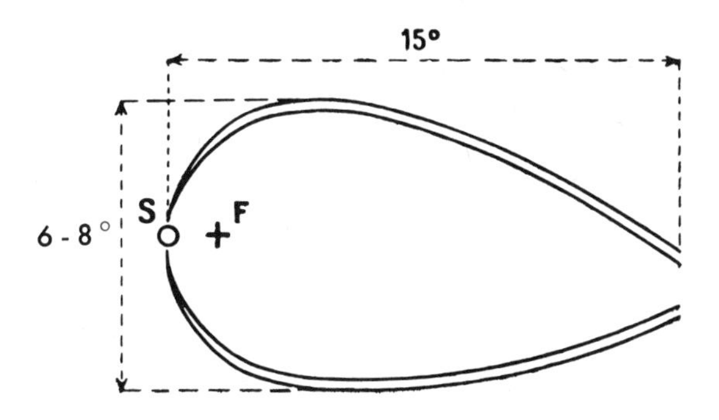

Fig. 48. Purkinje's arcs. *F*, fixation point; S, source (right eye).

a color of the arcs a purple complementary to 564–570 mμ with an excitation purity between 0.67 and 0.78. The color of the stimulating light hardly affects the color of the arcs and any source can produce them but long wavelengths are favorable. The luminance of the source must be at least 0.1 cd/m^2 and that of the arcs seems about 500 times lower. It is for a luminance of the source between 10 and 100 cd/m^2 that the brightness of the arcs (several thousand times less than the luminance of the source) passes through a maximum. If the luminance of the source increases beyond this value, the brightness of the arcs decreases. There are two principal theories. First, the bioluminescence theory suggested by Druault (1914) and supported by Ladd-Franklin (1924), Judd (1929), and Piéron (1939), according to which the nervous action of the retinal fibers would emit light (visible or ultraviolet) which would stimulate contiguous visual cells. Second, the theory of electrical stimulation (a secondary stimulation of the receptors by the action currents in the optic nerve) proposed by Gertz (1905) and developed by Friedman (1931). The fact that light preadaptation facilitates the phenomenon whereas dark adaptation impairs it (at least after 2 minutes of darkness, which is an optimum condition) recalls the laws of the electrical phosphene and would be in favor of the second hypothesis. Moreland (1965) has given two other arguments in favor of this second hypothesis: The threshold change in a simulated arc measured under closely controlled conditions during dark adaptation was found to exceed that for an actual arc; and this simulated arc may give an image by ophthalmoscopic photography, whereas the actual arc gives no image.

Light Chaos

The appearance of the visual field in total darkness is referred to as *light chaos* or "intrinsic light of the retina," but the latter term may be confused with an eventual bioluminescence. Purkinje described broad streamers, more or less curved with dark intervals between them, turning slowly in spirals around the papilla. It takes about 8 seconds for a streamer to complete its movement and vanish out of sight. With movements of the eyes, changes in accommodation, and respiratory rate, there is an accompanying variation of the light chaos. It is probable that part of this phenomenon is cortical, since J. Müller (1826) noted it on a subject whose eyes were enucleated and the optic nerves atrophied. London (1904) caused similar phantasms by applying radioactive materials to the surface of the tegmentum near the visual cortex. In cases of intense migraine, the name *scintillating scotomata* has been applied to star-shaped patterns, white or colored, flickering at a rate of about 10 per second. These scotomata first appear in the parafoveal region, then spread toward the periphery, where they disappear. The attack lasts about 20 minutes. This is the most common case of visual hallucinations of cortical origin.

It is also probable, as Marshall (1935) assumed, that part of the chaotic light originates from a spontaneous breakdown of the photosensitive pigments of the retina. Measurements in vitro on thermic decomposition of visual purple at 37°C gave a fraction destroyed which, for 1 hour, would be of the order of 5%, according to Lythgoe and Quilliam (1938), and 1%, according to St. George (1952). It has even been deduced that thermic energy in the red end of the spectrum may provide appreciable evidence for photochemical action and that the threshold for long wavelengths would reduce by increasing the temperature of the observer with a hot bath! But the experiments and calculations of Denton and Pirenne (1954) on the sensitivity and stability of the scotopic retina reduce to 5×10^{-6} the fraction of rhodopsin, which in the human retina can be spontaneously bleached during 1 hour, and it does not seem very likely that this process, if it exists, plays any visual role. In fact, the necessity that one functional unit should absorb several quanta (at least two for the most optimistic hypotheses) before a nervous discharge is produced is one more guarantee against visual perception of these spontaneous breakdowns.

Another possibility of retinal origin is the existence of spontaneous activity of nervous elements of the retina, particularly the ganglion cells. This phenomenon, first discovered by Adrian and Matthews (1928) on the eel, has been noted on numerous other animals, in particular on the cat by Kuffler (1953). It was found that light either stimulates, inhibits, or remains ineffective upon this spontaneous activity; in general, dark adaptation increases spontaneous activity. It is probable that the aim of these "dark messages" is, on one hand, to maintain the brain in a state of attentive alertness and, on the other hand, to enhance the effects of contrasts, owing to inhibition of spontaneous activity of the nonilluminated receptors near the image of a source (Granit, 1955).

CHAPTER TEN

Movements of the Eye

Having studied the static monocular visual field, with a fixation point occupying a fixed position in space, we will complete this study by removing this restriction; that is, the eye will move in its orbit.

Experimental Methods

The analysis and the recording of eye movements are possible with various techniques, of which we will review the principles. For more details, one should consult the excellent accounts by Carmichael and Dearborn [9], Lord and Wright [31], and Alpern [1a].

1. Subjective method. This method consists of observations of the projection of after-images (which are stable on the retina) onto immobile objects in the visual field. This method can give only qualitative results. The number of perceived images during a movement of the gaze, using an intermittent light source, has also been utilized by Lamansky (1869), Erdmann and Dodge (1898), and P. W. Cobb and Moss (1925) for the measurement of the speed of displacement of the gaze.

2. Mechanical technique. Ohm (1914–1928) used a very light lever, one end of which was in contact with the eye and the other end on a smoked drum moving with a helicoid movement. In the *nystagmograph* of Cords (1927), the tip of the lever was placed on the anesthetized corneal limbus. Buys and Coppez (1913), then Yourevitch (1929), used a pneumatic intermediary. Marey capsules are placed on the eyelids of the subject and the variations of pressure are transmitted to other capsules which control the recording lever. These systems are not very sensitive, and are uncomfortable for the subject.

3. Electrical method. The movements of the globe of the eye modify the distribution of the electrical potentials in its vicinity. Thus the variation of potentials (of the order of a few millivolts, at the most) between two fixed electrodes can be recorded and amplified. This method is reasonably easy to perform with modern electronic techniques, but its interpretation is difficult because the intraocular potentials and those of the extrinsic muscles intermingle. Schott (1922) placed the electrodes in the fornices. Meyers (1929) placed the electrodes merely on his subject's temples and found that the variation of potential was maximum if the displacement of the fixation point is horizontal. Jacobson (1930) placed one electrode on the edge of the orbit and the other behind the ear. Other positions of the electrodes have been used by Mowrer et al. (1936) and Miles (1939). This method is convenient for prolonged recording of large movements, such as those made during reading [9], but it lacks precision if the movements are small (see Baudoin et al., 1939). An interesting review of the possibilities of the electrical method has been prepared by Gemelli et al. (1952). It seems that the response is linear up to more than 15° displacement (Law and DeValois, 1957). Ford et al. (1959) attempted an analysis in two dimensions from this principle.

4. *Optical method, with mirror.* This method consists of fixing a lightweight mirror to the eye and optically recording the oscillations by photography on a moving film. The precision is excellent, but obviously the mirror must not be too uncomfortable for the subject, yet must follow faithfully the small eye movements, which is difficult. Marx and Trendelenburg (1911) placed the mirror on a corneal shell of aluminum; Dohlman (1925) used a rubber shell adhering to the cornea by suction, a method repeated by Yarbus (1957). Adler and Fliegelman (1934) placed a small galvanometer mirror 3 mm in diameter on the anesthetized temporal limbus of the eye. The subject kept his finger on his eyelid to avoid blinking. These are very artificial conditions and the eye-mirror relationship is doubtful. Recent studies utilize a contact lens on which a small facet is cut. It is, of course, the simplest solution but it has been criticized, as it makes the eye a bit heavier and therefore modifies the natural balance, and also the contact lens may slip on the eye. It seems, however, that these criticisms are not supported. According to Byford (1962), there is no slippage of the contact lens on the eye for deflections less than 30′ and only small differences for 9°. It is possible to record both vertical and horizontal movements at one time and even to measure torsion (Matin, 1964). The movement of closed eyes can be recorded in a similar manner, by placing a mirror on the eyelid (Dodge, 1921). A contact lens marked with lead, of which the motion is recorded on an X-ray instrument, can also be used (Rodin and Newell, 1934).

5. *Direct optical method.* This is theoretically the best, because the eye remains in its natural state, but it is difficult to obtain good accuracy. The observation of the subject's eye through a telescope (Javal, 1879) or with a special optical system (Newhall, 1928) provides only qualitative information. Direct photography of the eyes was used by Dodge and Cline (1901); here also the precision is low, but it is the most convenient method to record eye movements of large amplitude. Thus, to record eye movements during reading, Tinker (1931) constructed a large camera which records the horizontal movements of the two eyes. If one wishes to record the vertical movements at the same time, two moving films at right angles are needed, or one only with the artifice of a mirror at 45°, which transposes one of the movements perpendicularly (Weaver, 1931; Clark, 1934). A portable *ophthalmograph* of low precision (1 to 2°) has been designed by E. A. Taylor (1937). Karslake (1940) studied with it the eye movements of a subject reading the advertisements in a periodical. One can also measure the torsion of the eyes by photography to an accuracy of about 4′ (Howard and Evans, 1963). Instead of photographic methods, photoelectric methods have been investigated (W. M. Smith and Warter, 1960; Rashbass, 1960; Nelson et al., 1962).

To increase the precision of this method one can place a small mark on the eye: a small white spot on the cornea (C. H. Judd et al., 1907) or a droplet of metallic mercury (Barlow, 1952). It is even preferable to use a natural mark such as the small blood vessels of the sclera. This latter technique was suggested by J. Rösch (1943) and performed by Higgins and Stultz (1953). The eye of the subject was illuminated by a powerful mercury vapor lamp and the image of a small scleral blood vessel, appropriately chosen, was projected through the objective of a microscope magnifying 26 times, onto a slit perpendicular to the vessel and behind which a film moved at the speed of 13 cm/second. The precision obtained was 0.5′. The movements of the head could be recorded at the same time (with a mark on the bridge of the nose).

A special mark is the *corneal reflex*, the image of a light source reflected from the anterior face of the cornea. This reflex is displaced with the eye and by recording it photographically, the movement of the eye can be deduced. This method was applied particularly by Vernon (1930) and by Hartridge and Thomson (1948). The latter investigators have photographed, at a speed of 60 images/second, the corneal reflex of a light and its optical apparatus which were attached to the head in order to be independent of head movements. In all these photographic methods, one can use ultraviolet or infrared, provided the emulsion is sensitive to it. This has the advantage of not disturbing the subject. The principal disadvantage of the corneal reflex method is the considerable importance of head movements. According to Ditchburn and Ginsborg (1953), a lateral translation of the head of only 0.01 mm, produces a deviation of the beam of light as important as a rotation of the eye of 8′. Even by holding the head firmly, movements of that order occur constantly.

An ingenious optical method has been devised by Lord (1948). A beam of ultraviolet light (365 mμ) is incident on the cornea, in such a way that the fraction of the light transmitted through the cornea falls on the blind spot. Therefore, the subject does not see anything. The beam of light reflected from the cornea is split into two beams by a semi-reflecting mirror *L* (Fig. 49). One of the beams falls on a horizontal straight edge *B*, the other falls on a vertical straight edge *B*′. Eye movements modify the proportion of the beam blocked by each straight edge, producing variation of the current output of electron multiplier photocells *C* and *C*′. These outputs are amplified and fed to cathode-ray oscillographs *O* and *O*′ and the traces are photographed on a continuously moving film *F*. The film also includes a scale of time, 50 dots/second. The precision of this system is about 1′ of arc. Riggs et al (1954) improved this to 5″, but the disadvantage of this method, as with the

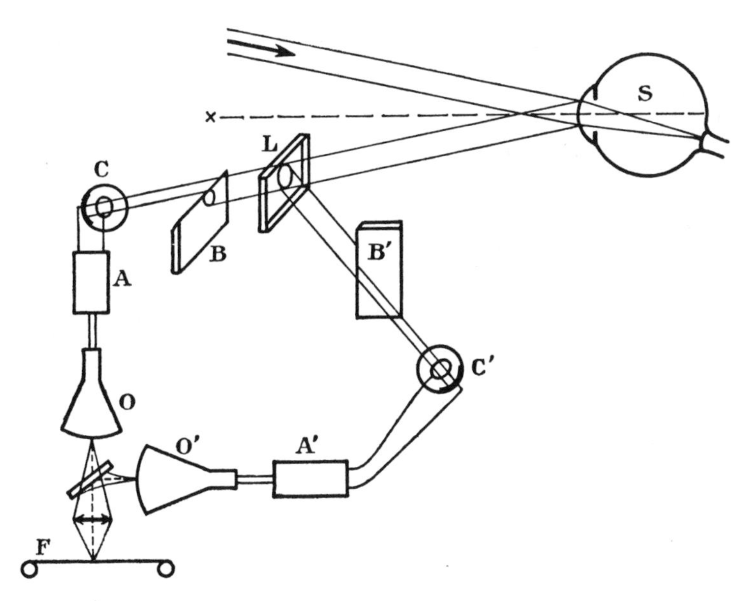

Fig. 49. Photocell recording of eye movements (Lord). *AA*′, amplifiers. *BB*′, straightedges. *CC*′, photocells. *F*, film. *L*, half-aluminized mirror. *OO*′, oscillographs. *S*, subject's eye (purposely much too large as compared to the scale of the figure).

corneal reflex method, is that even if the head is held by a dental impression, its movements are recorded superimposed on those of the eyes (Lord and Wright, 1948). A variation of Lord's method consists of locating the blind spot by a beam of light sweeping over the retina, the light which is reflected out of the eye falling on a photomultiplier plugged into an amplifier feeding into an oscilloscope (Cornsweet, 1958).

Geometry of Ocular Movements

Since J. Müller (1826), it has been agreed that the eyeball turns in its orbit around one point, fixed with respect to both the eye and the head; this point is referred to as the *center of rotation*. The measurements of Donders (1862) arrived at a distance of about 13.5 mm between the apex of the cornea and the center of rotation, but this value may vary a great deal; in myopes it is often larger. Actually the steadiness of the center of rotation is only an approximation, but its displacement hardly exceeds 1 mm.

The *fixation axis* is the line connecting the center of rotation to the point of fixation. The *field of fixation* is the total angular range of rotatory excursions of the eye with the head fixed, represented by a plot of the limits of fixation, on a spherical surface whose center coincides with the center of rotation of the eye, or on a tangent screen. This field rarely exceeds 45° from the straight-forward position. Each visual field shifts with the point of fixation; the total extent of physical space visible to an eye is the limit of the visual fields when the point of fixation moves along the limits of the field of fixation.

The geometry of eye movements is ruled by two laws. The first is Donder's law (1847). In general, the position of a solid which can turn around a fixed point is given by three independent parameters. Donder's law states that there are only two for the eye; that is, if the line of fixation is known (which sets two parameters), the position of the eye around this line is determined. This law implies a precise link between the six extraocular muscles which act upon the eye to rotate it within the orbit. This link enables verticals and horizontals projected onto the retina to return to an identical position when the eye fixates the same point on a different occasion; it is therefore an important factor in the sability of visual space.

The second law is Listing's law (1854), which states: For a given position of the head, when the line of fixation is brought from its primary position (in general this corresponds to the natural position when the head is straight) to any other position, the torsional rotation of the eyeball in this second position will be the same as if the eye had been turned around a fixed axis perpendicular to the initial and final direction of the line of fixation. Listing's law is satisfied with much less precision than Donder's law, which is usually exact within a few minutes of arc.

Movements of the Fixation Point

When a subject is requested to fixate a given point in the visual field, he has the impression that his gaze is motionless, but in fact this is not so. A subjective method can demonstrate this. If one fixates a mark in the center of a green circle on a white background one can observe easily at the edge of the circle a red fringe caused by successive contrast, due to small movements of the gaze. Dodge (1907) noted that in spite of efforts to maintain a stable fixation, the afterimage of a contour projected on a fixed background moves with an amplitude of 10′, sometimes increasing to 30′. These amplitudes increase with the fatigue of prolonged fixation.

An observation by O'Brien and Dickerman (1948) seems contradictory to the existence of these movements. They looked at a luminous point in the center of a dark circle surrounded by a bright ring. This point emitted flashes of light with a frequency of 18 to 240 per second. Each exposure lasted between 0.01 and 0.2 second. They assumed that if the eye moved during that time, the subject would see a series of points in different locations, but this does not happen. This hypothesis is incorrect, and we will return in Chapter 11 to this matter of a rigid localization in spite of ocular movements. The reasoning of Hartridge [19d], that the precision of a trained rifleman who hits the center of a circle of 2′ diameter would be impossible if the eye moved, is not convincing either, for the same reason as in the above experiment.

Observation through a telescope of an ocular detail on a subject (scleral vessel or corneal reflex) led Ohrwall (1912) and Sundberg (1918) to conclude that there existed some "rests" of fixation lasting 1 to 2.5 seconds and separated by jerks of about 30′ amplitude. But the precision of this type of measurement is too low to allow one to be sure that during the "rest" the eye is motionless. Park and Park (1940) did, in fact, find great individual variations. Some subjects appear to keep their gaze motionless, others move it slowly, and others rapidly.

Only the optical methods of recording eye movements possess sufficient precision to analyze in detail small ocular displacements. In general, researchers are all in accord on the permanent state of agitation of gaze but differ in the quantitative evaluation of these displacements, which shows that the parasite influence of the recording technique is important. Dodge and Cline (1901), who used a method of photography of the corneal reflex, described the *micronystagmus* of fixation (sometimes called *pseudonystagmus* to differentiate it from pathological nystagmus) which was later specified by Dohlman (1925). The tremors have an amplitude of about 50″ and a frequency of 5 per second. Every 2 seconds a larger movement occurs (4′), caused by pulse, respiration, or perhaps by compensation of small movements of the head (Verwey, 1918) or by the necessity to excite new receptors on the retina (Kestenbaum, 1921–1925). After a prolonged fixation, a true nystagmus of larger amplitude appears (Luhr and Eckel, 1933–1934). The results obtained by Adler and Fliegelman (1934) are frequently cited: micronystagmus of 2′ with a frequency of 50 to 100 per second, movements of greater amplitude (12 to 30′) with low frequency. But their technique lent itself to criticism, as we mentioned previously. Furthermore, they seem to have made a mistake in their calculations; the amplitudes they give should have been divided by 2.

The very thorough study of Lord and Wright (1948) reveals the importance of posture. If the subject is lying on his back, his head held still, the eye remains motionless for periods of about 0.5 second. This rest is interrupted by rapid flicks of 3 to 14′ amplitude which last 0.02 to 0.03 second. Following the rapid flick, fixation does not recover exactly its original position; there is an error of a few minutes of arc. The lateral flicks are more important than the vertical ones. If the subject is seated, the head movements are multiplied by a factor of 2.5. The saccades of

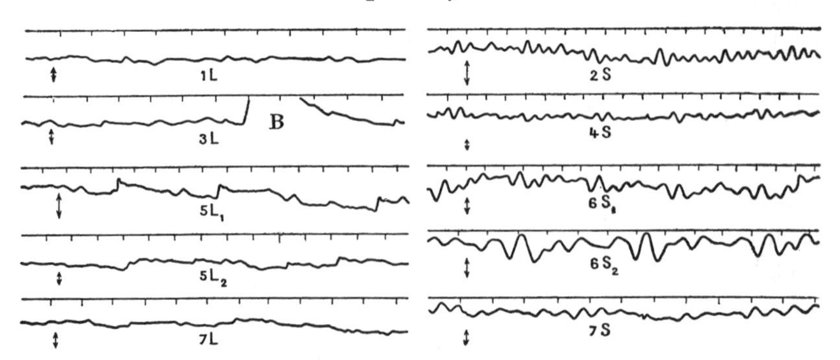

FIG. 50. Recordings of ocular movements of five subjects (Lord). The time axis is scaled from left to right and graduated in fifths of seconds. Amplitude of eye movements is measured by the arrows, which represent a rotation of 20′. Vertical movements are indicated by *L* (a movement upward appears as an upward deflection), lateral movements by S (movement to the left is shown as an upward deflection). Curves 1 and 2 represent movements of the head alone, recorded on an artificial cornea held between the teeth of the subject. Curves 3 and 4 represent eye movements with the same subjects as curves 1 and 2, respectively; 5 and 7, the eye movements of other subjects. At *B*, the subject has blinked.

the eye are less frequent (on the average one every 4 seconds on seven subjects). Some typical recordings are shown in Figure 50.

With a plane mirror attached to a contact lens, Riggs and Ratliff (1949–1950) recorded a micronystagmus of low amplitude (10 to 20″) with a frequency of 30 to 70 per second, superimposed on a slower and irregular movement of an amplitude of several minutes of arc and on a slow drift and compensating saccades. Ginsborg (1951) obtained similar results.

Higgins and Stultz (1953), who used a slit camera photographing a scleral blood vessel, have recorded horizontal and vertical eye movements on four subjects and found them to be very similar. Upon the micronystagmus (Fig. 51*a*) are superimposed saccades of larger amplitude which increase when the detail to be perceived is at the threshold of visibility (Fig. 51*b*), probably because the subject moves his gaze more or less consciously in order to see better. The statistical distribution of the mean period of the micronystagmus (Fig. 51*c*) shows a maximum around 0.01 second (that is, a frequency of 100 per second), but the curve is very asymmetrical and the average is about 0.02 second. The distribution of amplitudes (Fig. 51*d*) is represented by a similar curve, with a maximum of 1′, and an average of 1.2′ if we consider only the micronystagmus and about 2.4′ if we take into account all the movements up to an amplitude of 10′.

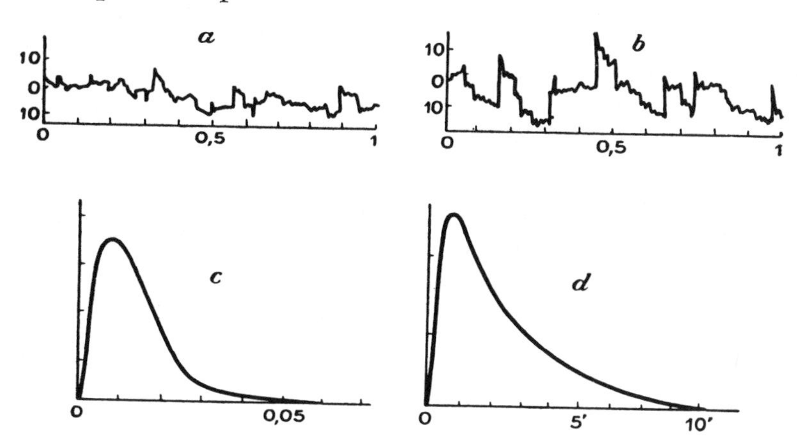

Fig. 51. Recording of ocular movements (Higgins and Stultz). (*a*) Normal micronystagmus. Abscissa: Time in seconds. Ordinate: The rotation of the eye in minutes of arc. (*b*) Eye movements when the fixation target is at the threshold of visibility (same conventions along the coordinate axes). (*c*) Frequency distribution of eye movements as a function of the period. Abscissa: Periods, in seconds. Ordinate: Relative frequency of occurrence. (*d*) Frequency distribution as a function of the amplitude. Abscissa: Rotation, in minutes of arc. Ordinate: Relative frequency of occurrence.

According to Riggs, Armington, and Ratliff (1954), during an exposure which lasts t seconds ($0.01 < t < 1$), the average amplitude ω of the eye movement is given by the following expression, in seconds of arc:

$$\log \frac{\omega}{200} = 0.9 \log t$$

For example, if $t = 0.1$ second, we find this formula $\omega = 25''$.

Besides the rotation, the micronystagmus includes some torsions of amplitude betyeen 5 and 45″ but which can sometimes reach 5′ (Fender, 1955). Finally, let us indicate that besides the micronystagmus (or tremor) and the saccades (5′ on the average), the slow drifts of several minutes of arc seem to be influenced by several visual factors (Nachmias, 1961).

Consequences

In spite of the disagreements among investigators, we can assert that the eye practically never remains immobile. Micronystagmus is usually interpreted as a nervous fluctuation in the oculomotor system due to simultaneous and antagonistic contraction of the extraocular muscles. According to Chiba (1926), there are approximately 3000 motor nerve fibers in the sixth cranial nerve which supplies the lateral rectus, and according to the all-or-none law each of these fibers is responsible for a contraction (total or nil) of a group of muscle fibers. The variation of only one of the nerve fibers, from rest to innervation, would suffice to produce a rotation of the eye of about 1′. According to Nordmann [45], this type of tetanic contraction would be exceptional in the muscles of man. It can, however, be revealed by several methods. By applying a stethoscope on the eyelids one can hear a rumbling sound from the muscles (Hering, 1879). In recording the action currents of the extraocular muscles, one observes spikes at a rate of about 70 impulses per second (Hoffmann, 1913; Perez-Cirera, 1932; Heijirmans, 1934, and others).

It is not completely obvious that these very small eye movements displace the image on the retina. Park (1936) found some tremblings of the lens, synchronous to those of the globe, such that the image remains motionless. This does not seem very likely. But Lord and Wright [31] have pointed out that the center of rotation of the eye seems different for small and large movements. In the classical position where this center is very near that of the eyeball, the image and the retina move in opposite directions, which doubles the movement of the image with respect to the retina. If the center of rotation were close to the retina this movement would be reduced by half, and if the center were at infinity (that is, if the rotations became translations) this movement would be nil. Conversely, if the center of rotation coincided with the entrance pupil the image would be stabilized in space and the retina would scan the image.

If the retinal image of the fixated point oscillates constantly, should the notion of the fixation point, defined on the retina with a precision on the order of a minute of arc, be abandoned and replaced by a *fixation area* of several minutes of arc diameter? This was the idea of Landolt (1891), who attempted to measure this diameter by the minimum separation between points on a line, such that they can just be counted. He obtained 5′ and by the same method Gertz (1908) found 3 to 4′. But this applies to a particular type of acuity which is difficult to interpret. Dodge (1907) was also in favor of a fixation area. If, as we mentioned previously, the fixation point is defined only by a maximum of physiological acuity and not by a privileged anatomical receptor, this area is merely the equivalent of the region of maximum efficiency, and the problem left is to find whether this maximum is very sharp and can be defined within the order of 1 minute of arc (that is, a few foveal cones) or rounded. In favor of the precise definition, we can cite Javal's remark that one knows which point source he is fixating when two sources are presented at the limit of resolution. But this may be due to a psychological effect of attention rather than to a movement of gaze.

We discussed in Chapter 6 the fact that the dynamic theory of acuity assumed that micronystagmus is necessary to the sharpness of vision. This hypothesis is contradicted by the results of experiments with a stabilized retinal image by means of a mirror placed on a contact lens worn by the subject. The beam of light reflected on the mirror produces a test object whose image does not move across the retina in response to small eye movements. This method was devised independently by Ditchburn and Ginsborg (1952) and by Riggs, Ratliff, and Cornsweet (1953). Through these systems it is not the acuity which is altered but the differential threshold. Ditchburn and Ginsborg (1952), then Ditchburn and Fender (1955), used a test of 1° diameter with a vertical line dividing the circle in half. Differences of luminance, between the two halves of the field, of 30 to 60% disappeared with the stabilizing system, from time to time, for several seconds even in foveal vision. But the subject does not feel comfortable and compensates by large involuntary movements or perhaps displacements of the crystalline lens, and the contrast reappears. Riggs and his collaborators (1953) have observed with a similar system that the acuity was better with than without stabilization as long as the exposure was short (<0.2 second); then the acuity becomes better if the eye moves, probably by refreshing the contrast. Similar results have been obtained by Fiorentini (1954) and by Yarbus (1956). Stabilized image techniques have been improved by Clowes and Ditchburn (1959), Ditchburn (1963), and Millodot (1966). Clowes (1962) and Krauskopf (1963) studied color discrimination with stabilized images and found it similar to that of the tritanope. Therefore it seems that the usefulness of micronystagmus is essentially to transform spatial differences of retinal illumination into temporal variations to make them less erasable (Ditchburn, 1955; Gaarder, 1966), but the effect of movement would be rather to impair visual acuity, at least for a short exposure of the test. This method of stabilized retinal images does, however, provide a very interesting technique for studying temporal retinal factors. For example, vision returns with flickering lights between 4 and 10 cycles/second, or by oscillating a mirror at this same frequency with an amplitude of 0.4 to 1 minute of arc.

Binocular Fixation

Lord (1952) demonstrated that the flicks of large amplitude retain their same characteristics in monocular and binocular vision, for a given subject at any fixation distance (170 or 45 cm). The eyes move very little in the vertical meridian and in the horizontal they move at the same time, in the same direction, and almost at the same angle (Fig. 52). These rapid flicks do not seem to be governed by a local or tetanic origin, and the synchronization of the two eyes reveals the activity of a motor center (for the micronystagmus, it seems that there is no correlation between the two eyes, according to Krauskopf et al., 1960). With one of the subjects the frequency of the flicks increased in near vision. Lord attributed this to a habit of relaxing the accommodation periodically in close work.

We will see, in Chapter 12, that binocular movements are determined by the *fusion* of the two retinal images. Dodge (1903) had already noted that *vergences* (that is, movements of both eyes in opposite directions which occur when an object is moved closer or further away from the eyes) were much slower than *versions* (identical movements due to a lateral displacement of the point of fixation). Subsequent studies confirmed this observation; the maximum velocity of vergences is of the order of 20°/second [1a]. According to Rashbass and Westheimer (1961), there is an almost linear relationship between the initial velocity of the vergence movement and the amount of disparity (see the beginning of Chapter 13) which is the stimulus to fusion. An increase of convergence is a little more rapid than a decrease in convergence.

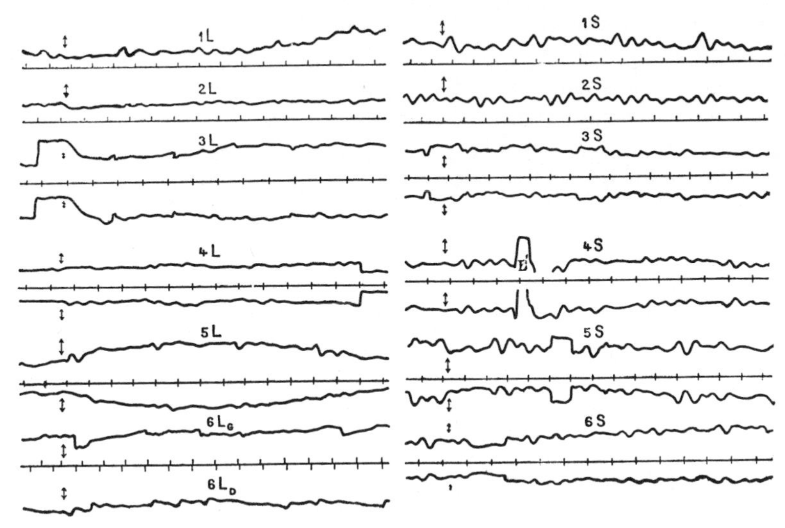

FIG. 52. Recording of monocular and binocular eye movements (Lord). The conventions are the same as in Figure 50. Curves 1 and 2 are head movements, 3 to 6, eye movements. In the curves of binocular eye movements, the curve corresponding to the left eye is above the time axis and that of the right eye below.

Absence of Fixation

In the absence of a fixation point the eye cannot stay immobile. An afterimage projected on the sky seems to move constantly. Even if the field is structured, the eye stops only if it fixates a detail. Furthermore, rotations of the head of about 3° occur slowly and unconsciously. In total darkness or after closure of the eyelids the eyes slowly move back and forth. During sleep the eyes of an adult usually stay motionless (Slotopolsky, 1931), whereas children's move about every 10 seconds (de Toni, 1934). However, it has been found (Aserinsky and Kleitman, 1955; Dement and Kleitman, 1957) that dreaming is associated with intermittent bursts of conjugate rapid eye movements, and that this pattern recurs four or five times during a typical night's sleep.

Change of Fixation

When a subject is presented with two fixation points A and B separated by an angle u, and he is requested to turn his gaze from A to B, the total duration of the movement increases with u, but not proportionately. It is 0.015 to 0.03 second for $u = 5°$; 0.02 to 0.04 for 10°; 0.05

to 0.10 for 40° (Erdmann and Dodge, 1898; Dodge and Cline, 1901). The speed is maximum in the middle of the displacement, which is sometimes performed in one sweep, sometimes in several flicks even if $u =$ 4° (Lord and Wright, 1949). If the movement is large it is fractioned (into 4 or 5 parts for $u = 150°$), but according to Westheimer and Conover (1954) a regular movement can sometimes occur for $u = 30°$. The eye rarely follows a straight line between the points, except when they are on a horizontal line. The speed is a little greater in the horizontal meridian than in the vertical and it diminishes with age. A mathematical study of these movements was made by Westheimer (1954). According to Hyde (1959), and contrary to common belief, the maximum velocity would occur at the beginning of the movement, and the deceleration phase would be double that of the acceleration phase. The maximum velocity can reach values of the order of 8°/0.01 second.

At the end of the sweep one often notes small rectifying movements made to place the fixation point on *B*. This is caused either by an imbalance between the muscles (frequently the medial recti are stronger than the lateral recti) or, as Sundberg (1917) assumed, the approximate amplitude would be determined at the start of the movement by the position of *B* in the visual field, but the final adjustment would take place at the terminal stage when *B* arrives on the fovea.

An interesting case studied by Lord (1952) is as follows. *A* and *B* are aligned in front of one eye, say the right, and at the distances 170 and 60 cm. When the subject is requested to fixate from *A* to *B*, the right eye, which in principle should remain immobile, accomplishes a saccade of about 1° in an oblique direction, the angle of which depends upon the subject but the horizontal component of which is larger than the vertical. This phenomenon is constant whether the left eye is occluded or not (except if the left eye is amblyopic). On the other hand, the left eye accomplishes a saccadic movement of which the displacement is less (by about 25%) than would be expected from the calculation. According to Lord (1953) this discrepancy would be due either to a change of the center of rotation or to a displacement of the lens following the variation of accommodation.

The fixation reflex, which is initiated by the appearance of a stimulus in the periphery (after a latency of about 0.2 second, according to Diefendorf and Dodge, 1908; this latency is about 0.2 second for an almost central stimulus and increases with eccentricity), is one of the most important phenomena of vision. Because of its poor visual acuity and lack of psychological training, the peripheral retina does not recognize much, but it signals anything new, and the immediate movement of the gaze places the image of the discovered object on the fovea, which analyses it in detail. Thus the functions of peripheral and central vision complement each other harmoniously. Some pathological cases in which one or the other disappears completely prove, in fact, that peripheral vision is the more necessary. A central scotoma (similar to the one which occurs at scotopic levels) is certainly a handicap and greatly diminishes the fineness of visual perception, but the subject can still live and manage alone. On the other hand, if only foveal vision is left, the subject is almost blind and incapable of any activity although he can read very fine characters, which is small consolation. Finally, it must be pointed out that the precision of fixation movements seems essential to assess visual space correctly. In pathological cases such as those studied by Paterson and Zangwill (1944) and by MacFie et al. (1950), the incoordination of ocular movements was accompanied by serious errors in the visual organization of the external world.

Pursuit Movements

When the fixation point is in motion, the eye movements can be regular or saccadic. To be regular the motion must not be too slow (Gertz, 1914), otherwise the micronystagmus of fixation reappears; nor must it be too rapid (less than 30°/second), otherwise the eye cannot follow the fixation point (Ohrwall, 1912; Zeeman, 1929; Roelofs and van den Bend, 1930). The size of the object and the level of illumination are also of some importance. There are large individual variations, some of which can be attributed to the ametropia of the subject (Borries, 1926; Nordmann and Lieou, 1928) but which, in general, are psychological in nature. According to Westheimer (1954), the fact that the subject is able, or not, to predict the movement which is about to happen plays an important role (see also Bennett-Clark, 1964). According to Rashbass (1959), the stimulus which initiates a pursuit movement is the movement of the target rather than its position. The reaction times of pursuit movements would be of several possible amounts at regular intervals, as found by Latour and Bouman [60b].

A particular case of pursuit movement is that in which the target moves sinusoidally through, for example, about 1°; then tracking movements of the eye occur. These movements are regular, but are smaller in amplitude than the target movements and show a phase lag (Fender and Nye, 1961).

If the nature of the movement is such that the subject cannot follow the objects—as, for instance, a series of white and black stripes on a rotating drum—an *optokinetic nystagmus* [45] occurs, which has been of much interest to medical people but is beyond the scope of our subject.

Voluntary Eye Movements

A voluntary eye movement is nearly always saccadic, even when one tries to make it as regular as possible. Stratton (1906) showed that when the subject believes that his eyes are following a circle, his gaze describes either a triangle or an irregular polygon, depending upon the size of the circle, with a few stops.

The most investigated topic is that of reading, which, following Javal (1878), Landolt (1881), and Lamare (1892), has given rise to innumerable studies [52a]. The reader thinks that he is following the line regularly but in reality the eye is still for 90 to 95% of the total time. Tinker (1946), who wrote a very thorough review of this question, points out that a trained adult only gazes at about 80% of the length of each line, with three to five fixation pauses per line (the lines are about the same length as those of the present text and the book is assumed to be held at 30 to 40 cm). Each pause lasts 0.2 to 0.3 second. The saccadic movement between two fixation pauses takes about 0.02 second and the return of the eyes to the next line takes twice as long. Evidently these figures apply to favorable conditions of illumination and legibility. The reader interested in the typographic factors which affect reading may consult the studies of Paterson and Tinker (1929–1946) and Burt et al. (1955).

According to Carmichael and Dearborn [9], the eyes tend to diverge at the end of each line and to converge when refixating the next line. During each fixation pause a slight divergence tends to occur, particularly for the first pause of the line. A small movement of the head generally accompanies that of the eyes.

Of course, the learning of reading skill is accomplished somewhat differently in children. Although regressions (that is, a movement of the eye returning to a previously scanned part of the material) are rare in good adult readers, where they only occasionally occur at the beginning of a line, they are frequent in children, as is shown in Table 22, which I have borrowed from Buswell (1922). As the child improves his reading, the number of fixations per line decreases and so does their mean duration.

Even in an adult, the difficulty of the text affects the number of pauses but not their duration. According to Erdmann and Dodge (1898), the number increases from a mean of 4.5 per line for one's native tongue to 5.3 for a known foreign language and 15 to correct papers; thus the corrector must read like a child of 7 years of age.

There does not seem to be any rule regarding the choice of words and letters on which the eye stops, and the individual variations among readers are considerable. Not all the letters of a line are read, only four

TABLE 22

The Learning of Reading

Age, years	Mean number of regressions per line	Mean number of pauses per line	Mean duration of each fixation pause, sec
6	5.1	18.6	0.66
7	4.0	15.5	0.43
8	2.3	10.7	0.36
9	1.8	8.9	0.32
10–14	1.4	7.1	0.25
15–18	0.7	5.9	0.24

or five groups of words are read and the rest is guessed. Therefore, peripheral vision plays an important role in reading, which may be surprising when one thinks how poor peripheral vision of form is, but it must be kept in mind that in this case the recognition is familiar and very special. Cattell (1885) demonstrated that with an exposure of 0.01 second the average reader perceives only three or four isolated letters, whereas he reads two words (without any meaning) made up of 12 letters or a complete phrase of four words made up of as many as 20 letters. Another proof of the insignificant role played by visual acuity is provided by the experiments of J. Wagner (1918) and Thorner (1929) on the brief presentation of an eight-letter word deprived of any meaning, with a fixation spot in the center of the word; the letters in the periphery, particularly the first few on the left, are recognized more easily than the one in the center. In the reading of newspapers, where the lines are very short, it seems that the lines interfere with one another and that the following one helps the eye to choose its pauses in the preceding one in the same way as blind people reading Braille often follow with the other hand the next line to read.

Since 1937, Brandt [7] and others have devoted a great many studies to the eye movements occurring when viewing periodicals, advertisement boards, classified ads, etc. Here again there is a mean of four fixation pauses per second. If no detail attracts the gaze particularly, it first aims in the center, then drifts upward and to the left, probably because of our habits in reading. They have also studied eye movements when looking at geometric figures. The more gifted subjects use many fewer movements than do others. As an example, we have reproduced from Brandt the eye movements of a subject looking at the Müller-Lyer illusion when requested to say which line appears the longer (Fig. 53). The important problem of eye movements was also studied during radar watchkeeping by White and Ford, 1960, and Michon and Kirk, 1962.

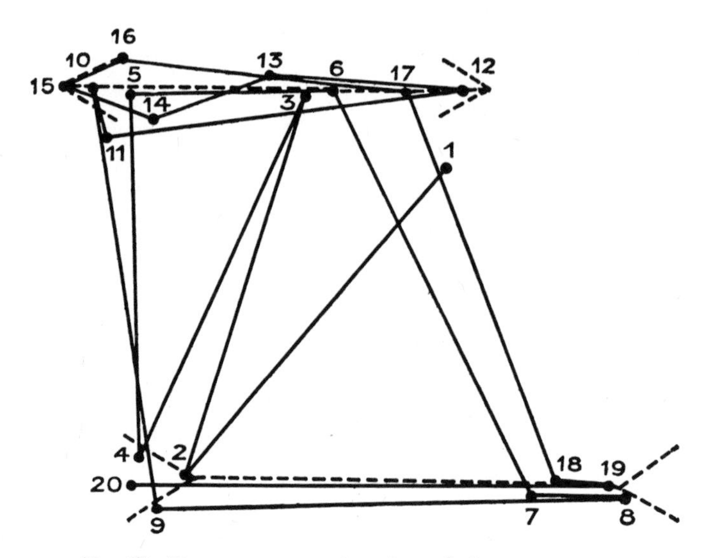

FIG. 53. Eye movements of a subject looking at the Müller-Lyer illusion (Brandt). The numbered points indicate the successive fixations; the total duration was 5 seconds.

Search time (the time required to find a specified target in a television display) was studied by Erickson (1966) in the presence of visual noise; neither foveal nor peripheral acuity correlated significantly with search time, which seems to depend more on mental processes than on retinal ones.

Movements of the Eyelids

The regular blinking of the eyelids is useful to vision, as it cleans and moistens the cornea. There exists a *reflex* blinking which is caused by a light touch of the eyelids or cilia, a bump on the head, a loud noise, the sudden approach of an object toward the eyes, or a strong light. Its latent period is short (<0.1 second), which explains why explosions rarely produce corneal burns. Spontaneous blinking is bilateral and synchronized for both eyes, but its amplitude is often less in one eye. According to Miles (1931), the duration of the closing of the eyelids is about 0.05 second, the eyelids remain closed for 0.15 second, and the opening of the eyelids takes 0.20 second. During each blink the darkness period is about 0.3 second (subjectively much less), and if a subject blinks every 3 seconds he spends more than 10% of his time in the dark (Lawson, 1948). Other authors arrive at lower values (3% according to

Gordon, 1951). The darkness time is increased by Bell's phenomenon (1829), the upward movement of the eyes which occurs with blinking. This displacement was believed to be of large amplitude (10 to 15°, Miles, 1925), but according to Ginsborg (1952) it is hardly greater than 1° and is accompanied by a convergence of the same order of magnitude. The upward movement lasts from 0.04 to 0.09 second and the downward movement from 0.07 to 0.15 second.

The curve of frequency of spontaneous blinking is typical for each subject (Ponder and Kennedy, 1927; see also Fig. 54), but the mean frequency varies a little with the state of nervous tension and increases when the conditions of vision are defective (Luckiesh and Moss, 1940). The approach of sleep does not modify this frequency, which remains

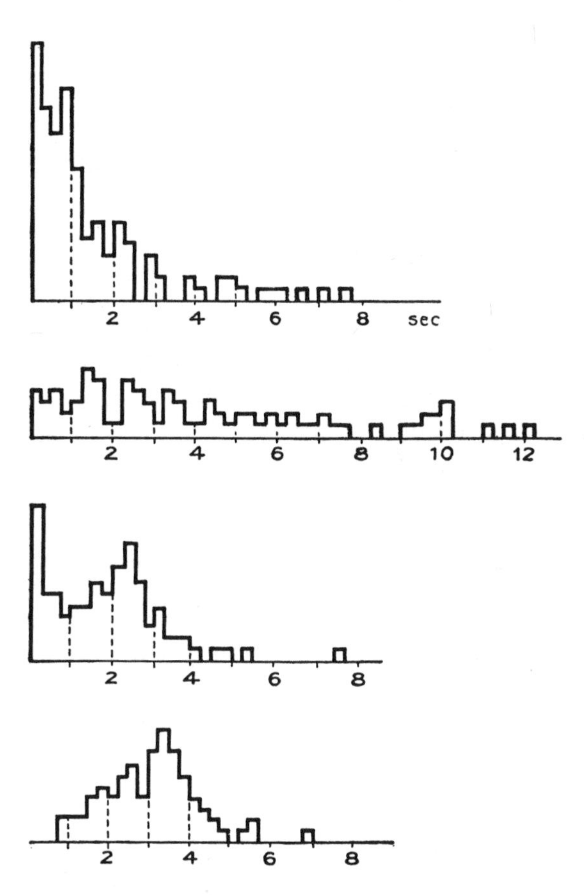

Fig. 54. Histograms of the interblink periods of four subjects. The top one is the most common (Ponder and Kennedy).

unchanged even when the eyes are closed in the state of drowsiness (Lawson, 1950). During sleep the blinking stops.

The origin of spontaneous blinking is not retinal, as it continues in the dark, nor corneal, as it is retained after anesthesia of the cornea. However, the absence of blinking produces a drying of the cornea in time. It seems that the principal cause of blinking is merely a fatigue of the levator palpebrae muscle, which relaxes periodically after a certain static tonus.

Speed of Accommodation

The duration of changes of accommodation is a problem linked to that of eye movements because of the relationship between accommodation and convergence. Convergence is more rapid because it is performed by the striated extraocular muscles, whereas the ciliary muscle which acts on the lens is smooth muscle.

Early measures gave longer durations in looking from the punctum remotum to the punctum proximum than for the contrary. According to Aeby (1861) the interval is 0.5 to 1.9 seconds in the former direction and 0.2 to 1.2 seconds in the latter direction. Schmidt-Rimpler (1879) maintained a steady convergence at 25 cm and found 2.7 seconds from the remotum to the proximum (in relation to this convergence) and 2.4 seconds in the opposite direction. By converging at 6 m he found 1.5 and 0.9 second, respectively. These results were obtained subjectively, by measuring the time needed for the test to be seen clearly by the subject. Angelucci and Aubert (1880) showed that the objective method based on observing the variations of the curvature of the anterior surface of the lens by means of the Purkinje images arrived at noticeably shorter times (0.2 to 0.5 second) and they were almost independent of the direction and of the amplitude of accommodation. According to these authors, in this manner, a gross adjustment is measured, which would be completed by a fine movement, especially necessary in near vision.

Following the research of Seashore (1893), it was admitted that, on the contrary, a slightly longer time is necessary to go from the proximum to the remotum than the reverse. By a method of brief presentation, Ferree and Rand (1918) obtained 0.4 to 0.8 second to go from far to near and 0.5 to 1.2 seconds from near to far. According to Banister and Pollock (1928), these durations increase with age but diminish with training in subjects who are used to handling moving objects. Fatigue increases the duration of accommodation (Robertson, 1936).

Kirchhof (1940) measured the durations on three subjects by an objective method, photographing the Purkinje images on a moving film, and obtained 0.43 second from near to far and 0.50 second from far to near. The difference is small but in the same direction as the former measurements. Fatigue increases the difference because the first result does not change, whereas the second can increase up to 0.7 second. Travis (1949) used a subjective method on 50 subjects with Landolt rings situated at 56 cm and at 13 m. Each recognition of the position of the ring takes on the average 0.9 second (whether at near or at distance), whereas with alternated recognition (from one distance to the other) it takes 1.2 seconds. The difference would be the time of accommodation.

Finally, let us cite the measurements of Allen (1953), who used at the same time an objective method photographing the Purkinje images to record convergence in one eye, and a subjective method based on Scheiner's principle to record accommodation in the other eye. For a variation of accommodation of 2 diopters there is latency of 0.2 second for the convergence and of 0.3 second for the accommodation. The total duration of the phenomenon is about 1 second. According to Campbell and Westheimer (1960), the lag is longer by 0.02 second on the average to look from distance to near than vice versa.

Finally, let us point out that when an object is brought progressively closer to the eyes, the changes of accommodation are discontinuous (Westheimer, 1954), which is the opposite for pursuit movements.

CHAPTER ELEVEN

Vision of Movements

In this chapter we will review the principal problems of the vision of movements, and we will present the difficulties encountered in the perception of true and apparent movements. The reader interested in the psychology of movement perception may consult [19*a*] and [61*b*].

Threshold of Displacement

The vision of movements can be produced in two ways. Either the eye remains steady and the image of the mobile object scans the retina, or the eye follows the displacement of the object as the subject fixates it.

The first method is more precise. In foveal vision the minimum displacement perceptible is of the same order as vernier acuity, that is, a few seconds of arc in the best experimental conditions and about 20″ in the most common measurements. In peripheral vision the minimum displacement perceptible increases toward the periphery but not so quickly as for vernier acuity. Several investigations in this area, starting with Dodge (1904) and Basler (1906), give a threshold of 3′ at eccentricity $\eta = 20°$ (whereas it is between 10 and 20′ for vernier acuity) and 5′ at $\eta = 40°$ (25 to 50′ for vernier acuity). The comparison with the limit of resolution would be even more advantageous to the threshold of displacement. This can easily be verified by looking peripherally at one's hand held at arm's length at the edge of the visual field. The separate fingers cannot even be discriminated, but a little movement of one of them is easily perceived. Stratton (1902) insisted on the biological role of this remarkable finesse which is better in the horizontal than in the vertical meridian. The least displacement in the periphery attracts the attention of the animal and brings the movements of the eye to place the image on the macula (or the fovea if the animal has one), where it is analyzed in detail. Therefore, the best tactic to avoid being noticed is to remain immobile and animals that act as if they were dead apply it, no doubt instinctively. The visual perception of movement seems to appear at the early age of a few weeks and is one of the earliest visual functions.

Physiologists have found it difficult to account for this keenness of the vision of movement. It can either be assumed, as Fleischl (1883) did, that it is produced by the excitation of the neighboring cones on other ganglion cells, the connections being such that a movement causes variations of excitation without perceptible separation, or, as Helmholtz suggested, that a particular ganglion cell can receive different excitations when the image moves in a receptive field. Electrophysiological experiments seem to support the latter hypothesis. Moreover, Warden [14] noted that an immobile object disappears easily in the periphery because of local adaptation, and movement of the object, like movement of the eyes, reinstates the perception.

The values mentioned above applied to a high luminance (according to Basler, the threshold is multiplied by a factor of 10 when the illumination is changed from 1000 to 1 lux) and with an almost instanteous displacement. The results do not alter as long as the speed of the displacement is greater than 1°/second (Blackburn, 1937). Below this level the problem changes and we will study the question of the speed threshold later.

If the object is placed in an unstructured field, such as a light in the dark, Basler determined that the displacement threshold was at least 1′ in foveal vision.

Clemmesen [12] studied the vision of a light oscillating periodically, using two subjects. The optimum frequency is from 2 to 3 swings per second. If the amplitude is 30″, the light appears to be stationary and for 1′ it moves slightly in foveal vision; for $\eta = 5°$ these results are, respectively, 2′ and 3′.

Speed Threshold

With an immobile fixation point the minimum speed perceptible is about 1′/second in foveal vision and 3′ at $\eta = 5°$, 14′ at $\eta = 10°$, and 34′ at $\eta = 20°$ (Bourdon [6]). Actually, the curves of equal threshold are not circles centered on the fixation point but ellipses of which the long horizontal axis is almost twice the length of the short vertical axis (McColgin, 1960). The duration of the exposure is obviously important and Dimmick and Karl (1930) found the threshold to increase from 44″ to 2.5′/second when the exposure decreases from 4 to 0.5 second. According to Crook (1937) it is around 1 second exposure that the limit of resolution of an object is equivalent to its speed threshold. Gordon (1947) studied the influence of the level of luminance; at 3×10^{-4} cd/m^2, he found that after complete dark adaptation the static and dynamic thresholds agree throughout the periphery for 2 to 3 seconds exposure (see also Leibowitz, 1955). In an unstructured field the minimum perceptible speed is 15′/second (Aubert, 1886).

A test often utilized since Klein (1942) is a type of Landolt ring with several gaps of different widths and which turns around its center. Warden et al. (1945) utilized this system on 28 subjects at 3×10^{-4} cd/m^2 and obtained a threshold of 20′/second at $\eta = 10°$, 31′/second at $\eta = 35°$, and 50′ at $\eta = 55°$. There are considerable differences among subjects (a factor of 40 between the best and the worst) without any correlation to the photopic acuity.

Bourdon made some measurements on the differential threshold of speed with a rotating disk at average speed and found a threshold of about 10%. J. F. Brown and Mize (1932) studied parallel translatory movements. The best differential threshold (2%) was obtained at a speed of 3°/second. R. H. Brown (1955) attempted to relate these differential thresholds to the Bunsen-Roscoe law. Finally, Hick (1950) studied the successive differential threshold for acceleration and deceleration.

Perception of Speeds

The impression of speed of a mobile object depends upon many factors. Fleischl (1882) observed that the velocity appears half as great when following a moving object with the eye as when fixating an immobile point. This phenomenon, sometimes referred to as the *Aubert-Fleischl paradox*, is due to the fact that the pursuit of the object shares the sensation of movement equally between the object and the background which seems to move in the opposite direction (Filhene [in 3]). According to J. F. Brown (1931), the apparent speeds are, on the average, in the ratio 1.4:1 for the conditions of steady versus moving eyes. A movement seen by the peripheral retina appears slower than when it is seen centrally, if the eye fixates a steady point (Czermak, 1854). Furthermore, Brown determined that if two movements are of the same velocity a vertical one will appear 10% faster than a horizontal movement and also that an increase in the size or the distance of a mobile object will diminish its apparent velocity. If two identical objects move across the field of view at 6 and 20 m from the subject, the speed of the farther object must be 1.6 times that of the near one to appear equal. A very heterogeneous background makes the objects appear to have a greater velocity. Metzger (1926) has brought attention to the effect caused by the structure of the field, first, on changeover of the perception of an object which appears to shift from one position to another to that of movement at low velocity, and, second, the changeover of perception from that of rapid velocity to the impression of fusion.

It must be noted that the perception of a movement cannot be considered as the ratio of distance to time. Piaget (1946) presented to children of 8 or 9 years of age, objects moving on two concentric circles. The children judged the speeds of the objects to be equal when the periods of rotation were the same. If it were pointed out to the children that the circles were unequal, they answered that the object on the smaller circle moves faster "because it has less to cover." Michotte (1946) demonstrated with very interesting experiments how remote the abstractions of kinematics are from the actual perception of movement where ideas of cause and intent play an essential part.

Appearance of Moving Objects

Visual thresholds for moving objects have attracted very little research. We can cite those of Bouman and van den Brink (1953, 1957), who investigated the absolute and contrast threshold of a mobile object on a dark background; Ludvigh (1947–1948), on the foveal acuity of a moving object (the results are represented in Fig. 55); and finally Low (1947), on the visual acuity of an object in motion seen in the periphery. The last investigator found that a Landolt ring moving at a velocity of 15°/second, at $\eta = 30°$, is recognized with a dynamic visual acuity which is 0.6 the static acuity at the same point. According to Ludvigh and Miller (1953) the visual acuity of a mobile test that is gazed at by the subject remains constant up to a velocity of 30°/second and even up to 80° for some very keen observers. In the case of a grating which is vibrating, the acuity was measured on three subjects by Fiorentini (1955). It would be interesting to have more data on the vision of mobile objects, as it is an important practical subject (see van den Brink, 1958; Bhatia and Verghese, 1963, 1964).

We have, however, a wealth of observations from psychologists on the particulars of the appearance of a source or an object in motion and we will review them briefly. De Silva (1929) presented to a subject a luminous slit moving horizontally in the dark. Up to a speed of 10°/second the subject noted that the contours were still sharply defined but they became blurred between 10 and 14°. Between 14 and 20° the slit looked like a bar with a tail, between 20 and 60° like a sort of luminous sheet unfolding itself, between 60 and 120°, a slightly vibrating sheet, and finally beyond 120°/second it appeared perfectly stable. Charpentier (1888) had observed the tail that appears behind a small moving light source and he explained it as a rhythmical diffusion of the excitation. Moreover, the object in motion seems to contract in the direction of the movement (Ansbacher, 1944), and a complicated figure is less deformed than a geometric figure.

When a source which emits periodic flashes illuminates a moving pattern, a *stroboscopic effect* is produced, which was first noted by Plateau (1836). This phenomenon has no physiological basis. A wheel appears immobile if the period between two successive flashes is equal to that needed for a spoke to replace the one in front. This effect is familiar to cinema audiences. Sometimes, even with constant illumination, one momentarily sees blurred stripes on a rotating disk. Some authors have been of the opinion that this effect may be due to some "retinal stroboscopy," but it is more probable that movements of the eyes are concerned, or it may be due to a cycloidal effect (in the case of the wheels of a car) caused by the trajectory of the wheels and which would only be seen in the lower part of the wheels (Gardner, 1941).

In some experiments a natural rhythmic effect of vision may occur. C. A. Young (1872) observed that an intense and brief electrical spark

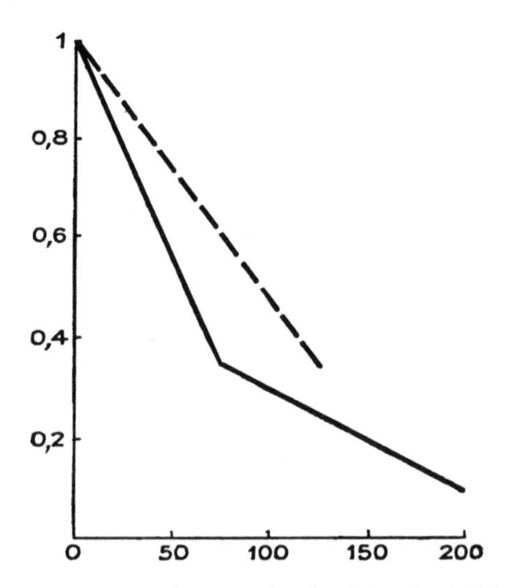

FIG. 55. Dynamic visual acuity (Ludvigh). The full line represents results with a Landolt ring; the dashed line, with Snellen letters. Abscissa: Speed in degrees per second. Ordinate: Relative dynamic visual acuity.

sometimes shows objects at several points of their trajectory. Young called this phenomenon *"recurrent vision."* MacDougall (1904) rotated a disk with a thin radial slit and observed up to seven images arranged fanlike. Richardson (1930) attempted a theory of this effect but it seems complex and it is probable that some inhibitions (similar to those that we will describe later) play a role in this phenomenon, which makes it difficult to relate it to simpler and known effects such as, for instance, the Broca-Sulzer phenomenon.* According to Sweet and Bartlett (1948), recurrent vision can account for the illusory sweep appearing under certain conditions, behind the real sweep, on the cathode-ray-tube screens used in radar. Viewing a television screen presents similar problems (van den Brink, 1957).

Anorthoscopic Illusions

Plateau (1836) described the illusion caused by the vision of an object in motion observed through a slit which also is moving. There are many variables in this effect and, therefore, a lot of work for psychologists. Zöllner (1862) got excited about the phenomenon and so did the theorists of the Gestalt school, particularly Wenzel (1926). These effects can be related to the deformation noted when a carriage wheel goes past some railings (Roget, 1825). All these effects and many others of the same kind do not in my opinion offer much interest but merely use up a lot of paper with unimportant material. To entertain and rest the reader, I invite him to look at the classical illusion in Figure 56 that can be seen by rotating the book, and if he is intrigued by the attempted explanations, to read the paper (not very convincing) of Bowditch and Hall (1880), or better, the first scientific note of Piéron (1901).

Metacontrast

Many authors have proposed the utilization of moving objects to measure the persistence of the luminous sensation (persistence is caused by a delay in disappearance which is greater than the delay in perception). Fröhlich (1929) and Monjé (1934) attempted to measure the total persistence by the apparent broadening of a moving slit, but the method is inaccurate because the eye tends to follow the objects. Piéron (1935), who used a slit with a less bright central part, observed an hourglass appearance which shows that at the center there is a greater delay in

* The initial brief enhancement of brightness that is observed at the onset of a luminous stimulus.

FIG. 56. Illusion of movement. By rotating the book and tilting it slightly forward the outside wheels are seen to rotate in one direction and the central cogwheel in the other. One will note grey sectors which appear in the outside wheels.

perception and a lesser delay in disappearance. Thus this method can measure differences in latencies.

Hazelhoff and Wiersma (1924–1926) tried to measure the absolute latency of a brief light by the following method. The subject follows the horizontal and uniform movement of a luminous point on a dark background. A bright flash occurs from a stationary point near the line of movement. This point is adjusted by the subject so that the two sources appear vertically aligned when the flash occurs. In fact, they are not vertical because between the flash and its perception the eye has traversed a certain distance which is a measurement of the latency. As van den Horst and van Essen (1933) have pointed out, it is very unlikely that an absolute latency can be thus measured because there is no reason for the movement of the eyes which follow the moving object not to be itself perceived with a certain delay. The results obtained range between 0.07 and 0.14 second, depending upon the subjects. Monjé (1934) utilized a similar method to measure the latency of the differential sensation, but, unlike the above experiment, the fixed source changes its intensity very suddenly instead of producing a flash of light. At threshold the latency is about 0.16 second and can be even as low as 0.04 second.

Fröhlich (1923) had proposed a method which he thought would measure the latency of sensation. If a moving luminous slit suddenly emerges from behind a screen which masks it, the beginning of its appearance is delayed because of latency and, therefore, a spatial difference occurs which can be measured with some fixed points of reference. But, here a phenomenon discovered by Stigler (1910), called *metacontrast*, intervenes. If the two halves of a photometric field are illuminated briefly and successively with equal luminance and exposure, the half which appears first seems darker, particularly near the diameter which separates the two halves. Piéron (1935) demonstrated that metacontrast could inhibit completely a previous luminous sensation. In fact, in the experiment of Fröhlich there is an inhibition of the slit at the very beginning of its appearance by the following positions which are perceived. Moreover, geometric forms also influence the phenomenon (Werner, 1937).

The laws of metacontrast have been studied by Fry (1934), Baumgardt and Segal (1947), and especially Alpern (1953). Alpern used a method of binocular photometry (Fig. 57). The comparison patches a and test b are illuminated simultaneously for 0.005 second, whereas the contrast-inducing patches c emit a flash having the same duration but occurring at a certain time t prior to a and b (asynchronous exposure). The maximum of the effect occurs for $t = 0.1$ second and the metacontrast disappears between 0.2 and 0.3 second exposure asynchrony. If the angular separation of the contrast-inducing patches and the test b is increased, the effect diminishes and becomes nil around $\theta = 2°$, and t diminishes as well and becomes 0.04 second when the effect disappears. For $\theta = 0.75°$ and $t = 0.1$ second, the results are represented approximately by the following formula:

$$\log \frac{L_b}{L_a} = \log \frac{L_c}{L_0}\left(0.6 - 0.13 \log \frac{L_a}{L_0}\right)$$

with $L_0 = 0.38$ cd/m^2, $0.38 \leqslant L_a \leqslant 380$ cd/m^2, and $0.38 \leqslant L_b \leqslant 10^4$ cd/m^2. Metacontrast disappears if the test itself is fixated. Alpern believes that the inhibition displayed by metacontrast is due to the fact that the arrival at the ganglion cells of the nervous impulses correspond-

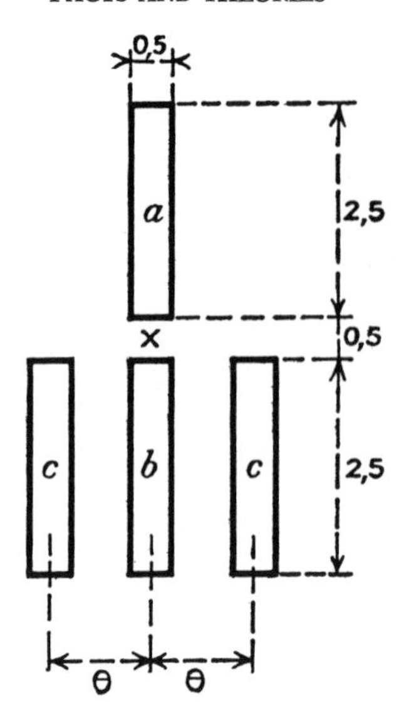

Fig. 57. Measurement of metacontrast (Alpern). The cross represents the fixation point; the four rectangles of equal dimension (indicated in degrees) are bright on a dark background; *a* is seen by one eye, *b* and *c* by the other.

ing to the inducing fields *c* blocks the synapses and prevents the normal transmission of the impulses from *b*.

If *t* is negative and of the order of 0.05 second, a slight effect of the same kind occurs; Stigler called this *paracontrast* and it was interpreted by Alpern as due to scattered light.

Subjective Colors

Motion produces several colored effects that have been known for a long time, which arise because the different colors have, at equal luminance, different latencies, shorter for red than for blue. Consequently, if a sector half red and half blue is attached to a rotating black disk, the red appears to lead and the blue to lag behind. The "fluttering heart" of Helmholtz is explained in the same manner, where hearts of red paper stuck on a blue paper background appear to dance about on the background when the card is moved to and fro.

These differences of latency can account for the appearance of subjective colors where there is only black and white. For instance, if black rectangles are drawn on a white disk at unequal distances from the center, one sees, when the disk is rotating slowly, rings which seem bluish (because of the predominance of the blue terminal fringes), and by contrast the intermediate ring seems yellow (Bidwell, 1896). It seems to us superfluous in this case to refer to chromatic aberration of the eye, as Hartridge (1947) suggested.

The famous "Benham top" of Fechner (1838) popularized by Benham (1894) gives pale red colors for the first group of lines (on the outside when rotating in the direction indicated on Fig. 58), reddish brown for the second, olive green for the third, blue green or dark blue for the last. We refer to Piéron [51] for the explanation of the appearances due to unequal speed of the development of the chromatic sensation. White light is not necessary since colors are subjective; they are seen even in monochromatic light, sodium, for example (Gehrcke, 1948).

Burnham (1954) described certain subjective colors that would appear with flickering light in binocular vision and that he interprets as due to a response shifted in time between the two eyes.

Phi Movement

Wertheimer (1912) has given this name to a very old illusion. If two identical images appear successively (with a time interval Δt reasonably short) at two points close to each other in the visual field, the subject has the impression that a single object moves from the first to the second position. Mibai (1931) found that the time interval must be $0.06 < \Delta t < 0.2$ second. The optimum time depends upon the configuration. According to de Silva (1926), the movement can always be recognized as apparent (because it is more saccadic) if the speed remains below 10°/second. The movements of the eyes do not account for this effect; they

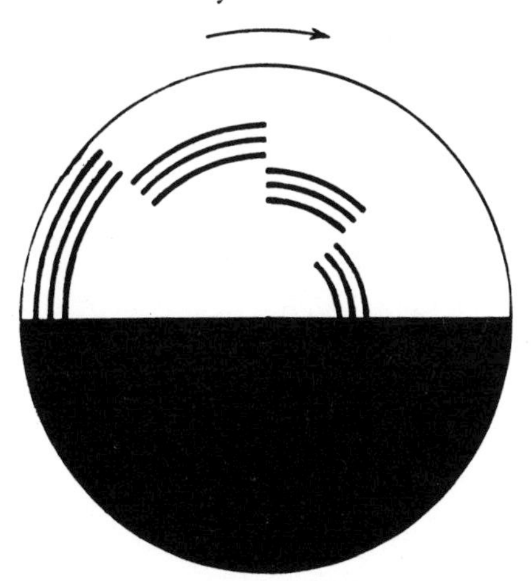

Fig. 58. Benham-Fechner top.

somewhat impair it (Guilford and Helson, 1929). In fact, the psychological attitude of the subject seems important, as was demonstrated by Kelly (1935). He showed this experiment to 400 students who had never seen it, and only half of them perceived a movement; but once they were told what to see, 90% of them saw it. Attention has sometimes been brought to φ movement in the cinema, but it is probable that for a greater number of successive images, the impression of saccadic or steady movement follows laws other than those applying to two images only. Interaction between flicker and apparent movement in motion pictures has been studied by Chatterjee (1954) and Braunstein (1966). Let us cite at last the experiments of Koffka (1924), who produced φ movement between points situated on each side of the blind spot or even in the blind spot itself. In the latter case it appears as a big blurred source which moves along a curved line. Koffka deduced from it that the blind spot possesses a small sensitivity, but the diffused light suffices to account for this effect, as we have mentioned earlier.

Psychologists have described many other illusions of movement and nearly the whole greek alphabet has been used up [3]. Benussi (1914) calls α movement that which is seen when showing successively one of the components of the Müller-Lyer illusion, then the other, at the same point in the visual field. Kenkel (1919) calls β movement a similar phenomenon produced by two stimuli of which the dimensions are actually different. Lindemann (1922) described a γ movement which occurs from the center to the periphery when a light is suddenly switched on in a dark room, and from the periphery to the center when the light is switched off. This movement is often accompanied by an apparent movement toward the observer, if the illumination is gradually increased (lasting 0.2 to 0.8 second) instead of being a spark. This phenomenon is seen just as well with a black object on a bright background and cannot be explained by irradiation. The δ movement deals with a pair of objects of different illumination, the movement appearing in the direction from the brighter to the darker. Finally, many apparent movements of afterimages have been described when the source is very bright (such as a photographic flash which is fixated) and particularly if it is red (R. L. Ives, 1942). Analogies and differences between real and apparent movements have been analyzed by Graham (1963) and Kolers (1963).

Autokinetic Sensations

This name was given by Aubert to apparent movements of an immobile luminous point on a dark background. One notes small displacements of the light, described by Schweizer (1858), on which, after sufficient fixation, are superimposed slow oscillations of large amplitude described by Charpentier (1886). According to Graybiel and Clark (1945) this latter phenomenon begins after a fixation of about 10 seconds and the amplitude of the oscillations is 3 to 4°. The movement changes its direction every 10 seconds. The illusion is seen even if the luminous point is not fixated, but it is then less pronounced. Several parameters, such as low intensity of the source, monocular observation, fatigue, and somnolence, enhance the autokinetic sensations, which are quite a serious drawback in piloting at night. It is recommended, to avoid these sensations, to increase the number of lights that may be used as a point of reference and to move the eyes often (Horsten and Winkelman, 1954).

It was assumed formerly that this illusion was due to eye movements. We will see later that for any voluntary gaze, a compensation occurs which immobilizes the object in spite of the displacement of the retinal image. This compensatory system is relaxed in darkness. But Guilford and Dallenbach (1928) photographed eye movements while they took note of the autokinetic sensations of the subject and found no correlation. There occurs a sort of entoptic streaming that can, in fact, be altered by pressing on the eye. Anyhow, the absence of perceptive stimuli in the visual field is essential and to have fixated previously a light object reduces the illusion a great deal (Crutchfield and Edwards, 1949). Group suggestion enhances the effect of the illusion very much, as Sherif (1936) found out by using a group of three or four subjects communicating their impressions to one another. Emotional influence also bears some importance (Haggard and Rose, 1944).

If the observer is subjected to acceleration as it occurs in an airplane, the phenomena become complex and the source takes up diverse apparent movements (Graybiel et al., 1946).

If the point of light is actually displaced, its apparent trace is the resultant of the true movement and the autokinetic movement, and the latter can thus be measured (Bridges and Bitterman, 1954).

If two identical light sources, one steady, the other in motion are presented in the dark to a subject, they seem to move symmetrically (Duncker, 1929). But if they are unequally bright, the more intense seems immobile (Oppenheimer, 1934). If the observer fixates one, it is this one which appears to move, while the other appears immobile. It is actually a common psychological observation that movement is attributed to the object fixated and immobility to the background, such as a train starting to move beside another remaining still, the moon running behind the clouds, the belfry which falls when it is windy, etc. (Thelin, 1927).

Motion Afterimages

After a brief observation (<1 second) of an object in motion, one sees, on closing his eyes, the object continuing its movement. This is more a memory than an afterimage. But the phenomenon perceived after having gazed at a regularly moving object for 20 to 30 seconds is very different, as, for example, a military parade, a waterfall, a train passing under the bridge where the observer is standing, etc. By looking elsewhere or after the movement is completed, one sees the background moving in the opposite direction for a few seconds with a speed which reduces gradually. Dark adaptation and rod vision impair the phenomenon (Granit, 1927). Durup (1931) has specified the role of fixation. He found that the motion afterimage is more prolonged if fixation was concentrated on the part which is moving. Motion afterimage has also been studied in a "Ganzfeld," that is, a homogenous visual field deprived of any detail (Miller and Ludvigh, 1961).

It could have been thought that eye movements were responsible for this phenomenon, but this is not so, as is shown by Plateau's spiral, which appears as concentric circles that seem to enlarge or contract according to the direction of the spiral's rotation. This is followed by an afterimage in which objects appear to reduce in size (therefore, go further away) or to increase (therefore, come nearer). Dvorak (1864) has made the experiment more striking by alternating on the same disk spirals in the opposite direction. The observer then sees an afterimage enlarging in some parts of the field and contracting in others, which proves a retinal or cortical origin but excludes eye movements. After Barlow et al. (1963) the origin of this phenomenon might be thought to come from the ganglion cells, but pressure on the eye which paralyzes these cells (Bornschein, 1958) leaves the motion afterimage intact, which therefore must be due to activity of the visual centers. This spiral aftereffect is fully studied in a monograph by Holland (1965); see also the paper by Thomas and Strong (1966). Entoptic phenomena may also give aftereffects of motion; for example, when the rotation of Haidinger's brushes is stopped, a brief apparent rotation in the opposite direction is observed (Anstis and Atkinson, 1966).

Other Apparent Movements

In spite of a displacement of the body, the head, or the eyes, objects usually appear to remain immobile, although their images move on the retina. The origin of this compensation has aroused many controversies.

Helmholtz thought that the principal cause was the sensation of the innervation producing the muscular movements, particularly of the extraocular muscles. He insisted on the beneficial role of Donder's law regarding ocular rotations which, by always bringing the images back into the same situation, facilitates the compensation. Wundt [75] placed the existence of these sensations under suspicion, and invoked the representation of movements which accompany voluntary displacements. Bourdon [6] insisted on the importance of the sensitivity of the eyelids and conjunctiva, but cocaine suppresses these sensitivities without altering the compensation. M. H. Fischer (1921) held the static tension of muscles to be responsible, but if the eyeball is displaced with the finger, an apparent movement is seen. For Hering [21] it was attention which plays the predominant role, and Hillebrand (1920) has completed this point of view by the consideration of the gradient of acuity, but these theories are difficult to reconcile with the localization of afterimages described by Purkinje [58] and Mach [38] and studied in detail by Lipps (1890). If the gaze is turned suddenly after having fixated a bright source of light, one sees a trail coming from the source, but it is in the opposite direction to the movement. Conversely, if the gaze is directed toward the source, the trail is in the direction of the movement. In both cases the trail is localized as if the gaze remained directed toward the initial point; the afterimage is evidently only perceived at the end of the movement, and yet it is localized in reference to the initial stimulus. Similarly, according to Kaïla (1923) a flash of light which occurs during a displacement of the gaze is localized according to the spatial reference points which precede the movement.

It is seen that the mechanism of the compensation is much more complicated than can be imagined a priori. The theory which seems at the present time the most satisfactory is that of Göthlin (1929), which involves complex synchronizations between the neuromotor systems of the eye and the head on the one hand, and the retina on the other. An apparent movement is seen when this synchronization is altered. One example is in some paralyses where the subject wants to turn his eye but cannot and the compensation occurs anyhow, which consequently produces an apparent movement although neither the eye nor the object has moved. Another example is a poisoning (drunkenness, for instance) which impairs the precision of ocular movements. Finally, an unconscious nystagmus which occurs after several rotations of a subject in a chair (acting on the labyrinth) makes objects and even afterimages appear to move in spite of their stability on the retina (M. H. Fischer and Kornmuller, 1931). It is necessary also to take into account the sense organs within the eye muscles themselves (proprioceptive muscle spindles), which provide a control and a knowledge of the position of the eyes, (Whitteridge, 1959; Bach-y-Rita, 1959; Donaldson, 1960).

Certain apparent movements have purely dioptric origin, as, for instance, a highly myopic or an aphakic subject corrected for the first time. Conversely, after he has adapted to his lenses, when he removes them he notes movements of objects which are due to a change in the size of the retinal images and a modification of the usual compensation. This effect becomes very striking when afocal magnification lenses are prescribed to an emmetrope; the walls of the room move about when the subject turns his eyes or his head.

A somewhat different problem of compensation consists in finding why, during the eye movements, the striae or blurred streaks of the images are not perceived. Holt (1903) had imagined that a "central anesthesia" occurred during the movement, but Dodge (1905) showed that with a little attention one can perceive appearances of this kind. As we mentioned in Chapter 10, the number of images (of a periodic source) seen during a displacement of the eye has been used to measure the speed of the displacement. This phenomenon can be easily observed by looking at a vertical tube of a neon sign which flashes on and off 120 times/second. If the gaze is displaced horizontally, one can perceive parallel stripes or not, depending on whether the attention is placed upon the mobile object that is gazed at or upon the neon tube itself, which proves the importance of central effects. For the same reason, the experiment of O'Brien and Dickerman (1948) that we discussed in Chapter 10 is not very convincing; one cannot see several points in spite of the micronystagmus, because it is upon the point itself that the attention is placed. This problem of vision of moving objects has not yet been resolved in spite of the attempts by Ditchburn (1955) and Hyde (1959). It seems most probable that there is no blankout during movement but a partial inhibition, the thresholds increasing by about three times (F. C. Volkmann, 1962); according to Latour (1962) this inhibition starts about 0.04 second before the beginning of the movement.

CHAPTER TWELVE

Binocular Vision

The existence of two eyes symmetrically situated with respect to the median plane of the body in man gives normal vision some remarkable properties, the importance of which is considerable in the synthesis of visual space. The simultaneous stimulation of the two retinas can, according to whether the images are similar or dissimilar, give rise to either an active cooperation, which we will see is the basis of the sense of depth and stereopsis, or to alternation and rivalry. We will begin by the study of the latter.

Binocular Visual Field

In normal vision the two monocular visual fields are superimposed because of the coincidence of the two points of fixation. As in each eye the field is wider on the temporal side, there remain two temporal half-moons of monocular vision in the binocular field (Fig. 59).

The protrusion of the nose also produces monocular fields in the inferior nasal regions. According to F. P. Fischer (1924), when binocular vision is induced artificially in this region by means of mirrors, it possesses all the characteristics that exist everywhere else in the retina, stereopsis in particular. Fischer deduces from this that this sense is innate. Actually this experiment is not very convincing, because when the eyes look directly up, this region can become binocular without artifice.

In the area of the monocular field of one eye which corresponds to the blind spot of the other, Ferree and Rand (1917) described a partial desaturation of colored vision. According to Wentworth (1931) this occurs only in individual cases. This also applies either to the diminution of the luminous sensitivity noted by Dubois-Poulsen [15] in a few rare

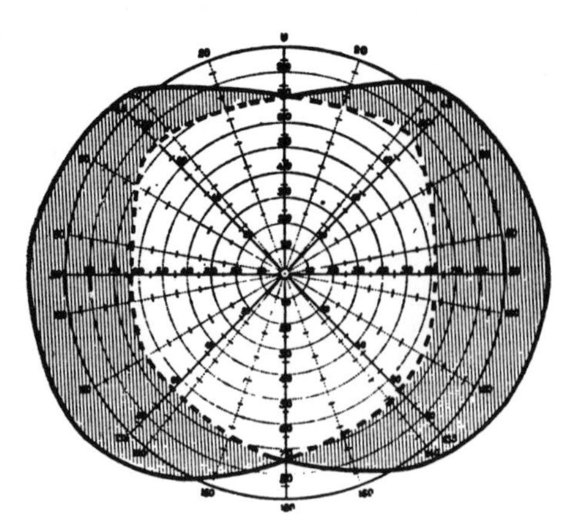

FIG. 59. Binocular visual field (after Dubois-Poulsen et al. [15]).

cases, or to the increase of sensitivity described by Wolf and Gardiner (1963).

On the contrary, a phenomenon which seems general is a slight enlargement of the blind spot which appears when one shifts from monocular to binocular vision. Dubois-Poulsen, Tibi, and Magis [15] have been able to demonstrate this effect by means of light polarized in opposite directions for each eye, so that when the test is inside the blind spot of one eye (in monocular vision) it remains unseen on opening the other eye. It is probable that the enlargement is due to an inhibition of the amblyopic zone which surrounds the blind spot.

Small scotomata for color, randomly situated but constant for a particular subject, are sometimes noted in binocular vision only, and instantly disappear in monocular field plotting.

Binocular Brightness

Generally speaking, the thresholds hardly differ whether one eye or both are used, provided, of course, both eyes are equally adapted and have the same sensitivity. Absolute threshold, differential threshold, and visual acuity are practically the same, or at least a superiority of binocular vision can only be demonstrated by some careful statistics and then is only a simple addition of independent probabilities. However, Matin (1962) found binocular vision slightly more advantageous than would be expected from a simple addition of independent probabilities. In peripheral vision and at very low intensities, some authors have also brought attention to the somewhat more pronounced effects of binocular summation than could be predicted by the simple addition of probabilities (Shaad, 1935; Collier, 1954). On the other hand, at photopic levels the summation is nil, according to Bartlett and Gagné (1939).

At supraliminal levels of intensity, the problem of the addition of monocular brightnesses is disputed. When one eye is closed, one perceives a slight temporary darkening at the moment of occlusion but then one does not feel, in general, that the brightness has reduced by a half; it is true that the pupil may have dilated a little. De Silva and Bartley (1930) performed the following experiment. Two neighboring patches are seen, one by both eyes, the other by one eye only, and their luminances are adjusted until they appear matched. To equalize the two patches, the luminance of the monocular test has to be multiplied by a factor of the order of 1.3 to 1.4. Fry and Bartley (1933) noted that this factor varied with the intensity, from unity at about 1 cd/m^2 to 2 at high luminances (this result does not prove, in fact, the summation of high luminosities and can be explained more easily by a difference of adaptation of the two retinas). According to Ivanoff (1947) the summation factor is 1.2 at 0.1 cd/m^2 and 1.4 around 50 cd/m^2.

The famous *Fechner's paradox* (1860) is, contrary to the preceding experiments, a binocular subtraction. If a fairly dense absorptive filter, transmitting only about 5% of the light, is placed in front of one eye, the luminosity seems to increase when shifting from binocular to monocular vision, as that eye is closed. It is advisable to utilize artificial pupils to eliminate the variations of pupil size with the retinal illumination. Hemholtz showed that this was due to a perceptive effect linked to the presence of objects in the visual field and it disappears if the whole field is of uniform luminance, as when a ground glass is placed in front of the eye. Schön and Mosso (1874) also described alternating rivalry and addition, that is, the monocular luminosity appearing successively brighter or darker than the binocular. According to Ivanoff (1947), if the more illuminated eye looks at a field of 25 cd/m^2 luminance, Fechner's paradox is most pronounced when the other eye looks at a field of 1.7 cd/m^2. The effect disappears when the more illuminated eye looks at a luminance below 6 cd/m^2.

If, in the above experiment demonstrating Fechner's paradox, the subject looks at a moving object, a curious illusion occurs (Pulfrich stereophenomenon), to which we will return in Chapter 13.

Another classical experiment relative to a difference of monocular luminances is *stereoscopic luster* (Dove, 1850). If the images presented to the two eyes are identical but one is white and the other black, the surface takes on a polished appearance. The most direct explanation (Oppel, 1854) is based on the fact that any point of a reflecting surface returns to each eye light coming from different objects. Actually, the same impression may occur monocularly with bright objects on slight movements of the head. Of course, psychologists have complicated the matter and introduced other factors (see Harrison [19c]).

Discrimination between the Sensations in the Two Eyes

The above phenomena show that the sensations which arise from each retina maintain their individuality until they reach the level of perception. Spatial *unification* is not a simple addition in which the monocular components become indistinguishable.

Some authors have avoided this difficulty by simply denying any unification. Porta (1593) maintained that only one of the two retinal images was perceived at any one time, and that they alternated rapidly in consciousness. This theory was improved by Bose [5], then by Verhoeff (1935), who applied it to parts of images, the perceived synthesis being a patchwork of pieces taken at the same time from each eye. Supported by some experiments of Neuhaus (1936) that Fleischer (1937) did not consider very convincing, this theory is sometimes still held nowadays, and Walls (1948) was able to ask seriously if vision is ever binocular.

Helmholtz thought justifiably that the best denial of such ideas is given by experiments made with the instantaneous light of an electric spark (or even better now, the "flash" lamps of photography) with which the duration is too short for any alternation to be possible. Stereoscopic luster, for example, occurs in this case (at least for some subjects and with a little practice). This would prove that at any moment the whole of each of the retinal images is, at least in principle, at the disposal of the mind.

But it must then be assumed that in the unified final image, what arises from each eye is identifiable without ambiguity (otherwise the stereopsis perceived with momentary illumination could, with equal chances, be correct or reversed). This problem of the distinction between the sensations of the two eyes is not yet resolved. Schön (1876–1878) explained it by local variations of retinal excitability and Bourdon [6] explained it by muscular proprioception related to the stimulated eye. This specificity of monocular sensations varies a great deal from one subject to another (Van de Geer and Moraal, 1963) and depends much on their training. I recognize easily on which lens of my spectacles there is a spot, whereas others, such as Brückner and von Brücke (1905), cannot. But all subjects recognize whether they see with both eyes or only one eye, as monocular vision induces a certain feeling of discomfort somewhat similar to that felt when the two eyes are in a very different state of adaptation.

Binocular Rivalry

If the images presented to the two eyes differ so much that unification is impossible, there occur curious alternating effects referred to as *rivalry*. For example, present vertical lines to one eye and horizontal lines to the other eye; either one or the other is perceived, or sometimes both together, but in the latter case it looks very different from a pattern of perpendicular lines—the lines seem to appear at different depths. From time to time one of the patterns is suppressed momentarily. The rhythm of the alternations as a function of the luminance of the two fields, of the movements of the eyes, and of other factors, voluntary or not, have been studied by Breese (1909), Devries and Washburn (1909), Bose [5], Caneja (1928), Peckham (1936), Zanen (1948) and Levelt (1965).

The preceding example is very schematic and artificial. In real cases one of the images or certain parts of the images resist the suppression, particularly *contours*, which are difficult to suppress if they are projected on a uniform background of the other retinal image. A classical experiment (Rogers, 1855) is the "hole in the hand": The hand is placed at about 20 cm in front of one eye and in front of the other is placed a cardboard tube through which one looks at the landscape. One sees a circle of the landscape through a hole in the hand. What is called *binocular contrast* results in the same effect: a black board on a white background seen in diplopia produces an intermediate band which is black against white and then white against black (Piéron, 1929).

The suppression may sometimes only affect part of each of the images in order to reconstitute a coherent figure. This is shown as an example by the figures of Caneja (1927) in which the half-fields of each eye alternate, which proves the importance of psychological factors of form (Fig. 60). This was also shown by Alexander (1951).

To end this, let us mention some curious experiments of Barany and Hallden (1948), in which a luminous source does or does not induce a contraction of the pupil, depending upon whether it is present in the perceived or the suppressed image, and another study by the same authors on the effect of different drugs on the rhythm of the alternations (Barany, 1947).

Binocular Mixtures of Colors

If, in the unified image, the corresponding regions of the monocular fields are of different colors, there is usually alternation and rivalry or

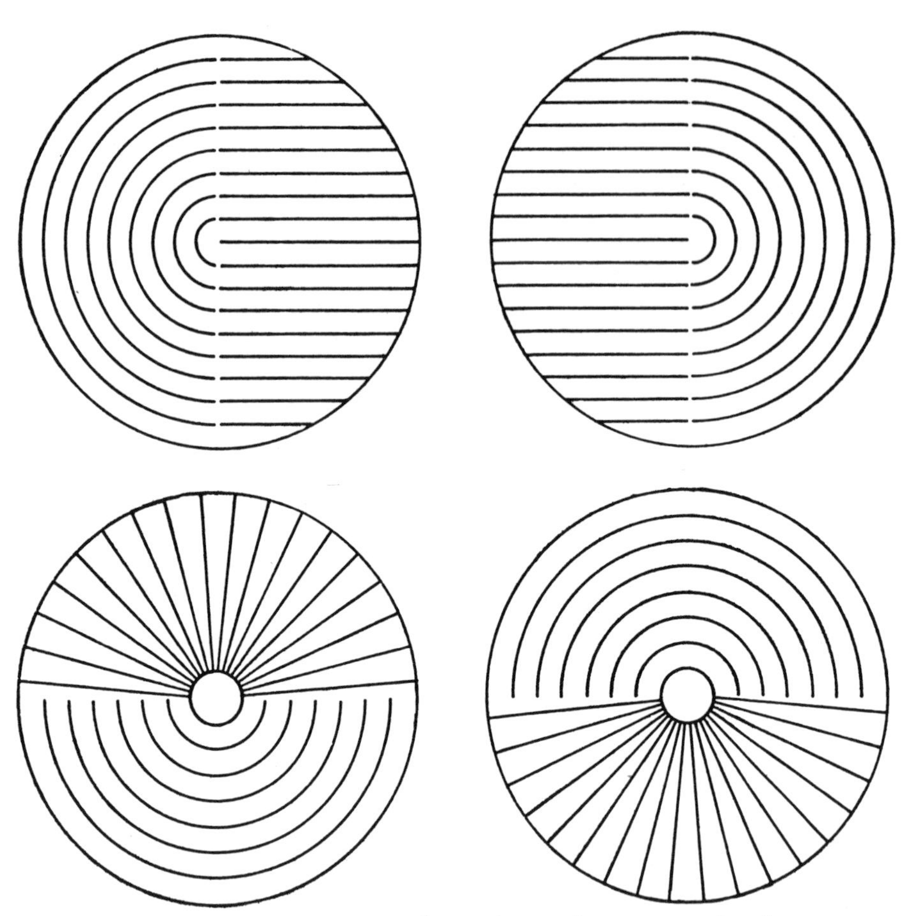

Fig. 60. Stereoscopic patterns of Diaz Caneja, demonstrating that the suppression reconstitutes a whole figure (circles, lines, or radii) and neither a monocular image nor one related to a particular cerebral hemisphere.

stereoscopic luster. But it has sometimes been perceived as a resulting color, which would bear resemblance to the laws of monocular addition. The question is still debated. Helmholtz and many excellent investigators have never noted a mixture but only a desaturation of each image due to color adaptation which reduces their difference and may therefore simulate an addition. Others, following Foucault and Regnault (1849), claim to have obtained white by binocular mixture of complementaries. According to Panum (1861), it is with small areas, and with colors not too saturated or too different in hue, that mixture occurs best. Trendelenburg (1913) and Rochat (1922) found that the result is about the same as it would be in a monocular mixture, differing mainly in the proportions of the components. Hecht (1928) saw yellow by mixing green and red binocularly and so did Prentice (1948), but others cannot. Pickford (1947) investigated the optimum conditions to obtain a binocular mixture. He found that the complementaries are the most difficult to unite, and green and red yield yellow with some difficulty. According to Hartridge [19d], the mixture is easier with a high intensity, whereas Johannsen (1930) recommends low luminances. Precise comparisons between monocular and binocular mixtures were carried out by Livshitz (1940), Thomas et al. (1961), and Hoffman (1962); the laws are very different.

We end this summary with a curious observation made by Creed [in 17]. He noted that when presenting two stamps of slightly different colors and patterns, one to each eye, the dominance of one color during the alternation was not necessarily associated with the pattern of the same stamp.

Other Binocular Interactions

It is known that, generally speaking, the adaptations of the two retinas are independent. However, according to Geldard (1928), the illumination of one eye will slightly affect the absolute sensitivity of the other. Crawford (1940) and Ivanoff (1947) deny any effect of the state of adaptation or glare of one eye on the absolute or differential thresholds of the other, whereas Bouman (1952) did note such an effect and so did Solis et al. (1953). According to very thorough research by Bouman (1955), it seems that some interactions of central origin are possible, which are similar to rivalry with alternations (see also Mitchell and Liaudansky, 1955; Fiorentini and Radici, 1961; and Fiorentini and Bittini, 1963). It is possible that such interactions are due to the centrifugal fibers (Shortess, 1963).

A type of interaction was described by Sherrington (1904). The critical fusion frequency, i.e., the number of flashes per second beyond which the source appears steady, is slightly higher when both eyes are excited at the same time than when they are out of phase (when the dark phase in one eye corresponds to the light phase in the other eye). The monocular c.f.f. is usually a value in between the two binocular c.f.f.'s (O'Brien and Perrin, 1937). A much more marked effect of interaction exists in peripheral vision with flickering light (Le Grand, 1963). J. R. Smith (1936) also pointed out that if one eye looks at a field of constant luminance and the other eye at a field of flickering illumination slightly below the critical fusion frequency, it is possible to produce an alternation of constant and flickering illumination. But this is more an effect of rivalry than interaction, although, according to Lipkin (1962), there is a slight inhibition from one eye to the other.

The existence of motion afterimages in binocular vision is unsettled. Wohlgmuth [in 47] fixated a mobile object with one eye, then opened the other eye while closing the former and perceived a motion afterimage, less marked than with both eyes. According to Exner (1875), movements in opposite directions induced on the two retinas would not give rise to a motion afterimage, whereas H. K. Müller (1928) perceived a motion afterimage with a sensation of depth.

Corresponding Points

We will leave these phenomena of rivalry, which are amusing curiosities without much practical importance, to embark upon the fundamental problem of binocular vision, the *unification* of the two images. We see objects single (in general), although we have two eyes. This process is not confined to vision. We have two ears and hear only one sound and, even simpler, we only feel one marble when it is held between two fingers. But because of the remarkable finesse of the perception of details by the retina, the unification of two retinal point images into one perceived point source is much more perfect than the "unification" of other senses, and poses much more difficult problems.

These questions have been pondered for a long time and Huygens wrote three centuries ago, "Nature has provided for us in a very particular manner so that our two eyes would not show us objects double. She has done so, that each point of the fundus of the eye has its corresponding point in the fundus of the other, in order that if a point of the object falls on any two of these corresponding points, then it appears single as it is." Pursuing this idea of Huygens, J. Müller (1826) has defined *corresponding points* of the two retinas as two points whose simultaneous stimulation gives rise to the sensation of a single external point. The points which are not corresponding points have been referred to as *disparate* by Fechner.

This definition seems clear but it causes a difficulty. Consider two points seen simultaneously in foveal vision situated close to each other but first, at the same distance from the observer and, second, at different distances. In both instances they are seen single, but in the second case the observer perceives the difference in depth because of the very precise sense of stereopsis that we will study in Chapter 13. However, it is impossible that in both cases the image falls on corresponding points of the retinas since the separation between the images is not the same in both cases. Wheatstone (1838) deduced from this, with a certain wisdom, that the excitation of two disparate points could in some cases produce the sensation of a unique object, which removes all rigor from the definition of corresponding points!

Attempts have been made to avoid this difficulty in several ways:

1. It could be assumed that single vision results from a shift of gaze to another point of regard. The subject would fixate *successively* the two objects with a slight variation of the convergence of the axes of the two eyes when fixating one object, then the other, at different distances. The images would still fall on the foveas of the two eyes, which would then be the only two corresponding points, strictly speaking, and it would be this variation of convergence which would enable one to perceive the variation of distances. We will see later that this explanation is correct for widely separated points in the visual field, but for points which are seen at the same time by each fovea, stereopsis occurs with momentary illumination which excludes any movement of the eyes, and the points are seen single whether or not they are at the same distance from the observer (provided the separation remains small).

2. There could be suppression of one of the retinal images, which would prevent diplopia. This was the opinion of Porterfield (1759), who regarded the perception of double images as abnormal or at least artificial (in the same way as two marbles are felt if you cross the forefinger and the middle finger). With a little practice double images are easily seen, as we will soon see, and the hypothesis of constant suppression is unfeasible.

3. Another possibility would be to extend the notion of corresponding points by substituting for it an *area of fusion* (Panum, 1858). With two images, one falling on a point of one retina, the other inside an area of fusion in the other eye corresponding to this point, vision is single. But it is obvious that the position of this image in the area of fusion affects the apparent distance and direction of the object (this is what Werner in 1937 called "functional displacement"), and within this area the receptors are not interchangeable. This idea of fusion is therefore an empirical concept which only evades the problem, without having any explicative value. This notion seems to me superfluous.

4. In my opinion the best solution is the one offered by Bourdon [6]. The corresponding points would exist, just as there exist cones in each retina. But, as two points are seen in monocular vision if their separation is of several minutes of arc, whereas the vernier or displacement acuity is much finer (several seconds of arc), likewise in binocular vision diplopia is compatible with a very fine stereoscopic acuity. The difficulty is therefore only apparent. The stimulation of disparate points very close to each other does not give rise to any perceived diplopia, and consequently a precise experimental definition of corresponding points must include both single vision and the notion of apparent distance.

An inverse paradox has also been put forth against the notion of rigid corresponding points. Double vision has been noticed in strabismic subjects in whom the images did fall on normally corresponding points. We will not study these pathological cases but will point out that Javal [26], who had pondered at length on those problems, regarded them either as alternations of attention or to projections falsified by habit. Tschermak [65], who squinted himself, is of the same opinion, and Burian (1947), who studied this whole topic, concluded that all cases of strabismus are compatible with the hypothesis of normal retinal correspondence (see Le Grand, 1955).

Diplopia

As opposed to the pathological diplopia of strabismic subjects, *physiological diplopia* is the perception of double images when the retinal images fall on sufficiently separated disparate points. This phenomenon was described very precisely by Alhazen in the eleventh century. Huygens also described it in these terms: "It is easy to see why distant objects must appear double when the two eyes are positioned to see another, closer object, and why, conversely, the near object is seen double when the distant one is seen single," which depicts *uncrossed* and *crossed* diplopia, respectively, so called because the right eye sees the diplopic image either to the right or to the left of the fixation point. Some subjects perceive these double images very easily by looking into the distance and interposing their finger. Others attain it with difficulty because they suppress one of the images. For these subjects the following experiment is recommended: Place a piece of string horizontally between a wall about 1 m away and the nose of the subject. The subject places his finger on the string halfway between the wall and his nose, and fixates his finger; then he sees two strings that intersect at his finger. Helmholtz remarked: "When the attention of a person is brought for the first time to the perception of these double images in binocular vision, he is usually very surprised to realize that he had never noticed them before and, what is more, the only objects that he ever saw single (very few of them) were those located at the same distance as the point of fixation." Parinaud [50] pointed out that double images were more easily seen with artificial objects such as threads or needles than with the contours of usual objects, for which the true stereoscopic effect remains even if there is diplopia, thereby masking the doubling.

TABLE 23
Thresholds of Separation with Eccentricity

	η, deg						
	1	2	4	5	8	10	13
Binocular separation	6–12′	7–15′	8–20′		13–28′		19–38′
Monocular separation	6–7′	8–11′		12–23′		15–54′	

Diplopia is of course more easily seen in the fovea than in the periphery. The threshold of diplopia has been measured by several authors. J. Rösch (1943) found a foveal threshold varying between 7 and 33 minutes of arc using two rods separated by 14′ and aligned vertically. Table 23 shows the results Ogle [46] obtained on three subjects using vertical lines. It is worth noting the striking agreement between these binocular results and the monocular limit of resolution measured by Clemmesen [12]. There is therefore no need to imagine, as Panum did, an "area of fusion," because the phenomena of binocular and monocular resolution are very similar.

Most authors who refer to Panum's areas describe them as elongated horizontal ellipses, the small vertical axis being about half the long horizontal axis (the values of Table 23 are equal to half the horizontal axes). This is due to the fact that the test is made up of two parallel lines. When these lines are vertical it is easier to determine the depth between the two lines by slight changes of convergence than it is when they are horizontal. It may also be possible that the physiological nystagmus plays some part if it is not synchronized in the two eyes, as Gertz (1935) suggested. With a test which did not yield so easily to this change of convergence, Brecher (1942) noted that the vertical and horizontal thresholds of separation were equal.

As in all questions of peripheral acuity, the thresholds of separation decrease with training as Volkmann (1859) had observed. If Panum's area possessed a physiological basis, it would seem peculiar that its dimensions decrease by half with practice! In support of Panum's concept, Tschermak (1939) assumed that the areas of fusion are made up of an area which is fixed and another which can be reduced with practice. Is it not simpler to abandon this hypothesis of areas of fusion which expresses an empirical threshold in a fashion that can only distort the truth of the matter?

Retinal Correspondence

Once the existence of corresponding points is taken for granted, the first problem is to determine experimentally how these points are distributed throughout the retinas.

First, the two fixation points correspond. Actually, the fixation point is determined only by a maximum of visual acuity. It cannot be determined with precision, and it is therefore somewhat awkward to verify this correspondence, which is more a definition than an experimental property. However, under the name of *fixation disparity*, a phenomenon has been described that is sometimes interpreted as a deviation from this correspondence. We will show later that this is not so.

Second, coordinate axes intersecting at the point of fixation must be defined on each retina. Consider a subject standing up, his head erect, looking straight ahead to the horizon. We will call the *median* plane the vertical plane of symmetry of the head, and the *frontal* plane a vertical plane perpendicular to the median plane and therefore parallel to the *interpupillary line* which joins the centers of the two entrance pupils. Imagine a horizontal line H situated in a distant frontal plane and on the same horizontal plane as the interpupillary line, and a vertical line V also situated in the frontal plane and in the median plane. In monocular vision (one eye being closed) the *horizontal meridian, h,* which is the image of H, is determined on each retina and then the *vertical meridian, v,* which is the image of V. It was noted by Volkmann [69] that the v meridians are not perpendicular to h but converge upward on the retina, therefore downward in the visual field (Fig. 61). The angle ϵ which

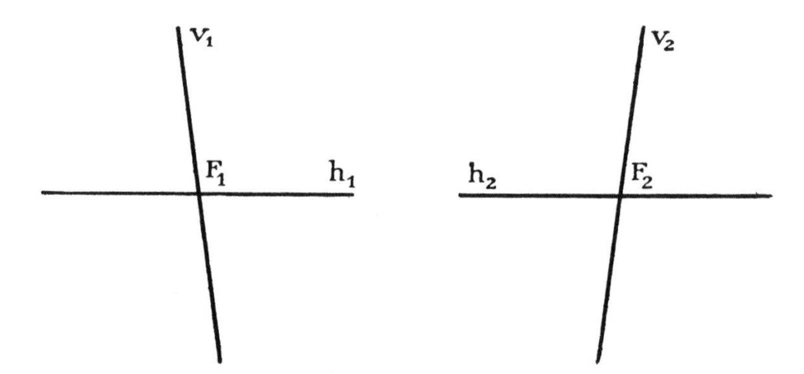

Fig. 61. Left and right visual fields (index 1 and 2); F, fixation point; h, hoizontal meridian; v, vertical meridian.

measures the deviation between a right angle and the angle made up by the lines v and h is the same for each eye but differs from one subject to another. It is, on the average, about 1° for an emmetrope and smaller for a myope.

Therefore each eye executes, in opposite directions, a cyclorotation ϵ (rotation around the visual axis) when going from observation of the horizontal H, presented alone in the field, to observation of the vertical V which replaces it. In most subjects the position of rest of the eyes is the one which corresponds to the eyes observing the horizontal and consequently there is no rotation of the eyes when shifting from monocular to binocular vision of H. If this is not so the subject has *cyclophoria*, which is considered to be a muscular imbalance. But many subjects have a small cyclophoria even when observing H and this cyclophoria increases by 2ϵ when looking at V. Thus the angle of convergence (downward) of the meridians v_1 and v_2 is greater than 2ϵ and may be as high as 5° or even more, as Helmholtz noted when looking at a uniform surface through two cardboard tubes, each with a wire across its outer end, adjusted so the lines appeared parallel and vertical. Fabry (1930) mentions the following experiment: If in a fixed telescope the vertical reticule is adjusted to appear parallel to a true vertical seen with the other eye outside the telescope, and the eyes are then changed, a considerable difference occurs (4ϵ plus twice the amount of cyclophoria, if any).

According to the above, if two perpendicular lines (vertical and horizontal) are fixated binocularly the vertical line should be seen in diplopia if the horizontal one is seen single. Actually, it is very difficult to notice because of a partial superimposition of the images. But it can be seen if a point closer (or further) than the perpendicular lines is fixated. The double images of the vertical line converge upward (or downward).

The origin of the divergence of the vertical meridians is uncertain. Helmholtz noted that with the usual separation of the eyes (60 to 65 mm) the meridians v_1 and v_2 intersect near the ground and consequently the details of the ground fall on corresponding points, which is advantageous for the appreciation of slight differences of level. It is true that the relief of the ground seems biologically important, and if the field is reversed (by means of mirrors or placing one's head upside down) the clouds give a sense of depth whereas the ground appears as a flat painting with bright colors. It could be argued, though, that one walks along hardly looking at the horizon but lowering the eyes from time to time. The other explanations proposed are not any more convincing.

We have now determined two coordinate axes (not exactly perpendicular) h_1v_1 and h_2v_2 on the retinas. Volkmann showed experimentally that the corresponding points were, as a first approximation, those for which the oblique coordinates were the same. It follows from this that the distance from two corresponding points to the fixation point is not equal, in general. For instance, along the bisector of the axes, the ratio of the corresponding lengths is $\cos(45° + \epsilon/2)/\cos(45° - \epsilon/2)$, that is, 0.97 for $\epsilon = 1.5°$. In fact, careful observation of equal circles in a stereoscope (white for one eye and black for the other, to facilitate the comparison) shows that the superimposition is exact along the vertical and the horizontal meridians only.

Mathematical Horopter

The *horopter* is the locus of points whose images fall on corresponding points of the retinas for a constant position of the eyes. The inventor of this concept is Aquilonius (1613), who thought that the horopter was the frontal plane passing through the fixation point. But the exact theory of the horopter was presented independently by Helmholtz (1864) and Hering (1864). If we accept the correspondence proposed by Volkmann, it is a mathematical problem without much difficulty (nor interest, as this correspondence is not strict). We will consider only the *longitudinal horopter*, that is, the curve which is located in the plane passing through the interpupillary line and the fixation point. If we assume that the horizontal meridians h_1 and h_2 remain in this plane, the horopter is the circle which passes through the fixation point and the center of the entrance pupils (Fig. 62), for the visual angles u_1 and u_2 are equal since they are inscribed in the same circle, and the retinal lengths are the same if the two eyes are identical. The images of M fall on correspond-

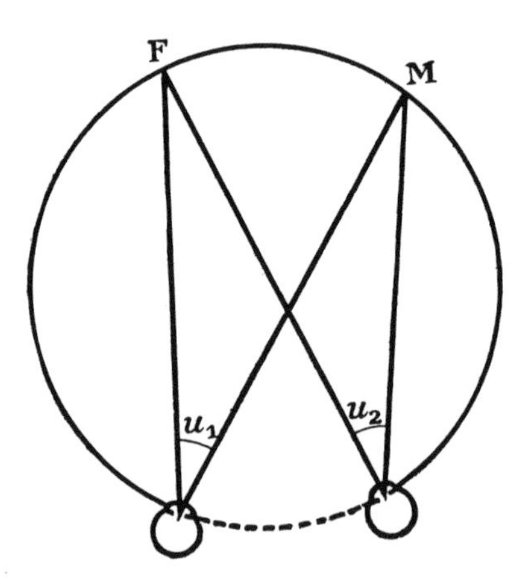

Fig. 62. Longitudinal mathematical horopter.

ing points according to the law of Volkmann. This theoretical longitudinal horopter is called the *Vieth-Müller circle* in honor of Vieth (1818), who was the first to consider it and J. Müller (1826), who made it well-known.

True Longitudinal Horopter

Very few data are available on the true horopter based on the definition of corresponding points, that is, the absence of diplopia. To determine it the subject fixates a point F binocularly, and a test (usually a vertical rod) seen in peripheral vision is moved back and forth until it is seen double. Two curves are thus determined which correspond to the thresholds of crossed and uncrossed diplopia. The mean is the true horopter. The measurements are, of course, not very precise (see Table 23). The data of F. P. Fischer (1924), obtained with fixation at 30 cm, and those of Ames et al. (1932) at 40 cm, gave in much the same way a curve nearly half-way between the Vieth-Müller circle and the tangent to this circle at F. The symmetry accepted by Volkmann between the nasal and the temporal region of each eye is thus not strict in the horizontal meridian.

Suppose the fixation point F is in the median plane. If the eyes are identical, the horopter is then symmetrical in relation to the median plane. Ogle [46] represents it by the conic section defined by the equation

$$\cot u_1 - \cot u_2 = H \tag{51}$$

The parameter H measures the nasal–temporal asymmetry of each eye since $H = 0$ for the Vieth-Müller circle ($u_1 = u_2$); H would be negative for a conic section of which the radius of curvature at point F is smaller than that of the circle. H is actually positive, because the radius of curvature at F is greater than that of the circle (Fig. 63). If we call ω the interpupillary distance O_1O_2 and d the distance FC, for $H < \omega/d$ the horopter is an ellipse, for $H = \omega/d$ it is a straight line, and for $H > \omega/d$, a hyperbola (see, for example, Lefèvre, 1955).

If the angles are small the tangents can be replaced by the angles expressed in radians and equation (51) can be written

$$1/u_1 - 1/u_2 = H$$

or, since η is the mean of u_1 and u_2,

$$u_2 - u_1 = H\eta^2$$

For instance, the data of Fischer lead to $H = 0.12$ and those of Ames to $H = 0.13$. If we assume $\eta = 10° = \pi/18$ radian, we arrive at

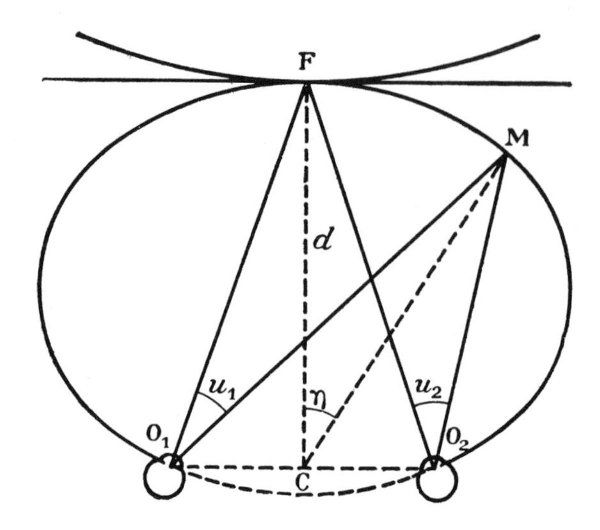

Fig. 63. True longitudinal horopter.

$u_2 - u_1 = 13'$, that is, considerably less than the threshold of diplopia. This method of measuring H is therefore not very precise.

The experimental fact that H is always positive means that in the right side of the field the visual angle u_2 of the right eye is greater than the visual angle u_1 of the left eye, the difference increasing with eccentricity. It is the opposite on the left side of the field. As Frank (1905) suggested, this asymmetry can be related to the phenomenon discovered by Kundt (1863) to which we have already referred (Chapter 7). A subject is asked to bisect a horizontal line in monocular vision by placing his fixation point in the apparent midpoint. He adjusts the right side larger if he looks with the right eye and the left side larger if he looks with the left eye. The temporal half is larger than the nasal (remember that these adjectives apply to the visual field and not to the retina, for which they must be reversed). In fact, the measurements of several authors [46] have confirmed the quantitative agreement between the value of H determined by the horopter and the monocular asymmetry, but the latter varies greatly from one subject to another. Its origin is actually uncertain. H. C. Stevens (1908) interprets it as a perceptive effect of central origin, whereas Ritter (1917) explains it by an unequal density of the retinal receptors in the nasal and temporal halves.

Besides these asymmetries, which are the same for both eyes, there may be differences between the eyes which give rise to *aniseikonia*, which we will discuss later.

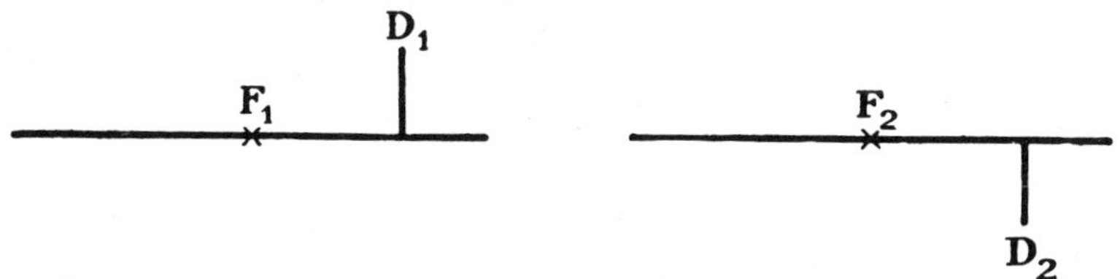

FIG. 64. Binocular vernier method (targets presented to the two eyes).

The Binocular Vernier Method

The determination of the horopter can be greatly improved by the method proposed by Volkmann (1859). The subject fixates a point binocularly and adjusts the monocularly seen lines D (Fig. 64) until they seem vernierly aligned. Van der Meulen (1873) modified the apparatus ingeniously. The subject sees in peripheral vision a vertical rod through screens specially shaped so that the left eye sees the lower part of the rod and the right eye sees the upper part of the rod. It is the distance of the rod from the observer which is adjusted until the upper and lower parts are aligned, while the fixation point remains immobile. The rod is then situated on the true longitudinal horopter.

Ames et al. (1932) perfected the method by replacing the screens with gratings so that alternate parts of the rod are seen by each eye. They have called this the *binocular vernier* or *nonius method* (in honor of the Portuguese mathematician Nuñes, who invented the vernier in the sixteenth century). Some practice is necessary because the alignment is difficult to perceive in the periphery and practically impossible if $\eta > 12°$. Precision is better than in the case of the diplopia method, particularly if the fixation point is not too close (otherwise the precision is impaired, either because of the uncertainties of fixation with strong convergence or because of instrumental uncertainties). Table 24 gives the measurements of Ogle [46], and its last line gives the results of Bourdon [6] and Hofmann [23] for monocular vernier alignment.

TABLE 24

Threshold of Vernier Alignment with Eccentricity

	η, deg						
	1	2	4	5	8	10	12
Binocular alignment							
$d = 76$ cm	20″	35″	50″		90″		2′
$d = 40$ cm	30″	40″	70″		2′		5′
$d = 20$ cm	2′	5′	6′		9′		11′
Monocular alignment	23″	54″		2–4′		4–7′	

If we take into account the fact that the data of Ogle are standard deviations of equality settings, whereas the others are minimum limits of resolution, there is satisfactory agreement and the binocular processes seem similar to the monocular.

Fixation Disparity

In determining the horopter by the nonius method, Lau (1921) noted that the horopter does not pass through the fixation point if the normal synergy of accommodation and convergence is impaired either artificially (by prisms) or naturally (heterophoria). The system behaves as if the subject looks a little behind the fixation point if he is exophoric and a little in front if he is esophoric. The deviation is usually less than 2′ (Riggs and Niehls, 1960; Palmer, 1962), and the supporters of Panum's areas interpret it as the value of the area for $\eta = 0$. It is simpler to assume it to be the limit of resolution, particularly if one takes into account the fact that the attention is placed upon the alignment in the periphery, which consequently lowers the central visual acuity. The phenomenon is called *fixation disparity*, but the word "slip" would be more exact. There is some kind of a struggle between the peripheral alignment and the central fixation. Peckham (1934) and Clark (1936) have found much greater disparities by a technique of recording eye movements which leaves much to be desired.

An opposite method was used by Ogle et al. (1949). Peripheral fusion is assured by a letter chart seen by both eyes, while alignment is measured at the fixation point by means of the two halves of a vertical line, each seen by one eye by use of polarized light. The slips are in this case more pronounced (up to 30′), although those above 5′ are rare. The correlation with heterophoria is rather clear, although it is not a fixed relationship (see Charnwood, 1951; Tani et al., 1956; Ogle and Martens, 1957; Verhoeff, 1959). In my opinion these results do not tell us much about the correspondence between the two retinas and can be explained by the increasing limits of resolution with eccentricity, in the same way as Panum's areas. In fact, the orders of magnitude are the same. Let us also mention the experiments of Mitchell and Ellerbrock (1955), who studied the effect of variations of convergence on fixation disparity by means of prisms, and those of Hebbard (1962), who found agreement between the objective methods of measuring disparity (by photography of the position of the eyes) and the subjective methods (binocular vernier).

The main interest of these experiments is to verify that binocular fusion is possible and stable when only the lateral retina is presented with fusional stimuli. Burian (1939) had already shown it with a square test of 30′ width, up to an eccentricity of 12°. Of course, the precision of the convergence of the eyes decreases as the eccentricity increases; this precision is about equal to the limit of resolution in the periphery. There are noticeable differences from one subject to another, which is a general rule in peripheral vision. But it is remarkable that a peripheral test incapable of immobilizing the eye in monocular vision can, in binocular vision, fix one eye relative to the other. It is without doubt the best proof of the existence of corresponding points throughout the retinas and the disproof of the hypothesis that the two foveas are the only corresponding points, the others being produced by variations of convergence.

A phenomenon similar to fixation disparity occurs in night vision. Scotopic convergence settles at a different value for each subject and is independent of the distance at which the test is fixated binocularly (Ball, 1951; Ivanoff and Bourdy, 1954). Figure 65 shows the results obtained for one subject as a function of luminance, according to whether the test is at 1.5 m (circles) or 20 cm (crosses). At high levels of luminance the convergence is about equal to the theoretical values (that is, 0.7 and 5 metric angles or diopters), but below 10^{-4} cd/m² the curves tend toward each other and reach the convergence of 2 (that is, 50 cm) for this subject. This *scotopic convergence*, which depends on ametropia and muscular imbalance, may be one of the causes of nocturnal myopia, the convergence inducing accommodation because of their well-known synergy. According to Fincham (1962), accommodation and convergence are active in the dark but fluctuate without relation to one another; the synergic relationship between them still holds near threshold, but when there is no retinal image the two mechanisms act independently.

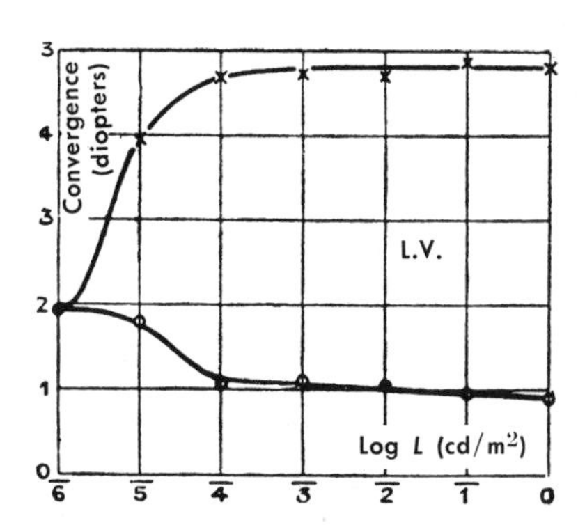

FIG. 65. Scotopic convergence (the luminances are represented on a logarithmic scale in the abscissa).

CHAPTER THIRTEEN

Binocular Spatial Localization

The monocular retinal image is a projection of a three-dimensional world on a single surface. Thus the monocular image is only a perspective and lacks one parameter for representing space completely. However, some cues exist to the appreciation of depth monocularly, as we will see in Chapter 14, but with incomparably less finesse than binocularly. The two retinal patterns are perspective images from two different points of view and their simultaneous utilization gives the sensation of stereopsis, the perception of depth, in other words, the third dimension.

Stereoscopic Acuity

Imagine two object points A and B (Fig. 66) close to the median plane and located at slightly different distances d and $d + \Delta d$. The segment AB subtends the angles u_1 and u_2, which are unequal from the centers of the entrance pupils O_1 and O_2 (unless A and B are on the same horopter), and consequently the images of points A and B fall on disparate points. Thus there exists a *binocular disparity* or *stereoscopic parallax* which is measured by the difference between the angles u_1 and u_2. It is this difference between the two retinal images which is the basis of the sensation of depth.

It can also be conceived that the subject does not gaze steadily but fixates binocularly first A, then B. The *convergence angle* (angle between the two lines of sight) changes from the value α to a different value β and the change in convergence $\beta - \alpha$ may also be perceived and interpreted as a difference in depth. In this second case the points O_1 and O_2 should be the centers of rotation of the eyes. If the difference between the entrance pupils and the centers of rotation is neglected, the following relationship exists:

$$u_1 - u_2 = \beta - \alpha = \omega\Delta\frac{1}{d} = \omega\,\frac{\Delta d}{d^2} \quad (52)$$

The *stereoscopic acuity* corresponds to the threshold of perception of binocular disparity $u_1 - u_2$. For a long time it has been believed, on the basis of rough observations and influenced by the value of the monocular limit of separation, that this threshold was of the order of 1 minute of arc. Stratton (1898) found much lower values; then Bourdon (1900) found a threshold of 5″ and von Kries (1909) found 3.3″. The "record" was gradually improved. Howard (1919) with a criterion of 75% correct answers found 10″ among 22% of his subjects and less than 2″ for 11%. Woodburne (1934) also obtained values of the order of 2″ while keeping the visual angle of the tests (slits) constant (see also Heinsius, 1944; Best, 1949).

For such low values the parameter Δd of equation (52) must be measured from the experimental horopter, and its curvature influences the results. This apparent systematic error was displayed by Fruböse and Jaensch (1923).

This finesse of stereopsis may seem prodigious. For instance, equation (52) shows that with a parallax threshold of 5″ a difference in depth of 0.4 mm can be discriminated at 1 m if ω = 63 mm, or that an object situated at 2600 m is seen in front of another which is at infinity. The clouds are seen in front of the moon (of course, they are known to be

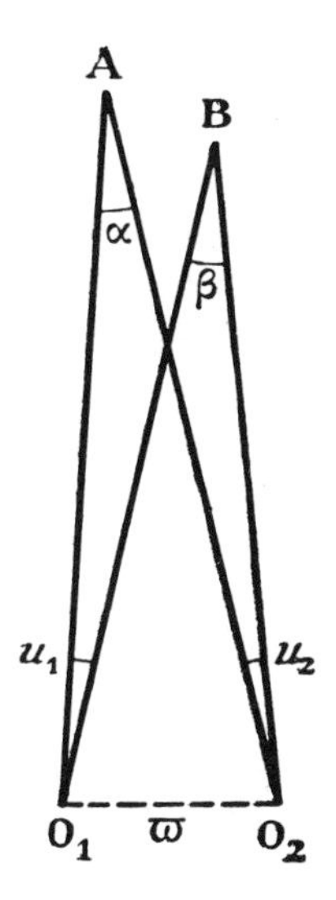

FIG. 66. Definition of stereoscopic acuity (O_1 and O_2 are the centers of the entrance pupils).

closer since they can hide it). This limiting fixation distance where stereopsis ceases occurs when ω subtends an angle of 5″.

Stereoscopic acuity can be measured with several instruments. In the apparatus of Galifret (1945), derived from that of Brooksbank (1908), the subject sees vertical lines through an aperture in a screen and must judge their relative distances. In the Howard-Dolman test (Howard, 1919), the subject manipulates two vertical rods by means of strings until they appear in the same plane. The rods have a diameter of 1 cm, are 5 cm apart, and are viewed from a distance of 6 m. With such thick rods the variation in apparent size of the rods may either help or hinder the measurement. In Verhoeff's apparatus the tests are black strips of paper of different widths on a white background, and for a fixed difference of depth it is the observation distance which is varied in order to determine the threshold. Another simple clinical test is the one by Davidson (1935) made up of a piece of black cardboard attached to a wall 6 m away from the subject on which are 5 white pin heads, one against the cardboard and the others 14, 28, 42, and 56 mm, respectively, in front of the cardboard (these correspond to parallaxes between 5 and 20″). Other stereoscopic tests are also utilized. The correlation between the different methods is satisfactory [76] (see also Beyne and Monnier, 1947).

There are very marked individual variations. Some persons are nearly monocular functionally, although unaware of it and with two normal eyes. Heterophoria and accommodation have no correlation with stereoscopic acuity, according to MacCulloch and Crush (1946), but there is a marked correlation between monocular visual acuity and stereoscopic acuity according to Heine (1900), Kirschberg (1946), and Hirsch and Weymouth (1948). However, some subjects with good monocular acuity have very poor binocular vision. Hofstetter (1965) has found that between 14 and 21 years of age there is a marked increase in stereoscopic acuity, owing to experience of the subject. The sharpness of contours is actually an important factor in the determination of stereopsis (Fruböse and Jaensch, 1923). On the other hand, if both eyes are slightly blurred to the same extent, stereoscopic acuity will decrease only by a small amount (Ogle [46a]). Stereoscopic acuity is also virtually independent of the contrast of the target (Ogle and Weil, 1958). Finally, according to Scott and Sumner (1949), a marked ocular dominance sometimes introduces a systematic error similar to the aniseikonia effect that will be discussed later.

Theories of Stereoscopic Acuity

Stereoscopic acuity is of the same order of magnitude as monocular or binocular vernier acuity. Berry (1948) compared these thresholds on several subjects and found that they were practically the same. Therefore, it seems logical to assume a similarity of the mechanisms. The perception of relative depth would only depend upon disparity of the two retinal images and the slight difference of perspective from each eye. This was the theory of Euclid and it is still the simplest and most satisfactory.

Opposed to this static theory of stereopsis is the dynamic theory supported by Brücke (1841) and others since then, according to which the variations of convergence are necessary for the perception of depth. In equation (52) it is the difference $\beta - \alpha$ which would be the important factor and not the disparity $u_1 - u_2$.

In favor of the static theory, one can mention experiments with "instantaneous vision," where the flash is so quick that variations of convergence do not have time to take place. Dove (1841) verified the existence of stereopsis with an electric spark. Hering (1865) obtained the same result with the falling-bead test. Langlands (1926–1927) verified that stereoscopic acuity remains constant for exposures between 10^{-7} and 0.1 second then increases rapidly to 0.5 second and thereafter very slowly; the gain is a factor of four or five times between 0.1 second and a steady exposure. Langlands interpreted this result as an improvement due to eye movements but we can also point out, in support of the static theory, a choice of the best fixation to perceive the disparity. Also, with other tests, acuity is the same no matter how long the exposure. It must be admitted, however, that some subjects seem unable to perceive stereopsis instantly: M. Tscherning [66], for instance. According to Karpinska (1910), Skubich (1925), and Bonaventura [4], some untrained observers find it difficult to perceive stereopsis on the basis of disparity alone. On the other hand, the experiment of Shortess and Krauskopf (1961) proved that stereoscopic acuity was unaltered when the images of both eyes were stabilized.

Another argument in favor of the static theory is the possibility, pointed out by Wheatstone (1838), of obtaining stereopsis by combining two afterimages which may even be induced one after the other in the two eyes (Rogers, 1860; Ewald, 1906). But here again, practice and steady attention are necessary; otherwise the afterimages are only projected onto the real objects of the visual field (Helmholtz). According to Ogle and Reiher (1962), stereopsis is better when images are induced simultaneously in both eyes rather than successively. This finding can be correlated with results of experiments by Ogle (1963) which show the possibility of fusion and stereopsis of two images each presented for a very short time but with a delay between the two, provided this delay does not exceed a value of 0.10 to 0.25 second.

Javal [26] thought that parallax provided the qualitative sense of depth but that variations of convergence were necessary for a quantitative perception. This is not correct either; the experiments of Bourdon [6], Le Grand (1932), S. Smith (1946), Wright (1951), and Rady and Ishak (1955) proved that stereoscopic acuity is nearly as precise with brief illumination as with movements of the eyes.

It is probable that everybody is right. The static theory would suffice, but by refreshing the retinal images by eye movements, stereopsis improves as does the perception of details. Poppelreuter (1911) and Jaensch (1922) noted that prolonged fixation attenuates the sense of depth and objects resume their relative positions when the eyes move. According to Holmes (1918), cerebral lesions which make eye movements impossible provoke severe troubles with the sense of depth, although form vision remains. Washburn (1933) thinks that some alternating rivalry between the two monocular images is a favorable factor in stereoscopic vision. Fleischer (1939) reconsidered the dynamic theory and attributed the origin of the perception of depth to the movements of convergence. Even if movements do not modify the retinal images (as with flash illumination or with afterimages), the stimulus (disparity) remains and the movement takes place, and therefore the stereoscopic depth is perceived. A discussion on the latest works regarding this eternal problem may be found in an article by Amigo (1963).

Factors Affecting Stereoscopic Acuity

Luminance level and contrast influence stereoscopic acuity in the same way as they affect vernier acuity (Hirsch and Weymouth, 1948). Stereopsis remains in scotopic vision and down to the absolute threshold although acuity is obviously poor (Dichman et al., 1944; Güggenbühl, 1948; Mueller and Lloyd, 1948; Berry et al., 1950; Lit, 1959). As an example, Figure 67 shows the results of Mueller and Lloyd. We find the classical break between photopic and scotopic vision.

The separation between the two point objects used to determine relative depth is an important factor studied by F. P. Fischer (1924) and by other authors.

Matsubayashi [in 19a] studied the case of small separations between the two point objects. The threshold is two or three times higher for 2° separation than when the objects are in contact. The best results are obtained with lines 2.4′ in diameter and at least 30′ high. According to

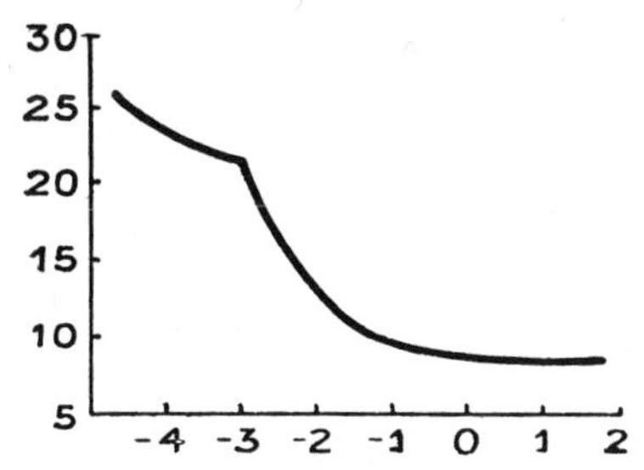

FIG. 67. Relationship between stereoscopic threshold (ordinates in seconds of arc) and the log of the luminance (in cd/m² along the abscissa), according to Mueller and Lloyd.

Hirsch and Weymouth (1948) it is not when the lines are in contact that the best threshold is obtained, but when their separation is about 20′.

Figure 68 represents the results obtained for stereoscopic acuity threshold in different conditions. In curve *a* (Le Grand, 1932) the sub-

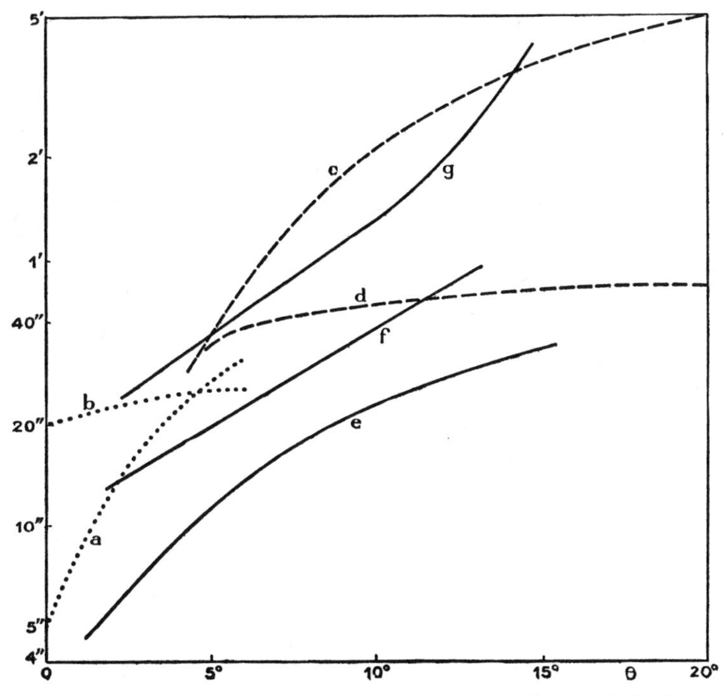

FIG. 68. Relationship between stereoscopic acuity threshold (ordinates, log scale) and the angle θ which separates the test objects (abscissa, linear scale). *a* and *b*, Le Grand; *c* and *d*. Wright; *e*, *f*, and g, Ogle.

ject fixates a luminous point situated 2.5 m away in the dark. A flash of light occurs at the same height as the fixated point but at an angular separation θ; therefore at eccentricity $\eta = \theta$. The subject must say whether he sees the flash nearer or farther than the point. The threshold is defined for 87.5% frequency of correct answers and 12.5% errors. Curve *c* (Wright, 1951) is obtained under similar conditions with circular tests of 25′ diameter seen at 1.8 m; the threshold was defined for 80% correct answers. These two curves characterize the disparity alone, since instantaneous illumination eliminates eye movements. In curve *d* (Wright) the eye moves freely from one test to the other and consequently the two mechanisms (disparity and convergence) function simultaneously. Convergence alone is responsible for curve *b* (Le Grand) obtained in the following way. The subject fixates for 1 second a point in the dark; this point is extinguished and after 1 second another point at a distance θ from the first is illuminated for 1 second and the subject must report if it is nearer or farther than the first point. While it is dark the subject is asked to slowly look from one point to another. Under these conditions the precision depends upon the exactitude with which the convergence of the eyes is maintained as they fixate one point then the other. If the eyes were left to move freely the results would become very poor, 14 to 40′, according to Bourdon. But then it is only a difference of two absolute localizations and we will see later that absolute localizations are very crude. We can, therefore, deduce that for small values of θ it is the disparity which accounts for the stereoscopic acuity, whereas for high values it is the variation of convergence. This conclusion has been directly verified by Wright in the case where one of the points falls on the blind spot of one eye ($\theta = 14°$), which of course eliminates the disparity.

Curve *e* is the mean of the values obtained by Ogle [46] on three subjects under the following conditions: a luminous point, 2.4 m away along the median plane, is fixated by the subject. It is switched off and then two symmetrical vertical lines appear, separated from the central point by $\eta = \theta/2$. The threshold was measured by the standard deviation (68% of correct answers) and the eye movements were left free. Curves *f* and *g* are the extreme acuities obtained in a similar experiment at 3 m, but the tests are also luminous points, which provide a lower acuity than the lines. Generally speaking, stereoscopic acuity depends upon the form of the objects, and Samsonova (1936) studied particularly the effect of the ratio between the vertical and horizontal dimensions of the test. When the object becomes purely horizontal, a telegraph wire, for example, stereopsis disappears because no reference marks exist for either disparity or convergence. Ebenholtz and Walchli (1965) have studied also the stereoscopic threshold as a function of head and object orientation. Acuity improves slightly with distance (for six subjects the threshold went from 3.9″ at 2.5 m to 2.5″ at 15 m, according to Matsubayashi; see also Teichner et al., 1956).

Another factor which influences the precision of stereoscopic acuity is the position of objects with respect to the horopter and the fixation point. The best results were found when objects were seen single. Tschermak and Kiribuchi (1900) have even attempted to determine the horopter experimentally by this method of minimum threshold. But the persistence of stereopsis with images seen double has often been demonstrated. Auerbach and von Kries (1877) flashed an electric spark in front or behind the fixation point; the flash was seen in diplopia but in its correct position. Wright (1951) repeated this experiment with considerable diplopia (1.5°) and found the same positive result. Tschermak and Höfer (1903) placed a pin 80 cm behind the fixation point and showed that another mobile pin can be adjusted at the same distance within a few centimeters accuracy. Let us also cite the experiments of Burian (1936) and those of Ogle (1953). The latter studied the perception of depth with double images up to a disparity of 25′ in the fovea and greater amounts at an eccentricity $\eta = 8°$. Finally, Westheimer and Tanzman (1956) have demonstrated on six observers the possibility of an exact localization for an exposure of 0.01 second up to 7° of diplopia. Uncrossed diplopia is somewhat more favorable.

Aniseikonia

So far we have assumed the eyes to be identical. But it may happen that one of the retinal images seems larger than the other, either because it is really larger if there is *anisometropia*, that is, a difference between the dioptric systems, or it may only appear so because of different distributions of the corresponding points; to this phenomenon Lancaster in 1932 gave the name *aniseikonia*. Aniseikonia can be induced artificially by placing in front of one eye a special optical system which is afocal but magnifies the image. More complicated systems may introduce different magnifications in different meridians.

In aniseikonia, the horopter is no longer symmetrical in relation to the median plane. If g is the ratio of the visual angles of the two eyes, near the fixation point, for images falling on corresponding points, we have

$$u_1 = gu_2$$

These angles are respectively proportional to the projections of AB on the perpendiculars to O_1F and O_2F (Fig. 69), which can be expressed

$$\cos(\varphi - \psi) = g \cos(\varphi + \psi)$$

and resolving this expression and taking into account the relation

$$\tan \varphi = \omega/2d$$

we arrive at

$$\tan \psi = \frac{g-1}{g+1} \frac{2d}{\omega} \tag{53}$$

The horopter is tangent to AB at the fixation point F in the median plane. For objects to be seen single they must be brought closer to the subject on the side of the eye with the smaller image, which is obvious. This is what Erggelet (1916) noted as he induced aniseikonia artificially; small objects were fused but bigger ones had double edges. This is very distressing if the subject has a marked aniseikonia. If the aniseikonia is small, fusion is still possible but spatial distortion occurs. If an afocal lens which magnifies in the horizontal meridian is placed in front of the right eye, the objects on the right side of the field appear further away and therefore larger; those on the left side of the field nearer and smaller.

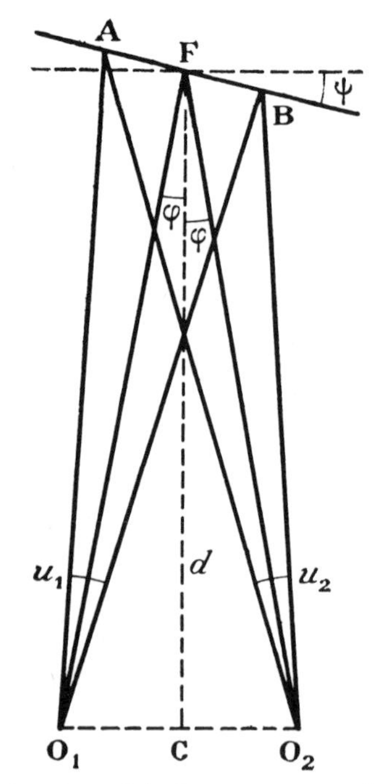

FIG. 69. Rotation of the horopter in aniseikonia.

Some subjects can appreciate this effect with only 0.25% disparity ($g = 1.0025$). This phenomenon of spatial distortion was first pointed out by Wadworth (1876) in the correction of patients with high astigmatism. It was explained correctly by J. Green (1889) and gave rise to much research by Ames and Ogle [46]. According to the latter, a magnification only in the horizontal produces generally a geometric distortion, as expected from equation (53). A magnification only vertical, which should not produce any stereopsis, usually causes a distortion as if the magnification were horizontal in the other eye. This is what Ogle calls the *induced effect*. Hence, an over-all magnification produces only a slight distortion or none, because the effects compensate for each other. Bourdy (1961) gives a good review of recent works on aniseikonia.

The induced effect varies a great deal from one subject to another. It is maximum around 8% ($g = 1.08$), then decreases slowly, whereas the geometric effect increases continually with g. Unlike the geometric effect, it does not occur instantly but within a few seconds. This effect is mysterious, although it is very likely to be of a psychological nature, with an impaired tendency for vertical fusion as a physiological basis. The effect disappears completely if no vertically separated details are present, for example, if the test is made up of vertical lines on a uniform background. Equation (53) then represents the phenomena exactly, even for an over-all magnification.

The artificial aniseikonia introduced by the wearing of lenses of different magnifications in the horizontal meridian for the two eyes produces immediate spatial distortion which diminishes slowly and disappears within 3 or 4 days. These distortions can, however, reappear if the subject is placed in unfamiliar surroundings where the empirical cues are insufficient to counterbalance the stereoscopic effect. These distortions do not, however, reduce stereoscopic acuity. When the subject removes his lenses after having adapted to them, a reverse distortion occurs, with a rapid return to normal. The induced effect produced by the wearing of lenses which only affect the vertical meridian does not show this progressive adaptation, nor is there any reverse phenomenon when the lenses are removed (Bourdy, 1960).

Asymmetrical Convergence

A special type of aniseikonia results from asymmetrical convergence in near vision (Fig. 70). A vertical segment placed at point F gives two retinal images of different sizes. For instance, if $CF = 20$ cm and $\omega =$ 63 mm, the ratio of the images is 1.11 for $\eta = 20°$ and 1.22 for $\eta = 40°$. Such aniseikonia would be intolerable in symmetrical fixation, whereas

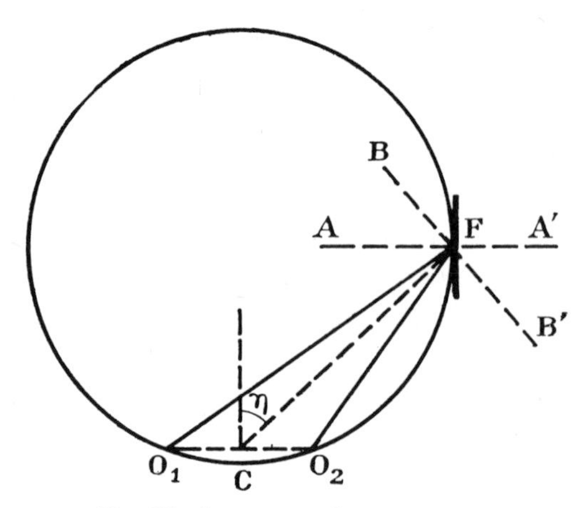

FIG. 70. Asymmetrical convergence.

it does not cause any difficulty in asymmetrical convergence. The images are, however, of the expected difference in sizes, as Desaguliers (1717) noted, and as can be verified by producing diplopia.

This difficulty was explained by Hering (1879), who assumed that because of a difference in accommodation the image would be clear only in the eye which had to accommodate less and the other image, being blurred, would play little part or would be suppressed. In fact, according to C. Hess (1899) the difference of accommodation cannot exceed 0.25 diopter or even 0.15 (Ogle [46]). According to Stoddard and Morgan (1942), in symmetrical convergence most subjects cannot induce 0.25 diopter of difference but some can, up to 0.5 diopter. But in asymmetrical convergence it seems that greater differences of accommodation occur (Rosenberg et al., 1953). Therefore the explanation of Hering is not supported.

Ames [46] discovered the existence of a compensatory mechanism later confirmed by Ogle. With a system of mirrors eliminating the difference of accommodation and enabling a direct comparison of the sizes of the images, it was noted that in the horizontal meridian no change in size appears, whereas in the vertical meridians it appears as if the eye turned toward the nose had its image enlarged by the amount necessary to make the two images equal. This compensation is neither dioptric nor retinal. It probably has a central origin related either to the movements of the extraocular muscles or to the disparities in the vertical meridian which normally do not exist. These disparities would produce a compensation to maintain the correct appreciation of distances and it may perhaps account for the induced effect.

Herzau and Ogle (1937) attempted unsuccessfully to see if equal afterimages would become unequal in asymmetrical convergence, which would have supported the notion of a muscular origin for the compensation.

Herzau (1929) determined the empirical horopter by the nonius method and verified that it was tangent at point F to the Vieth-Müller circle(Fig. 70).

Measurement of Aniseikonia

The measurement of aniseikonia (natural or artificial) is not directly possible by showing successively and rapidly to the subject his two monocular images of the same object. He detects a difference only if it is more than 4%. With a small prism base-up in front of one eye, which vertically separates the two images, a trained subject begins to appreciate a difference at about 2%. To obtain better precision, one can use stereoscopic tests consisting of a central fusion target and four pairs of opposing arrows, but it is preferable that the subject does not look into a stereoscope, which is an artificial situation. The best *direct comparison eikonometer* seems to be the one of Ames and Ogle [in 46]. On a screen is projected a cross having at its intersection a small white disk in the center of a black square (Fig. 71). Four pairs of opposing arrows (numbered on the figure) are placed at 4° away from the center. The even-numbered arrows are polarized in one direction, the odd-numbered

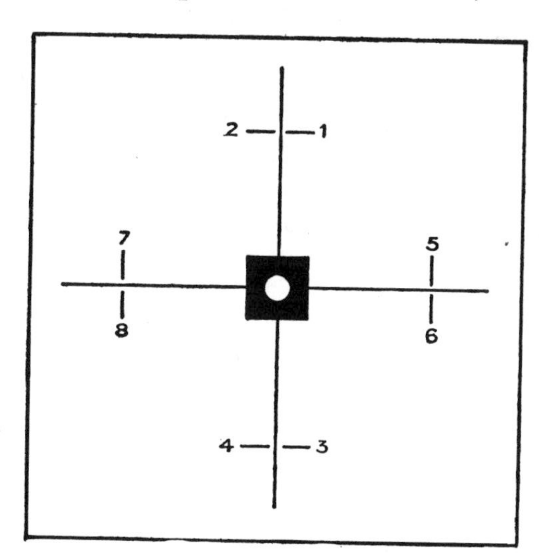

FIG. 71. Direct-comparison eikonometer.

arrows in the other direction by means of filters placed in the projector. The screen is aluminized so that it does not alter the polarization by diffusion. The subject wears polarizing lenses so that he sees the whole pattern with both eyes, while the even-numbered arrows are seen by only one eye and the odd-numbered arrows by the other. The arrows are actually aligned, but if the subject has aniseikonia, they are not seen in alignment, the fusion being maintained by the central figure (within the limits permitted by fixation disparity). By means of an adjustable magnifying unit in front of one eye, apparent alignment of the arrows is produced. The precision can be as good as 0.25% in subjects with good acuity.

An indirect way to measure aniseikonia is to use equation (53), dealing with spatial distortion, but to do so it is essential to eliminate the induced effect by not having any vertically separated details. For instance, in the usual form of the Howard-Dolman test the vertically separated details, although few, suffice to reduce by one third or one half the effect of aniseikonia (Sloan and Altman, 1953) with important variations from one subject to another and even from one day to the next for a particular subject. By improving the test to eliminate the vertical details and empirical cues, the errors can be made equal to the theoretical value. The *space eikonometer* of Ames (1945) consists of an oblique cross made up of two cords stretched at right angles to each other situated in a frontal plane 3 m from the observer, a vertical line suspended through the center of the cross, and two pairs of vertical plumb lines separated by 50 cm placed in two frontal planes 60 cm in front of and behind the cross. The cords are of different colors to enhance their identification and are seen on a uniform black background. I refer to Ogle [46] the reader who would like to know all the possibilities of the instrument which measures separately the aniseikonia in the vertical and horizontal meridian. A simplified model was described by Malin (1955). The results obtained with the space eikonometer agree well with those of the direct comparison eikonometer. For 75% of the subjects there is at the most 0.5% difference between the aniseikonia measured by the two instruments.

Asymmetry of Luminance

Some phenomena mimic aniseikonia and appear as spatial distortions. as such, a difference in retinal illumination does not produce any effect, and Szily (1921) even demonstrated that stereopsis is perceived in a stereoscope with figures in which the white and black parts were permuted for each eye. Julesz (1963) analyzed with a great deal of ingenuity the role of contours in such stereopsis with complementary figures. Sometimes, however, irradiation can create distortions, as Emsley (1944) and Cibis and Haber (1951) have observed. If a neutral density filter is placed in front of one eye, the rotation ψ of the horopter increases nearly linearly with the density up to N.D. 1.00 (transmission 10%) corresponding to a rotation between 20 and 30°, depending upon the subject. Then the rotation levels off and approaches an asymptotic limit of 30 to 40° at a density of about 2.5 (transmission 0.3%). An unequal adaptation of the eyes produces the same effect.

An amusing phenomenon based on the asymmetry of luminance is the *stereophenomenon* described in 1922 by Pulfrich [56]. The stimulus object is in motion, e.g., the bob of a pendulum in a frontal plane. Its path is represented by the segment *AB* in Figure 72. When a neutral density filter is placed in front of one eye, the bob appears to move in an elliptical path, partly in front and partly behind the frontal plane *AB*. This phenomenon is due to a difference in the latency periods for the two eyes; as the right eye sees the bob at M_2, the left eye sees it at M_1, slightly behind because of the greater latency. The bob is thus localized at M'. Kahn (1931) found that absence of eye movements was favorable for observation of the effect, the magnitude of the ellipse decreasing greatly if the bob is followed with the eyes.

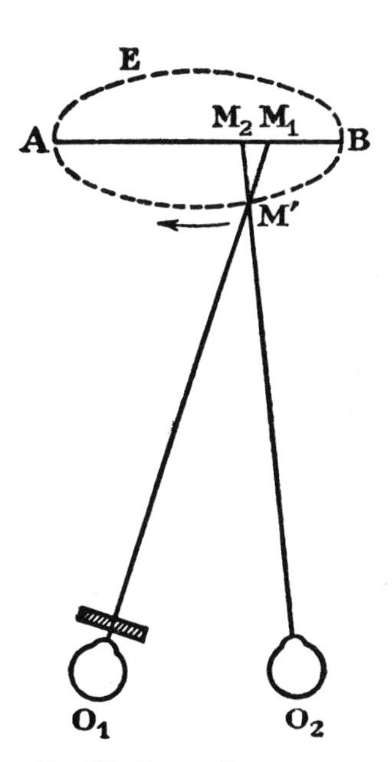

FIG. 72. Stereophenomenon.

The Pulfrich stereophenomenon provoked a great many researches after its discovery. Lit (1949) summarizes them. According to Banister (1932) it is principally the ratio of the luminances between the two eyes which is important, that is, the density of the filter placed over one eye. A ratio of 2 causes a latency difference of 0.004 second and a ratio of 20 causes a difference of 0.018 second. Fusion becomes difficult if the ratio is too great. Moreover, there is a certain range for which the effect does not occur. Granit (1932) thought that it was the difference in adaptation rather than the difference in retinal illumination which played a part. In fact, if one eye is adapted to daylight and the other to darkness, the direction of rotation reveals a longer latency for the latter although its image seems brighter (see Trincker, 1954). According to Holtz (1934) and Lythgoe (1938), for a given ratio of luminances the effect diminishes if the level of luminance increases greatly. The same results were also found by Liang and Piéron (1947) and then by Lit (1949, 1964). According to the latter, the latency period for each eye is proportional to the log of the retinal luminance. Lit (1960) also studied the effect of target thickness and its velocity. M. S. Katz and Schwartz (1955) have described curious experiments with polarized light which suggested certain monocular factors in the stereophenomenon. In some pathological cases the phenomenon occurs without filter and can be used to measure the loss of sensitivity of one of the two eyes (Dupuy-Dutemps, 1926).

Pulfrich had thought to utilize the stereophenomenon as a method of heterochromatic photometry by comparing two colored filters, one placed in front of each eye. Neutral filters would be superimposed on one side until the effect disappeared and the luminances would then be equal. But von Kries (1923), and then Engelking and Poos (1924), have shown that this was not correct; at equal luminances there are differences of latency inherent in the different colors, the red having a lower latency period than the blue. Moreover, a variation occurs with the luminance (reverse Purkinje effect).

Stability of Retinal Correspondence

Is the relationship between corresponding points innate as Hering [21] held it to be, or is it acquired, as in the conception of Wundt (1898), which was brilliantly defended by E. Jackson (1937)?

The first group of arguments comes from the well-known observations on congenitally blind individuals. When their sight is restored the organization of their visual space develops very slowly, but it seems, however, that the depth localization based on retinal disparity is somewhat intuitive. In support of the innateness theory, we can relate the case of Held [in 23], who was amblyopic in one eye since birth and succeeded one day in building a corrective system which restored to him at once the sensation of stereopsis with all its characteristics.

The observations gathered by Malbran [38a] on strabismic subjects also favor the notion of an innate correspondence which may appear during the treatment, while at the same time the abnormal localization of the deviated eye persists and, therefore, a curious "monocular diplopia" occurs.

A fixed and rigid correspondence would support the innate concept, whereas some plasticity would be favorable to the thesis of acquired correspondence. But there seems to be a small margin of adaptation, although it is very reduced. In fact, in subjects with good vision it is remarkable that aniseikonia is, in general, negligible. A study by Ogle [46] on 280 young pilots showed that 94% had an aniseikonia of less than 0.25% in the vertical meridian and 85% had this same amount in the horizontal meridian. This represents an extraordinary precision in the placement of corresponding points. If we assume that a 4° angle (between the center of the cross and the arrows of the eikonometer target) represents a distance of 1.2 mm on the retina, in order to obtain a precision of 0.25% the variation of length between corresponding points of the two retinas, within this distance, must be below 3 μ, that is, the size of a cone. This seems inconceivable for an innate anatomical mechanism and it seems therefore necessary to assume a small margin of adaptation.

The limits of this adaptation may be assessed by inducing artificial aniseikonia in a subject with normal eyes [46]. Assume a subject wears a 3% afocal magnification lens in front of the left eye. When measuring the aniseikonia a diurnal variation is noted (Fig. 73); the values taken at night are less than those taken in the morning because of some adaptation during the day. After the seventh day the rhythm disappears, and a stable value of the aniseikonia is reached which is about half what it was initially. If on the ninth day the lenses are removed, a 2% aniseikonia appears in the opposite direction but it subsides quickly and within 2 days the subject returns to his normal state. This compensating effect is more marked in the horizontal than in the vertical plane. This eliminates the purely physical hypothesis that the ocular dioptrics would play any major role. There would exist, therefore a mechanism, very limited in its effects, which would allow a precise adjustment of binocular vision. This mechanism is very likely of a cortical nature.

The fact that this adjustment is limited is proved by the high correlation between anisometropia and aniseikonia. It is also noted in cases

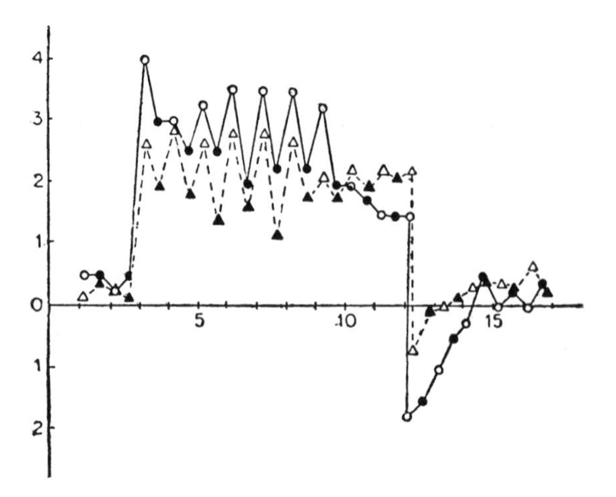

Fig. 73. Variation of artificial aniseikonia. Abscissa: Number of days. Ordinate: aniseikonia (in %) measured with the direct-comparison eikonometer (circles) and the space eikonometer (triangles). White symbols represent morning measurements and black ones evening measurements. On the third day a 3% meridional afocal magnification lens axis 90° was placed in front of the eye and was removed on the twelfth day (after Ogle).

of detached retina. When the retina is restored against the globe it does not always resume the original position, and subjects note a spatial distortion which subsides progressively but only within certain limits. All these facts seem to favor the hypothesis of the innate nature of binocular correspondence but with a small margin of adjustment.

Monocular Cues to Depth

If the perception of stereopsis—that is, of the relative depth of objects—is essentially binocular, monocular cues are not, however, negligible and have in fact given rise to a great deal of research. First, the importance of binocular vision diminishes as the distance increases, as we can see in Table 25, which was determined with an immobile object. The movement of an object diminishes the advantages of binocular vision very greatly (Walker, 1941). This comparison of binocular and monocular thresholds of depth was investigated a great deal in view of its importance in piloting an airplane (Pfaffman, 1948; Roscoe, 1948). The pilots wore goggles that eliminated binocular cues to distance. Piloting of the airplane remained possible but with reduced precision; in particular the pilot had a tendency to level off too high in the terminal stages of landing.

TABLE 25
Ratio of Binocular and Monocular Thresholds

Authors	Number of subjects	Distance, m	Ratio
Andersen and Weymouth (1923)	1	2	10
Deyo (1922)	100	6	6.35
Hirsch and Weymouth (1947)	4	61	2.8
	5	122	2.0
	5	183	1.8

The most important monocular cues that have been studied are the following:

1. *Accommodation* may provide some information if one perceives the effort of the ciliary muscle which acts on the lens, or if one senses changes of convergence (even when one eye is closed) which accompany variations of accommodation. Irvine and Ludvigh (1936) assumed that the extraocular muscles do not have a proprioceptive sense but mainly a sense or "knowledge" of the innervation sent to the extraocular and intraocular muscles together. Some authors found accommodation to be very precise. Wundt [75] looked through a hole in a screen at two lines close to each other and determined the depth difference between the two lines that could just be perceived. He found 12 cm at 2.5 m and 4.5 cm at 40 cm, that is, the changes of accommodation were $\Delta A = 0.02$ and 0.25 diopter, respectively. Arrer (1896) found 1 to 2 cm at 40 cm ($\Delta A = 0.06$ to 0.12 diopter); Piéron (1927), 5 cm at 1.5 m ($\Delta A = 0.02$ diopter). Grant (1942) held accommodation to be important up to 2 m. Some authors criticized the preceding experiments and inferred that factors other than accommodation play a role. By eliminating these factors, Hillebrand (1898) and Bourdon (1898) provided evidence that a change of accommodation of 1 diopter is necessary for the subject to notice the difference in depth, and Bappert (1922) found 5 diopters. The problem is as yet unsettled.

2. *Chromatic aberration* has sometimes been proposed as a cue to depth perception (Polack, 1900) but without many data. In binocular vision chromatic aberration creates a well-known illusion to which Goethe (1810) and others have referred—that on a blue and red surface the blue appears slightly behind the red. Helmholtz believed this to be due to a variation of accommodation, but Einthoven (1885) correctly explained this phenomenon by a displacement of the center of the entrance pupil with wavelength which produces a variable binocular disparity (we discussed in a similar manner in Chapter 8 the apparent

variations of the fixation point with wavelength). In fact, the effect is more marked when halves of both pupils are covered symmetrically. Hartridge [19d] gives a somewhat different interpretation, bringing attention to the dispersion of light as it enters the eye due to the angle of incidence (small but not nil) of the beam of light on the cornea. S. Rösch (1954), Vos (1960), and Kishto (1965) reviewed the history of this curious effect. It is possible that the eccentricity of the Stiles-Crawford effect in the pupil plays a role (Vos, 1963).

3. The preceding factors are doubtful; on the other hand, *monocular parallax* caused by movement is an important and strong cue. If the eye moves while the head remains immobile, there are theoretically some very slight movements which may be related to the distance, but it is very unlikely that they play an active role. However, if the head is moved, the differences in distance are easily perceived. If the whole body (or the object) moves, the effect is even more pronounced. For instance, if one looks up at the branches in the forest and closes one eye, the branches and leaves are incomprehensible, but as soon as one walks they seem to take their place in depth because of motion parallax. The relative movements of objects in motion pictures give a sense of depth which overshadows much of the importance of stereoscopic projections (to which we will come back later). Musatti (1924) gave evidence that in monocular vision an intense sensation of stereopsis could be obtained by rotating plane figures (stereokinetic phenomenon).

Tschermak (1939) established a list of the several effects of motion parallax. Graham et al. (1948) have studied thoroughly the principal factors of the phenomenon with the apparatus shown in Figure 74. Two vertical needles M and M' of $9'$ apparent diameter are situated almost along the same line (with a vertical separation of $3'$) and move at equal speed in two frontal parallel planes situated at distances r and $r + \Delta r$ from the observer. The needles cross the median plane at the same time, but when they are separated from this plane by the angle θ they have an angular parallax $\Delta\theta$ and the angular speed of this parallax (t = time) is

$$\omega = \frac{d(\Delta\theta)}{dt} = \frac{\Delta r}{r}\,\frac{d\theta}{dt}$$

A lens placed in front of the observer's eye, O, projects the images at infinity and the opening in the screen, E, reduces the field to the median plane and its proximity. On five subjects the threshold of ω is of the order of $40''$/second of time at high luminances and reaches $5'$/second at 10^{-3} cd/m^2 when $d\theta/dt = 7°$/second. The thresholds increase with $d\theta/dt$ and become about three times as great for angular speeds of the

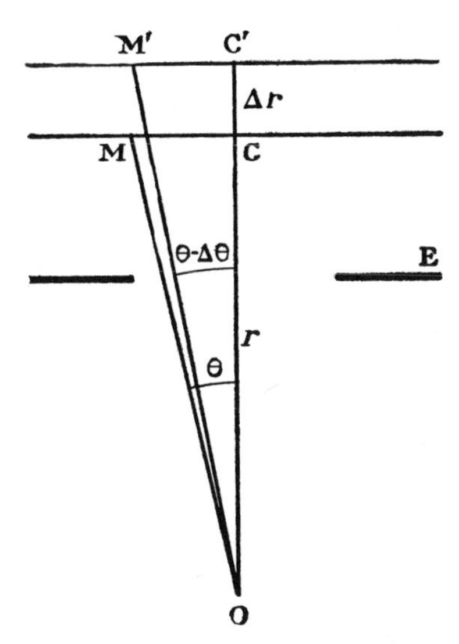

FIG. 74. Monocular motion parallax.

order of 20°/second. In the vertical direction the threshold is not so good (it is about twice that in the horizontal). In these experiments the subject follows the needles with his eye.

Zegers (1948) studied the effect of the total field in which the movement is visible and arrived at the simple law that for each value of $d\theta/dt$ it is the extreme value of the parallax $\Delta\theta$ which is responsible for the threshold. Gibson et al. (1953) investigated the role of different parallax factors in the landing of aircraft. It was probably also parallax which came into play in the experiments at great distances of Teichner et al. (1955). Let us point out also the research of E. J. Gibson et al. (1959) on the parallax of movement.

4. Finally, other monocular cues of depth perception are utilized (size of the object, linear and aerial perspective, light and shadow), which we will discuss briefly in Chapter 14. Another important monocular cue is *interposition*; that is, a nearer object may conceal parts of a farther object. The geometry of the phenomenon was studied by Ratoosh (1949).

CHAPTER FOURTEEN

Spatial Perception

It is with much caution that we have given some indication in Chapter 7 of two-dimensional perception, that is, the vision of forms. We will be even more cautious regarding the problem of three-dimensional perception. Very difficult problems of psychology risk making us stumble at each step. But these problems cannot be evaded and we shall try to present the reader with as objective a summary as possible.

One of the difficulties of these problems lies in the importance of the entire retina, even the more peripheral part, in the elaboration of spatial perception in spite of the rapid decrease in acuity away from the fovea.

Projection

The *projection* in the external world of the two retinal images has given rise to considerable literature. We will leave all the abnormal cases, particularly those caused by strabismus.

The first apparent difficulty is the pseudo-problem of "upright vision" in spite of the inversion of images on the retina. In reality, we call objects "upright" as we commonly see them and we do not see our retinal image. Moreover, a motor relationship is established, for, to quote Piéron [52], ". . . we see upwards the points in space that we fixate by an elevation of gaze, a movement brought about by the activation of certain cortical receptors, and the position of the retinal elements connected to these receptors is of no importance. . . ." Following Stratton (1896–1897) several psychologists, particularly I. Kohler (1951), have amused themselves by placing inverting optical systems in front of the eyes of a subject. Within a fortnight the new motor relationship became readjusted so that the subject could ride a bicycle or play football in the same way as shaving in front of a mirror becomes a habit enabling us to operate correctly in spite of the lateral inversion. Walls (1951) analyzed these experiments and showed that adaptation produced neither an upside down nor a right-to-left reversal but a new adjustment of movements.

Some anatomists have tried to link this reversal with the crossing of nerve fibers at the chiasma (which we will discuss later). They also related it to the fact that because the retina is concave, the cones are pointing practically toward the center of the exit pupil, and may be prolonged toward the object point in the same way that, in the convex eye of insects, each ommatidium (element of a compound eye) points directly toward the object. These facts may be meaningful in comparative biology but seem foreign to problems of perception.

Monocular projection was formerly interpreted by the theory of *local sign* that Lotze (1846) proposed for cutaneous localizations and that Hering (1864) applied to the retina. Each retinal receptor possesses an innate value of lateral and vertical localization. From recent observations of pathological cases (for instance, those of Teuber and Bender, 1949), it was noted that progressive spatial distortions of the visual field followed cerebral impairments. It is nowadays more commonly believed that the relationship between the retinal image and the perceived direction is functional and acquired and not only anatomical.

It is, of course, true that the spatial projection of a retinal image involves a nonretinal factor—the knowledge of the position of the eyes in the orbits. For instance, an entoptic or afterimage is fixed in relation to the retina but follows eye movements and is therefore projected in different directions in space. We have mentioned these effects in Chapter 11.

The localization of phosphenes poses curious problems. A pressure phosphene applied to the top of the globe appears in the bottom of the field and one applied to the temporal angle appears on the nasal side of the field, which seems logical, owing to the position of the stimulated retinal elements. But the subject, blind since birth, to whom Albertotti (1884) referred, had amused himself (before the operation which restored his sight) by producing pressure phosphenes with his fingers and had localized them exactly where he touched his eyelid, probably because normal projection was not yet developed. Similarly, electrical phosphenes are usually perceived on the side where the electrode is, which seems paradoxical; it is true that the lines of current in the eye are not known.

In binocular vision the projection of points seen single, that is, falling on corresponding retinal points, can be represented easily by using the notion of the *cyclopean eye* of Hering (an imaginary eye on which the visual directions of the binocular field of view are imagined to be superimposed). In the case of points seen in diplopia, if there is no suppression, each of the images is usually localized in a direction according to the law of *identical visual directions* (Hering). If one looks at a landscape for example, through a window, but fixates a small spot on the glass, each detail of the landscape is seen in diplopia, and two different details, although separated in space, are localized in the same visual direction* as that of the spot on the window on which they are both superimposed.

Distances

Generally speaking, an object fixated in space is localized at the point of convergence of the primary lines of sight of the two eyes. Parinaud [50] called this rule the *law of Giraud-Teulon*, in honor of the ophthalmologist who had described it (after Nagel [44]). In the appreciation of stereopsis, that is, the relative depth of objects, we know that two binocular mechanisms are involved (disparity and variation of convergence) and the monocular cues are secondary. On the other hand, in the evaluation of the distance of an object (*absolute* depth), the convergence of the visual axes is the only binocular factor that may play a role, and it is not impossible that accommodation may have some importance. However, by the interposition of prisms these two factors can be differentiated, and from the experiments of Swenson (1932) and Vernon (1937) it seems that convergence prevails with weight at least five times higher than accommodation. But these experiments made with prisms must be considered cautiously because they produce distortions which transform flat surfaces into curved surfaces. Thus one may explain some experiments that seem paradoxical, such as the distortion produced by identical small prisms placed base-in in front of each eye, which seems to bring the objects closer even though the convergence is relaxed. Ludvigh (1936) even thought that convergence is not a factor of spatial localization because it is not sensed by itself, but this argument could as well apply to the retinal disparity in the sensation of stereopsis, since most subjects cannot tell to which eye each image belongs!

In the absence of all empirical cues, as, for example, when a light source is seen in the dark, the absolute sense of depth is crude and subsides totally beyond 20 m. According to Bourdon [6] and Petermann (1924), the accuracy of convergence is only of the order of 20′.

* This is also referred to as *oculocentric localization.*

In the evaluation of absolute distance, empirical cues* become of primary importance. With well-trained subjects and in a surrounding offering many cues, the verbal evaluation of distance can be correct within 10 to 30% up to 4 km (2.5 miles) in daytime. A decrease in illumination is usually interpreted as an increase in distance. But Münster (1941) showed that a light gray piece of paper often appears farther away than a dark gray piece of paper, because aerial perspective, that is, the veil caused by the atmosphere in front of distant objects, tends to diminish contrasts and objects appear of a uniform less-saturated grey. The same applies for colors, and according to Johns and Sumner (1948) reduction in saturation and bluishness makes objects appear farther away. In mountainous countries or in polar regions where the atmosphere is very pure, the decrease in the effects of aerial perspective creates an illusion of nearness which often misleads explorers.

In near vision, perception of details of the surface enhances localization (D. Katz, 1935; Vernon, 1937). In monocular vision, for example, a white and lustrous piece of paper is localized less precisely than an uneven piece of paper. A similar distinction is made between "surface color" which is localized on the object itself and "film color" such as the blue of the sky, or any color indefinitely localized and not belonging to any particular surface or object. Two other cues are very important in the perception of distances—the size of the objects and linear perspective—to which we will return later.

The problem of the perception of distance is totally different from that of stereopsis. In the latter, binocular vision is so predominant that all other cues are negligible. This is not so in the perception of absolute depth. Consequently, the conclusion at which we arrived in Chapter 13 regarding the innateness (with a slight margin of adjustment) of the mechanism responsible for the stability of corresponding points cannot necessarily apply in this case. According to the Gestalt school, we would possess an innate tendency to perceive three-dimensional space, but precise localization would be acquired slowly during childhood. In fact, Cruikshank (1941) gave evidence that distance can be perceived by the child at about 6 months, convergence occurs around 8 months, and toward 1 year space acquires its definitive structure, but until 5 to 6 years of age many mistakes still happen. A study of subjects with congenital cataract who recover their sight in adulthood was carried out by Senden (1932) and Dennis (1934). Spatial localization remained uncertain for a long time.

An interesting practical question is that of the *bisection in depth* in

* Also referred to as secondary cues.

which both stereopsis and perception of distance are involved. In the median plane of the subject are placed two objects, one nearer, N, and the other farther, F. The subject is requested to place a third object M in the middle of NF. Issel (1907) and Filhene [in 47] found that NM is usually longer than MF if M is at a distance less than 15 m from the observers. Beyond this distance it is the opposite. In the case of near objects this overestimation of the half closer to the subject was interpreted by a greater number of differential increments of depth or, what amounts to the same thing, the disparity increases more rapidly than the proximity to the observer, as is shown by equation (52). Indeed, the apparent halves of the bisection are intermediate between the true halves and the "psychological" halves which would contain the same number of differential increments. This compensatory mechanism would function in the opposite way for distant objects.

Apparent Fronto-Parallel Plane

While a subject fixates a vertical rod in the median plane he is asked to place other vertical rods seen in peripheral vision so that they appear in the same frontal plane as the fixated rod. The subject must fixate the central rod continually while he adjusts the lateral rods. It demands some discipline to keep fixation steady, but this is made easier by having the observer adjust two rods at the same time which are equidistant from the median plane. Such measurements have often been repeated since Hering (1864) and the results have agreed satisfactorily. The more numerous data are those of Tschermak (1939) and those of Ames and Ogle [in 46].

When the distance d between the fixation rod and the observer is small, the apparent fronto-parallel plane (AFPP) is a cylinder with its axis vertical, and its horizontal section is the true longitudinal horopter. It is therefore concave toward the subject. This coincidence is seen in Table 26, where the parameter H [equation 51] has been measured for

TABLE 26

Comparison of the Values of H for the Apparent Fronto-Parallel Plane and the Longitudinal Horopter (Ogle)

	DISTANCE d FROM OBSERVER			
	20 cm	40 cm	76 cm	6 m
Apparent fronto-parallel plane	0.11–0.23	0.09–0.14	0.06–0.08	0.007–0.016
Longitudinal horopter	0.13–0.14	0.10–0.11	0.05–0.08	—

four subjects at several distances d with the apparent fronto-parallel plane and the longitudinal horopter (by the nonius method). Note that H does not remain constant but varies nearly in inverse proportion to d.

The precision of settings with the lateral rods at eccentricity η is measured, in seconds of arc, by the standard deviations as indicated in Table 27 from the experiments of Ogle on himself. These results are better than with the nonius method. The precision diminishes at near distance; this is due partly to experimental errors because of the mechanical apparatus, and also to diminution of acuity caused by accommodation.

If we assume that the apparent fronto-parallel plane can be identified with the longitudinal horopter and that the latter can be entirely explained by monocular asymmetry (H constant), the apparent fronto-parallel plane would be a real plane at a distance d such that $H = \omega/d$. This is the *abathic* distance (Liebermann, 1910). But since H decreases as d increases, experiment alone can give us the value of the abathic distance. Kröncke (1921) found a distance varying between 40 and 185 cm with 36 subjects. The mean was 1 m. Beyond that distance the apparent fronto-parallel plane is convex toward the subject. The same probably applies to the longitudinal horopter, but no direct proof is yet available.

These interesting results give rise to two different problems. The first concerns the confusion between the longitudinal horopter and the apparent fronto-parallel plane. This distinction was not obvious to Hering, so he took the apparent fronto-parallel plane as the definition of the longitudinal horopter, which is perhaps correct according to etymology but not to logic. The reasons advanced by Hering seem fragile and if a priori there had to be a confusion, it should rather have been between the longitudinal horopter and the curve of *equal apparent distance*,

TABLE 27

Accuracy of the Determination of the Apparent Fronto-Parallel Plane (standard deviations in seconds of arc)

d	η, deg					
	1	2	4	8	12	16
6 m	3.4	5	7	11	13	—
76 cm	7	10	14	35	52	67
40 cm	12	15	22	46	71	120
20 cm	—	32	62	89	170	320

that is, not a fronto-parallel plane, but a circle or sphere with its center at the pupil of the cyclopean eye. As Glanville (1933) demonstrated, this idea of Hering merely obscured the notion of the horopter. In fact, the confusion of the longitudinal horopter with the AFPP applies only to symmetrical convergence. In asymmetrical convergence (see Fig. 70) the apparent fronto-parallel plane seems to be located in an intermediate position between the objective fronto-parallel plane *AA′* and the plane *BB′* perpendicular to *CF* as far as we can assess, because these notions become somewhat vague.

The second problem concerns the variation of *H*. It is important because the first measurements of Helmholtz gave a constant value, nearly perfect, of the product *Hd* between the distances 13 and 75 cm. As *H* represents the monocular asymmetry (that is, the variation of separation between corresponding retinal points, whether they are in the nasal or temporal part of the field), it seems that this variation is in contradiction to the stability of corresponding retinal points. This difficulty is not resolved yet. Some authors, Ogle in particular, have suggested that a change in the optical distortion of the images caused by accommodation may be responsible. In fact, a change in wavelength of the focus with accommodation would be more likely. As we have seen, this phenomenon does exist and, in fact, is confirmed by some statements of Tschermak (1924), who demonstrated that the chromatic aberration of the eye plays an important part. The above results were obtained with black lines on a light background. With blue rods on a black background the AFPP and the longitudinal horopter are found closer to the Vieth-Müller circle, whereas they are farther away with red rods on a black background. The effect of luminance of the background was studied by Lefèvre (1955), who did not find great variations. With average luminance (between 10 and 60 cd/m^2), the curvature of the AFPP is at its greatest. Other factors may influence the curvature of the AFPP. For example, a very brief observation time brings the AFPP closer to the Vieth-Müller circle (Tschermak and Kiribuchi, 1900); if the slit in the screen which limits the parts of the rods seen by the subject is tilted, the AFPP rotates around the fixation point. Finally, some eidetic subjects (that is, capable of producing a visual image nearly as marked as an hallucination, although they themselves recognize that it is subjective) set the apparent fronto-parallel plane in the opposite direction; for them it is a convex surface at near and concave at distance (Jaensch and Reich, 1921; Hofe, 1926). Harker [in 64a] studied the presence of perceptual factors in addition to stereopsis in the determination of the apparent fronto-parallel plane.

Verticals and Horizontals

In the apparent fronto-parallel plane there exist two important directions, the vertical and the horizontal. In the absence of any visual cues, these directions are related to labyrinthine sensations (especially the semicircular canals). Jastrow (1893) and Pierce (1899) showed the precision obtained to set the vertical in these conditions to be better than 30′. In monocular vision there is often a systematic error for the apparent vertical, caused by the vertical meridians of the retinas (see Chapter 12). The localization of a vertical in the median plane of the head is not very precise. Dietzel (1924) found a mean error of 3° but it could be as great as 12°. Of course, the vertical and horizontal are defined by the apparent direction of gravity, that is, by the resultant of gravitation and of centrifugal force if the subject is moving. But if the subject perceives this movement by some visual frame of reference there is a partial correction of the vertical, with considerable variation from one subject to another (Witkin, 1950). In the absence of gravity, such as in space flights, there still remains a definite notion of up and down (down being in the direction of the subject's feet).

Aubert [2] noted that the tilt of the head toward one shoulder causes marked deviations of the vertical and the horizontal. For a tilt of 90° the vertical seems to incline in the opposite direction by an angle between 10 and 30°, which increases with time (G. E. Müller, 1917). It is possible that this effect results from an error that occurs on tilting the head and from cyclotorsional reflexes of the eye, which partially compensate for the inclination of the head. One will find the data reported by several authors in Ogle [46a].

Another illusion described by Hering (1862) happens when the head is tilted, not laterally this time, but upward or downward, the median plane remaining vertical. If the head is tilted forward, the lower part of a vertical line must be placed nearer the subject to appear vertical. The same effect occurs if the subject only gazes downward while his head remains erect. G. T. Stevens (1897) explained this appearance by the *fusional cyclotorsion* of the eyes. Suppose O_1O_2F to be the visual plane (Fig. 75) passing through the centers of the pupils and the fixation point. A line AB is inclined at an angle i with the perpendicular to the visual plane at point F, and is situated in the median plane. A_1 and A_2 are the projections of A on the primary lines of sight O_1F and O_2F. In the similar triangles FA_1A_2 and FO_1O_2 the ratio of the sides A_1A_2 and O_1O_2 is equal to the ratio of the heights $AF \sin i$, and FC. Letting $\omega = O_1O_2$, and $d = FC$, we have

$$\frac{A_1A_2}{\omega} = \frac{AF \sin i}{d}$$

The angle φ between AA_1 and AA_2, being small, can be expressed in radians:

$$\varphi = \frac{A_1A_2}{AF \cos i}$$

Therefore,

$$\varphi = \frac{\omega}{d} \tan i \qquad (54)$$

The angle φ represents the sum of two cyclotorsions necessary for the line AB to be seen fused; it remains small but when the visual plane is tilted the true cyclotorsions are in general less than φ, which consequently displaces the apparent vertical toward the perpendicular to the visual plane to reduce the angle i.

For the same reason, the existence of natural cyclotorsions in a subject (with cyclophoria) may induce a tilting of the vertical in the median plane (Hermans, 1943). The effect described by Nagel [44] and christened *binocular depth contrast* by Werner (1938) is thought to have the

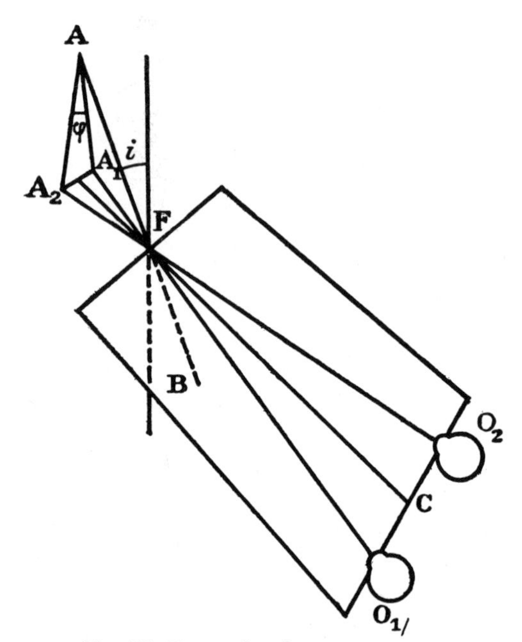

FIG. 75. Fusional cyclotorsions.

same origin. If three vertical lines are situated in an objective frontal plane where no other cues are present and the two lateral lines are tilted forward and backward, the central line seems to tilt in the opposite direction. Ogle attributes this effect to small cyclotorsional fusional movements compensating the true tilt and therefore creating an apparent tilt of the immobile line. Moreover, prolonged observation of lines tilted laterally can also produce reverse deviations of the vertical and horizontal meridians because of compensatory cyclotorsions, which in this case are in the same direction in both eyes (Noji, 1929; Vernon, 1934; J. J. Gibson, 1937); these "directional aftereffects" are greatly influenced by perceptive factors (Köhler and Wallach, 1944; Köhler and Emery, 1947; Hammer, 1949; Ancona, 1950; Howard and Templeton, 1964) and pose difficult problems of psychology [19a].

When a great number of empirical cues are present in the field, the notions of horizontal and vertical become very complex. For example, Asch and Witkin (1948) led a blindfolded subject into a room seen by reflection in a slightly tilted mirror. Depending upon the observer the vertical will be set either correctly or in relation to the walls seen by reflection. The same results can be obtained if the subject is placed in a tilted room (22°) or even if he is seated on a tilted chair. Some subjects seem more sensitive to visual cues, others to postural cues (Piéron, 1951).

Size Constancy

An important element in the appreciation of the distance of a familiar object is the size of the retinal image. The visual angle u is related to the dimension y (perpendicular to the line of sight), and to the distance d, by the relation

$$u = \frac{y}{d} \qquad (55)$$

From this expression, if y is known a priori, d can be inferred from the visual angle u (see, for example, Ittleston, 1951).

Actually, when a known object moves toward or away from us we perceive directly the variation of distance but it is very difficult to notice any change in the size of the retinal image. If one's hand is looked at when the arm is extended and then brought to about 20 cm from one's eyes, only a slight variation of apparent size is noted, especially in binocular vision, whereas in the cinema the same movement would have the effect of ridiculous enlargement of the hand (this is why the "foreshortenings" of paintings, as, for instance, the *Anatomy Lesson* by Rembrandt, are much reduced compared to photography). There exists, therefore, some sort of compensatory mechanism which attenuates the changes of size of the retinal image as distance varies. This compensatory system can sometimes produce errors, as when for some reason the subject judges the distance incorrectly. For instance, R. Smith (1738) placed a small object at the focal point of a convergent lens and noted that the image, although it subtends a constant visual angle, seems to increase in size as the observer moves away because the image is localized where the object actually is, that is, close behind the lens, and not at infinity. Similar experiments have been performed by L. Ronchi and Zoli (1955) with a clear glass reflector. Another example is the case of regular patterns on wallpaper; two adjacent patterns may be fused and, according to the way this was done, will be localized in front or behind the wallpaper and the pattern will therefore appear to shrink or grow larger (H. Meyer, 1842; Woodworth, 1938). Also along the same line of thought are the experiments of Donders (1871) on the apparent diminution of the size of distant objects in monocular vision when the subject accommodates, and also the pathological cases of *micropsia* caused by a partial paralysis of accommodation. When the subject looks at a near object he makes a greater effort of accommodation than he would otherwise, which is followed by an excessive convergence and the object is localized nearer than it actually is (see Trendelenburg, 1943; McCready, 1965).

The distance where the double images seem to be projected can be determined by equation (55) if the subject can evaluate their apparent size y. Giraud-Teulon localized the images seen in diplopia at the same place as the object, and this seems correct for very slight diplopia. But the images move progressively toward the longitudinal horopter, and Polliot (1921) demonstrated this by observing the marked variations in size with the amount of convergence. Actually, the longer the observation the more uncertain is the localization of the double images. Afterimages can also be projected on surfaces, and their apparent size changes proportionately at least up to a few meters away (Witte, 1919). This variation of afterimages has been called *Emmert's law* (1881). It has given rise to much research, particularly by Boring (1940) and W. Edwards (1953). Even with his eyes closed, if the subject imagines that he projects an afterimage on a piece of cardboard that he holds in his hand, he can see the image vary in size as he moves the cardboard closer or farther away.

We have said that the visual angle u subtended by a familiar object of size y is one of the important factors in the appreciation of the distance d. Conversely, if other empirical cues enable the observer to determine the distance d, the size of an unknown object y can be deduced and this size is correctly perceived in spite of the variations of u with the distance. This effect has been referred to by psychologists as *size constancy*. Holway and Boring (1941) have shown that size constancy occurs only if other empirical cues to distance are present. They placed a subject at the intersection of two hallways at right angles. In one of the hallways an illuminated disk was placed at various distances d between 3 and 36 m. The diameter y of the disk varied so that the angle was always 1°. In the other hallway a disk was placed at 3 m. Its diameter y' was adjustable. The subject was requested to equalize the sizes of the two disks. In binocular vision it was found that $y = y'$ whatever the value d was (size constancy). When eliminating all empirical cues by the use of an artificial pupil, monocular vision, and screens limiting the field to the disk only, the following relationship was found:

$$y' = 3y/d$$

which according to equation (55) equalized the two visual angles. If only the empirical cues to distance were hidden, the results were intermediate between the above two extremes. According to D. W. Taylor and Boring (1942), persons with only one eye do not appreciate size constancy as easily as people with binocular vision can, but better than normal people using only one eye. Gibson and Glaser (1947) carried out another experiment in which they placed a series of rods of different lengths 5 m away from a subject and another rod of unknown size at any distance up to 400 m. They found that size constancy is excellent but less precise at greater distances. With training, animals have been shown to have size constancy. In children it has been noted around 1 year of age and then it improves with time.

Even under the best conditions there are small deviations from size constancy in the sense that the farther object seems slightly smaller although actually of equal size (Martius, 1889), particularly when the objects are seen simultaneously instead of fixating them successively (Grabke, 1924). Conversely, if two objects are unequal and in the same plane, the larger appears nearer (Pouillard, 1933). The ingenious experiments of Bernyer et al. (1939) have demonstrated the effects of perceptive conflicts between size and convergence, the results varying from one subject to another. According to Gogel (1963) the apparent size and position of any object are determined by whatever size and distance cues occur between it and adjacent objects; this "adjacency principle" is the main organizing principle for the retinal stimuli. For seven observers, Leibowitz and Moore (1966) found that accommodation and convergence could mediate size constancy only at observation distances less than 1 m, and that other cues are necessary at greater distances.

To end this section let us say a few words about the famous problem: Why does the moon appear bigger when it rises, whereas in reality its apparent diameter is slightly smaller than at the zenith because its distance is greater by the radius of the earth? Since Ptolemy the classical explanation has been that the rising moon seems farther away because the celestial vault appears as a flattened arch, not a sphere, closer overhead than at the horizon, because of the perspective of clouds, the presence of distant objects toward the horizon, the interposed air, etc. Other hypotheses have been suggested, even aniseikonia in the horizontal meridian! Boring (1943) and Zwaan (1958) made a critical analysis of the phenomenon and its explanations, but none is really very convincing. Whether the eyes are lowered or elevated seems to be of some importance and there are notable differences among subjects.

Shape Constancy

A problem slightly similar to size constancy is the constancy of shape perceived in spite of perspective deformations of the retinal image. For instance, in a frontal plane we have a very precise notion of a circle, and an ellipse is recognized as such when its flattening is only 2 to 3%. But if an elongated ellipse is drawn on a paper and then observed obliquely, the long axis being directed toward the subject, the observer cannot say when the projected image of the ellipse is a circle, even in monocular vision. There are very marked differences among observers, but generally speaking, all observers tilt the paper too much and the retinal image is an ellipse flattened in the opposite direction, therefore intermediate between a truly circular image and the image of a circle situated in the inclined plane of the ellipse. Other interesting experiments of this nature have been done by Thouless (1931) and Stavrianos (1945). Thouless (1938) classified subjects in two principal groups according to whether the subjects perceive the object as it is or whether they perceive their retinal images. The first group is of course better adapted to practical life, but with fatigue they tend to shift to the second group. These problems are interesting for psychologists and important for the foundation of abstract ideas.

An afterimage projected on an oblique plane is also deformed; for instance, a square becomes trapezoidal. But often in these experiments the image detaches itself from the plane and regains its original shape while floating in the air.

Geometric Perspective

Perspective is one of our last considerations in the make up of visual space but one of the most important, especially when the play of light and shadows come to complete the "illusion." The inventors of perspective in the fifteenth century, Brunelleschi and Alberti in particular, had exhausted all the possible effects which have been rediscovered often since that time, as for example, the famous matchbox of Tastevin exhibited in 1937 at the "Palais de la Découverte" in Paris or in the amusing experiments conceived by Ames [24]. On the other hand, the total absence of shadows impairs spatial localization even in binocular vision, as shown with objects placed in a photometric sphere reflecting light in all directions (Wagner, 1941). The effect of the illumination on stereopsis was studied by Lauenstein (1938).

The fact that a flat object drawn in perspective gives nearly unmistakably the impression of an object in depth poses an interesting and difficult problem of psychology. Of course, classical perspective is not a convention; it is in fact the only one which is possible if the problem is correctly posed (see, for example, Pirenne, 1952). But this is not sufficient to explain why perspective enables us to see this spatial depth. Kopfermann (1930) showed, for example, that the structure in three dimensions is only spontaneously established if it corresponds to a simpler and more symmetrical construction than a plane figure. In Figure 76 both drawings represent the perspective of an hexagonal prism but it is difficult to see *A* in space because of its symmetry, whereas the more complicated figure *B* becomes simplified as soon as we see it in space, which is almost immediately.

In Figure 77, drawing *A* is seen in perspective, whereas *B*, which is the same with some striped areas, remains flat because of a symmetry which is not apparent in *A*.

Michotte (1948) showed that a plane figure could take on a very remarkable reality of volume if it could be made to spring out from the page. In the example of Figure 78 this can be done by looking at the

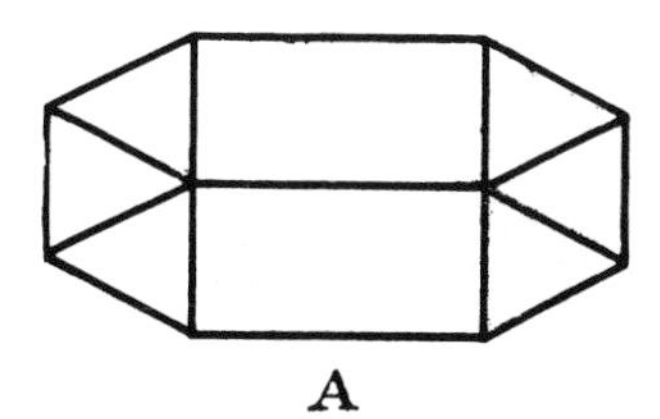

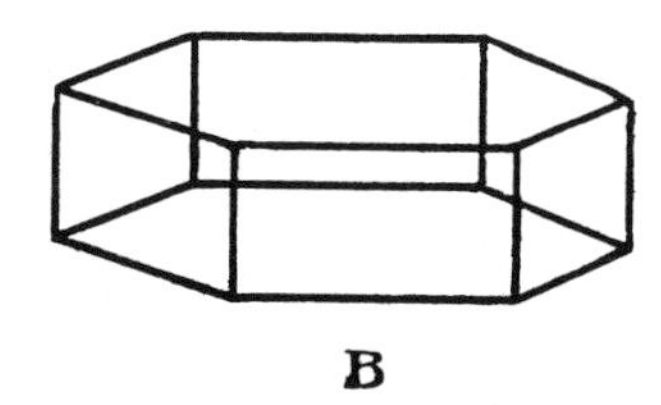

FIG. 76. Figures in perspective (Kopfermann).

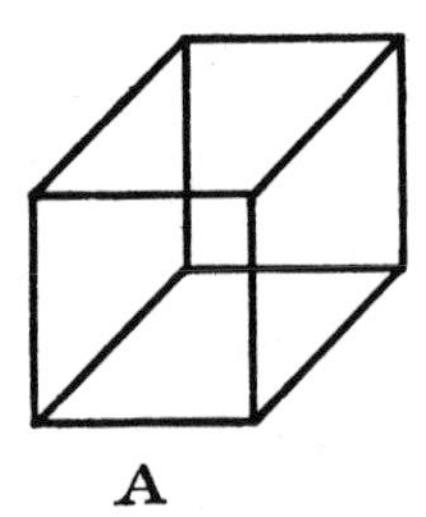

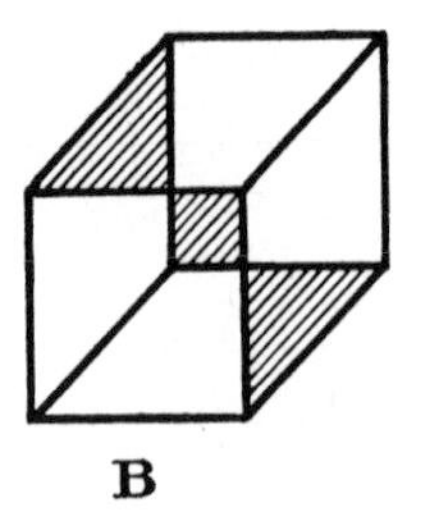

FIG. 77. Perspective drawings (Michotte).

diagram obliquely from the longer side of the page with monocular vision. Movements of the page produce curious deformations of the solid seen in space.

Some drawings in bold perspective can be interpreted as seen in two positions in space and this reversal appears periodically and very suddenly. The oldest example of "reversible perspective" is the Necker cube (1832) (Fig. 77, *A*) and the best known is the inverting staircase of Schröder (1858). In general, plane *A* (Fig. 79) appears in front and plane *B* behind and the staircase is seen from above, but *B* can appear in front and one sees a tiered ceiling. If the reversal from one perspective to another is difficult, it suffices to turn the book slowly 180°. Rubin (1921) studied a great number of similar ambiguous figures.

A related example is the windmill of Sinsteden (1860), its arms slowly turning against a bright sky are viewed obliquely as a dark silhouette on a bright background. It is observed that the arms of the mill seem to go around first in one direction and then in the other. Monocular vision facilitates the changes from one representation to the other and the alternations vary between 3 and 20 seconds, depending on the subject (MacDougall, 1903). An illusory rotation of the same kind is Ames' trapezoid (1951), recently worked out by Guastalla (1966).

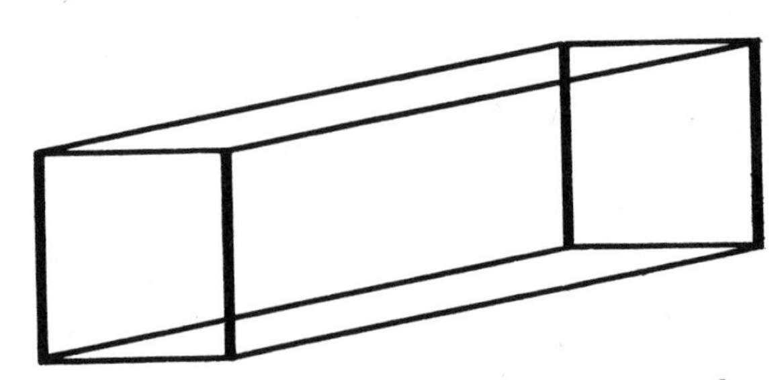

FIG. 78. Volume springing out from the paper when observing tangentially (Michotte).

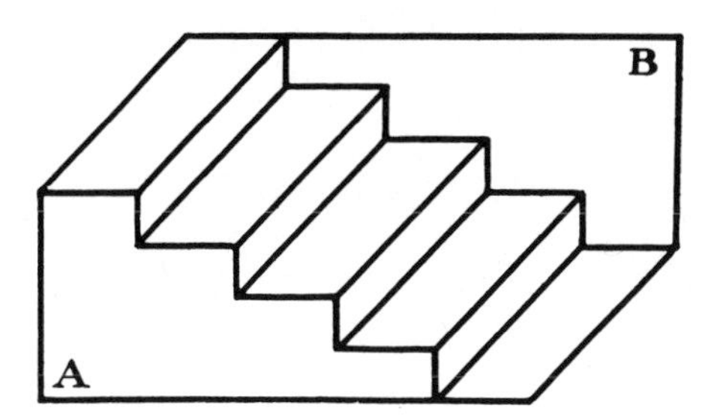

FIG. 79. Reversible perspective figure (Schröder).

Luneburg's Theory

The fundamentals of the geometry of physical space have been studied by many mathematicians, and Weyl [in 73] clearly presented the theory of this problem. With a precision sufficient for everyday life, physical space can be considered as "Euclidean," such that the distance between two points is equal to the square root of the sum of the squares of the differences between their rectangular coordinates.

There is no reason for the notion of subjective distance to coincide with its physical value either for the distance between the object and the observer or between two points. We could, of course, call this discrepancy between true and perceived distance an "illusion," but that is only a word and, for each illusion one must seek the origin of the discrepancy.

A more general method was proposed in 1947 by the mathematician Luneburg [36,37]. He assumed that visual space is not Euclidean and that visual distance is controlled by laws other than physical distance. All the discrepancies would be due solely to the curvature of this visual universe and would be accounted for by a geometric consideration alone.

The idea is attractive, but actually there are few good justifications to support the thesis that these rather vague entities—the "subjective distances"—could fit themselves into a uniform geometric framework. First, experiment must prove it. However, the results are not very conclusive. For instance, Luneburg assumes that points appearing equidistant from the subject are situated on a Vieth-Müller circle, whereas observations [19b] lead to a curve situated between the Vieth-Müller circle and another circle of which the subject is the center. Then Luneburg points out a remark of Helmholtz, who found that when the eyes are free to move, the apparent fronto-parallel planes are similar to those that we described earlier in the case of lateral vision. In fact, we have very few data on this phenomenon that Helmholtz explained by an underestimation of the absolute distance, but which would be simpler to relate to monocular asymmetry. According to Foley (1966), the locus of perceived equidistance is, between 1.2 and 4.2 m, always concave toward the observer and slightly asymmetric with respect to the median plane.

Hillebrand (1902) had remarked that parallel lines such as railways do seem to converge in the distance. This experiment of the "parallel alleys" that Luneburg reconsidered is done in the following manner. A subject is placed in the dark and fixates two light sources *A* and *B* (Fig. 80). Then he sets, two by two, a series of other light sources closer and closer to himself, leaving them all illuminated, until the two lines of lights appear straight and parallel. It is found that the lines converge behind the observer but it is possible that the height of the observer's eyes above the plane of the lights has some bearing on the phenomenon.

Another experiment is the "distance alleys" of Blumenfeld (1913). In this situation only the two lights *A* and *B* are switched on permanently and the observer sets a series of lights so that the separation of each pair appears equal to the distance *AB*. According to Hardy et al. (1949–1951), the curves found are always on the outside of the parallel alleys.

From the analysis of these experiments Luneburg concluded that visual space is Riemannian with constant negative curvature (hyperbolic space), each subject having a characteristic distance function in this space. In spite of the mathematical ingenuity of Luneburg and his

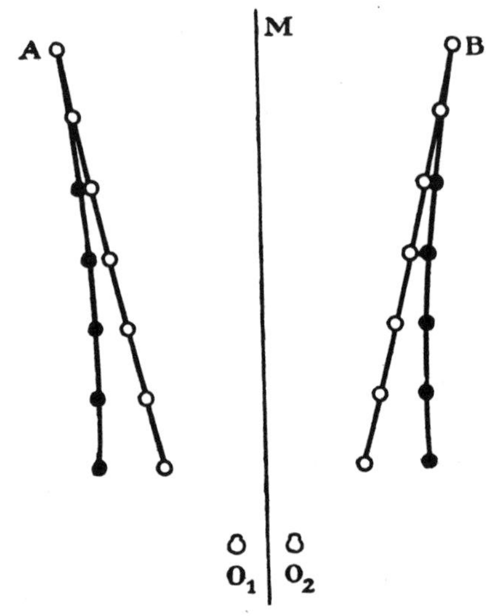

Fig. 80. Parallel alleys (open circles) and distance alleys (filled circles).

admirers [19b], this conclusion reaches beyond experimental observation. Stein (1947) applied the theory of Luneburg to binocularly equivalent configurations designed by Ames. Blank (1953) measured the characteristic distance functions on several subjects. (On the theory of Luneburg, see also Charnwood, 1951; Morrison, 1951, 1961; Bourdy, 1957; Blank, 1961; Gogel, 1964; Foley, 1964.)

CHAPTER FIFTEEN

The Visual Centers

In this chapter we will give succinct notions of anatomy and physiology which will enable the reader to have an over-all view—as much as is possible in our present state of knowledge—of the biological basis of visual perception. For more details, the reader can consult classical texts such as Brindley [7a] or the excellent monograph by Albe-Fessard (1957). To conform to medical usage we will make two modifications in this chapter to the terminology that we have adopted for the rest of this text. First, the terms "temporal, nasal, superior, and inferior" will apply to the retina and not to the visual field (because of the reversing of the retinal image there is double inversion from the retina to the visual field; for instance, the superior temporal quadrant of the retina corresponds to the inferior nasal quadrant of the visual field). Second, we will call "macula" what is in fact the fovea, that is, the central depression of the retina which is rod-free. The true macula, characterized by the presence of yellow pigment, is much wider and badly defined.

Retinal Fibers

We refer the reader to anatomy texts for the description of the retina. Let us recall only a few simple notions. Groups of visual photoreceptors (rods and cones) are linked to a bipolar cell which acts as a relay station, and several bipolars are connected to one ganglion cell. The axons of the ganglion cells constitute the *optic nerve fibers* which transmit the nervous impulses elicited by the absorbed photons in the sensory receptors. In the retina these fibers are very thin (0.5 to 5 μ diameter) and are collected in bundles which are separated from each other by the lateral filaments of the supporting cells.

The course of these retinal fibers is visible with an ophthalmoscope

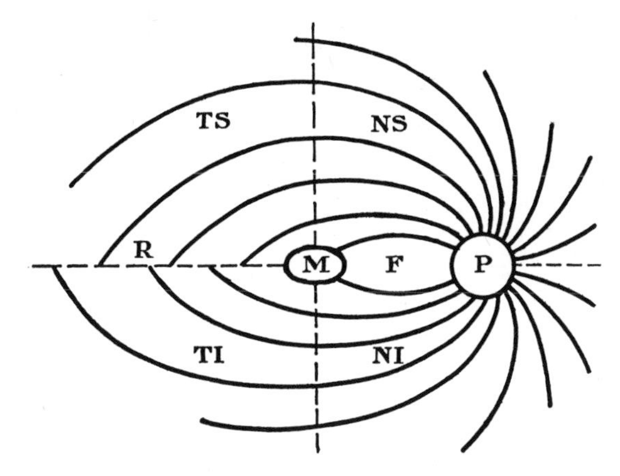

FIG. 81. Diagram of the optic nerve fibers of the right eye seen from the front. *M*, macula; *P*, papilla; *R*, raphé; *F*, papillo macular bundle.

(Fig. 81) in red-free light. The macula *M* is connected to the papilla *P* by the arcuate papillomacular bundle *F*, which entoptically appears as the blue arcs of Purkinje visible in the dark when a small red light is focused near the fovea. The remaining part of the retina is divided into four quadrants (superior temporal *TS*, superior nasal *NS*, inferior temporal *TI*, inferior nasal *NI*) separated by the vertical and horizontal meridians intersecting at the macula. The fibers of the superior and inferior halves meet at the horizontal, forming the retinal *raphé*, *R*, whereas there is nothing comparable in the vertical meridian.

Optic Nerve

The optic nerve is surrounded by three layers which are continuous with the three meninges of the brain (pia mater, arachnoid membrane, and dura mater). It has a diameter of 3.5 to 4 mm and a length of nearly 5 cm (of which 3 are intraorbital). The two optic nerves arising from the right and left eyes join to form the *chiasma* from which the two *optic tracts* emerge.

The fibers of the optic nerve are thin (on the average 2 μ), grouped in bundles, and organized in the same way as in the retina. The superior fibers remain in the upper part, the inferior in the lower part, the temporal and nasal on their corresponding sides (the temporal fibers are much more numerous). The macular fibers alone occupy one third of the section of the nerve and are situated on the temporal side as they enter the nerve, then assume a more central position along its course, and in the chiasma shift slightly medially among the nasal fibers. These localizations have been ascertained by *degeneration* of some areas of the retina which causes atrophy of the corresponding optic nerve fibers. Furthermore, some centrifugal fibers, conveying nerve impulses toward the retina, have also been described in the optic nerve, but their function is hardly known. It is possible that some of these fibers come directly from the optic nerve of the other eye through the chiasma, thus enabling a certain relationship between the two retinas. According to some experiments of Granit (1955), the centrifugal fibers may serve to regulate the spontaneous activity of the retina (see also [30a]). Ogden and Brown (1964) also put forward an argument in favor of the centrifugal fibers by showing that electrical potentials can be recorded in the retina of the monkey when the optic nerve is stimulated.

Chiasma

The crossing of the two optic nerves resembles the Greek letter χ, from which the name chiasma is derived. This crossing is general among vertebrates but with some variations. One of the nerves can cross over the other or through it or else the bundles cross each other like a checkerboard. In mammals there is partial *decussation*; some of the fibers form a *direct* bundle (as, for example, a bundle from the right optic nerve continues into the right optic tract), whereas other fibers constitute a *crossed* bundle which goes from the right optic nerve into the left optic tract. The direct bundle becomes more and more common as we pass up the vertebrate scale and constitutes over 40% of the total number of fibers in man.

The chiasma, which was already known by the Greek physician Galen, has aroused innumerable discussions. Newton (1704) saw in it the explanation of single vision with both eyes. He assumed that decussation existed in animals with binocular vision in which the two visual fields are at least partially superimposed, whereas in animals with panoramic vision, in which the two visual fields are adjacent to each other without intermingling, it would merely be an intercrossing of the fibers. This thesis is partly true but it is too rigid. Thus in some nocturnal rapacious animals there is an intercrossing of the fibers without decussation although the position of their eyes may provide them with binocular vision. Described by Hannover (1852), then denied, partial decussation was only established with certainty in man by some classical works on the chiasma, in particular that of Rönne (1910). He proved that the temporal fibers constitute part of the direct bundle and the nasal fibers part of the crossed bundle. The macular fibers are shared by both.

If we leave the macula aside for the moment, we can say that the transport of sensory information emerging from, for example, the *left* visual field of each eye (which corresponds to the nasal fibers of the left retina and the temporal fibers of the right retina) is entirely conducted along the *right* optic tract. As there is no more crossing along the pathway, the messages are received in the *right* cerebral hemisphere (Fig. 82). Cajal (1911) held this to account for the pseudo-problem of the

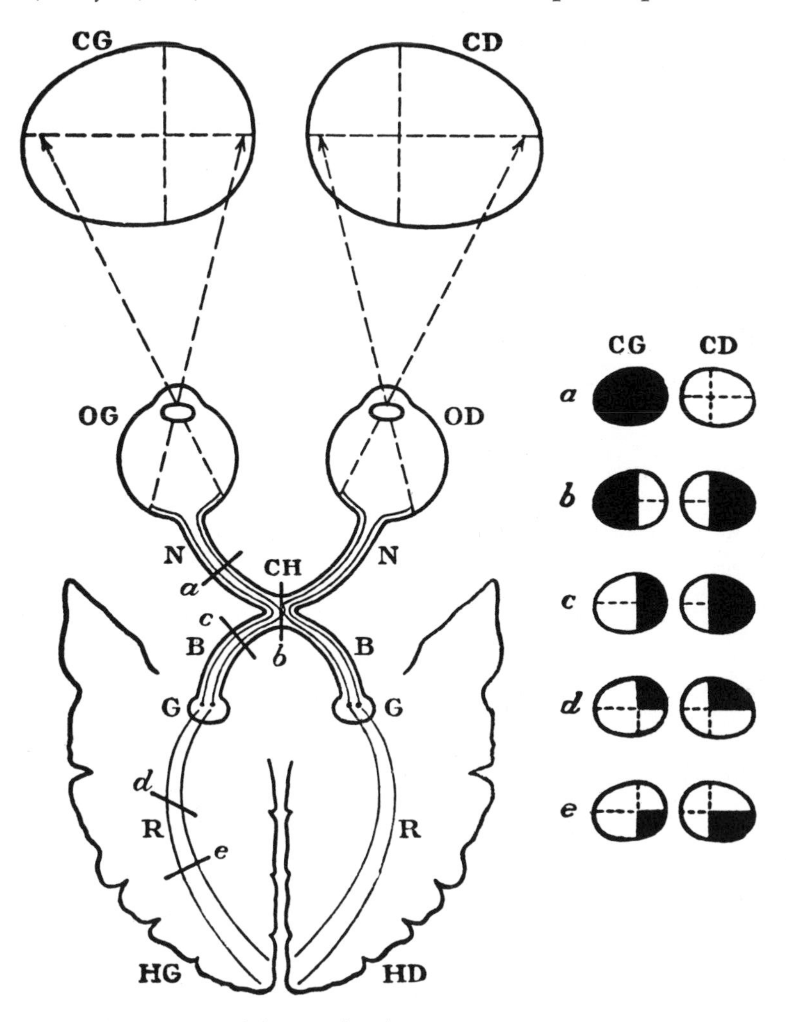

FIG. 82. Diagram of the visual pathway. *B*, optic tracts; *CD* and *CG*, right and left visual fields; *CH*, chiasma; *G*, lateral geniculate bodies; *HD* and *HG*, right and left cerebral hemispheres. *N*, optic nerves; *OD* and *OG*, right and left eye; *R*, optic radiations. *a* to *e*, several lesions along the visual pathway producing the scotomata indicated in black on the right side (at *d* and *e*, it has been assumed that the lower and upper part, respectively, of the radiations were sectioned).

"reinversion" of the inverted images of the retinas. As we have mentioned, this problem has not much meaning as far as visual perception is concerned, or at least one must take into consideration the motor responses arising from vision. Because of the decussation of motor nerve fibers, it is in the right cerebral hemisphere that the movements of objects situated on the left side of the body are elaborated. Therefore, the role of the partial decussation is not to invert an image but to gather in one cerebral hemisphere the visual messages relating to a given spatial area and the movements of reaction which are accomplished in this direction.

Geniculate Bodies

Emerging from the chiasma, the fibers form two flattened oval bands, the optic tracts. The course of the fibers is somewhat uncertain but it is generally agreed that there is a twist of 90° so that the macular fibers are in the upper part, the superior fibers on the medial side, and the inferior fibers on the lateral side. About 80% of the fibers of the optic tracts end in two flattened ganglia called the *external* or *lateral geniculate bodies*. The remaining fibers end in other organs called the *pretectum* and the anterior corpora quadrigemina* [26a]. Whereas the lateral geniculate bodies are very important relay stations in the conduction of visual messages between the retina and the brain, the other organs do not seem to intervene in this conduction. The corpora quadrigemina seems to be a center for the coordination of eye movements. Apter (1945–1946) and Hyde and Eason (1959) were able to map the whole of the corpora quadrigemina in terms of small retinal areas of the cat; the corpora may be a sort of calculating organ necessary to determine the contraction of the extraocular muscles required for the eyes to fixate a point stimulating the retina. Pitts and MacCulloch (1947) attempted to analyze the problem of constancy in visual form perception on the basis of the above findings.

Let us return to the geniculate bodies. The most precise knowledge on the structure of this organ has been provided by Minkowski (1920) and Le Gros Clark (1941) (see also [43a] and [55a]; Silva, 1956; Hayhow, 1958; Glees [in 26a]). In the monkey each geniculate body is made up of six layers, alternate grey and white lamina which we number from 1 to 6 starting from the outside. The crossed bundle of the optic tract terminates in layers 1, 4, and 6 and the direct bundle ends in layers 2, 3, and 5. Each fiber of the optic tract which is the axon of a

* These are also referred to as *superior colliculi.*

retinal ganglion cell terminates in a spray of five or six bulbous extremities each touching a cell of the geniculate body. The axons of these cells constitute the fibers of the *optic radiations*, which relate the geniculate bodies to the visual area of the cerebral cortex. The geniculate body would therefore be an amplifying relay station, since more fibers emerge from it than it receives. It is possible to map out precisely the whole of this relay station with the retina. The macular fibers end in the upper posterior two thirds of the nucleus. Furthermore, it has been attempted to attribute to the geniculate bodies their own visual functions. For instance, some have held it to be a center for binocular fusion because of the contiguity of the layers where the fibers coming from corresponding points of the retina terminate. But with this hypothesis it becomes difficult to understand the multiplications of fibers emerging from this relay station. However, after the work of several investigators—especially Bishop et al. (1959), who found in the lateral geniculate body of the rabbit about 10% "binocular" cells (responding electrically to stimuli in either eye) and many other cells which although they are "monocular" have their discharge modified by excitation in the other eye, and Fillenz [in 26a], who described two interlaminar layers in the cat containing synaptic terminations from both eyes—the hypothesis of the binocular role of the lateral geniculate body seemed to have gained ground. But it has lost ground since the important work of Hubel and Wiesel (1961–1962), who do not find any cell responding to both eyes in the geniculate body of the cat. According to DeValois [in 44a] the higher the organization of the animal, the smaller the role played by the geniculate body in subserving binocular vision. In the monkey he found practically no binocular cells.

Another hypothetical function of the geniculate body, suggested by Le Gros Clark (1942) is to consider the bodies as a trichromatic organ for a theory of colors: layers 1 and 2 would be the relay stations for blue, 3 and 4 for red, 5 and 6 for green. Some arguments have been advanced in favor of this concept; for instance, there are only four layers in some dichromat monkeys. Moreover, in the area of the geniculate body corresponding to the periphery of the retina, layers 3 and 6 are fused, which would be responsible for the experimental blue-yellow dichromatism. On the other hand, in the macular region, layers 1 and 2 grow thinner, accounting for the foveal dichromatism of tritanopic type. But this hypothesis still lacks evidence.

The best data on the role of the geniculate body have been provided by the research of DeValois [in 44a] on the monkey. In response to monochromatic stimuli by diffuse illumination of the retina, the cells can be grouped into two classes: the first consists of broad-band cells (either inhibitors which respond by a decrease of their spontaneous activity, or excitators which respond to light by an increase in firing) which seem to carry luminosity information, scotopic or photopic; the second consists of opponent cells which respond to some wavelengths with an increase and to others with a decrease in firing rate. In the macaque there is a considerable variability of these spectral-opponent cells with regard to which wavelengths produce inhibition and which produce excitation. In addition, changing the intensity of the light has a different effect on inhibition and excitation. This is probably the origin of the so-called Bezold-Brücke phenomenon, that is, the systematic effect of intensity on the colored appearance of different wavelengths. Chromatic adaptation has very little effect on broad-band cells and very great effect on opponent cells, which is also in accordance with the well-known colorimetric laws. In the squirrel monkey, which has been shown by Jacobs (1963) to possess color vision like that of the human protanomalous trichromat, opponent cells are less than 15% of the total number while they constitute more than 50% in the macaque, the vision of which is like a normal trichomat; in addition, the red mechanism is displaced spectrally toward shorter wavelengths (Jacobs and DeValois, 1965; DeValois et al., 1966).

These studies point to the fact that the lateral geniculate body acts as an organ "encoding" the retinal message, which becomes more easily assimilated by the visual cortex. DeValois suggests that this coding utilizes the spontaneous firing rate of neurons as a sort of carrier frequency around which signals are modulated, with decreases from this frequency as well as increases carrying information. This transformation would apply, on the one hand, to color information as we have just seen, and on the other hand to interactions between several parts of the visual field (Hubel, 1960) comprising, in particular, vision of movement (Grüsser [in 26a] and Arden and Söderberg [in 60b]) and that of flickering lights (Arden and Liu, 1960).

It is interesting also to note that the geniculate body has its own noise. By intraocular variations of pressure, Arden and Söderberg (1959) found in rabbits that silence in the optic nerve did not imply silence in the geniculate body where spikes were still recorded. This spontaneous activity seems to be largely determined by extraretinal factors, such as acoustic and tactual stimuli, or the state of wakefulness; it might also have an energizer action (Schlag, 1959).

Visual Cortex

After the geniculate bodies the sensory impulse is conducted by the fibers of the optic radiations, the upper part of each radiation containing the superior fibers of the retina, the lower part the inferior fibers of the retina, and the center of the radiations the fibers from the macula. These fibers terminate in the *cortex,* that is, the cerebral bark in the posterior region of the brain, referred to as *occipital.* This was first demonstrated by Henschen (1892). The visual area of the cerebral cortex is reasonably well defined by a stria, and this *striate area* lies symmetrically about the *calcarine fissure* (Fig. 83), a horizontal fold in this internal surface of the brain. Henschen also proved that the superior and inferior retinal quadrants are localized, respectively, on the upper and

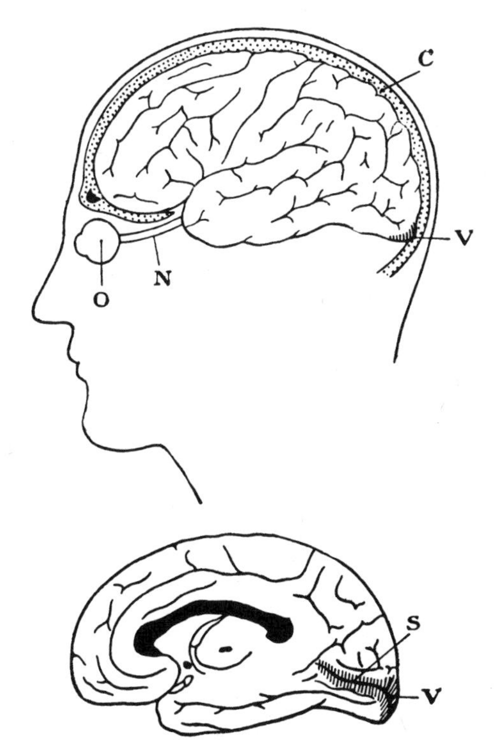

FIG. 83. Location (very diagrammatic) of the visual cortex *V* in man (indicated with stripes). The upper diagram shows an open section of the cranium *C* in order to see the brain in its position. *N*, optic nerve; *O*, eye. The lower diagram shows a median section of the brain and we are looking at the internal part of the right hemisphere. *S*, calcarine fissure.

lower lips of the calcarine fissure and spread over a width of some 5 mm on these lips. But he incorrectly localized the macula, the correct position of which was discovered by Lenz (1909) at the posterior pole of the occipital region on the internal side and extending a little on the external side. These localizations have been ascertained during the First World War by the study of injuries to the cortex causing scotomata in the visual field, and it is classical to represent them by a diagram designed by Holmes (1918 and 1945) that we reproduce in Figure 84.

It is often pointed out that the visual cortex is isolated in comparison to the other centers of the brain, which may attribute to it an exceptional character in the cerebral architecture. The striate area is some sort of *cortical retina* establishing a precise point-to-point localization with the retina of the eye. In this correspondence there are successively a di-

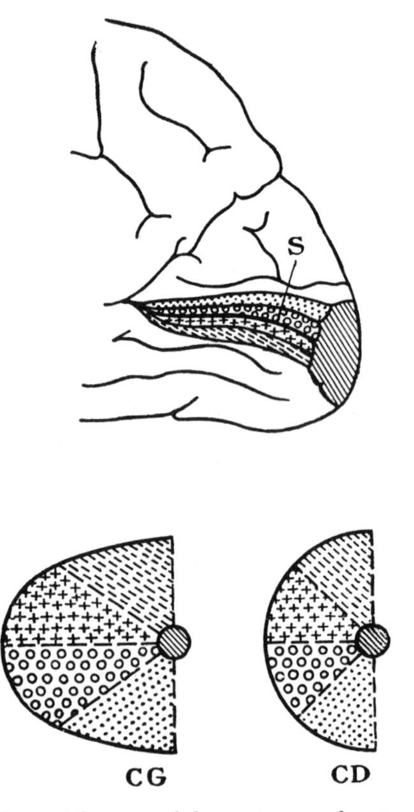

Fig. 84. Classical diagram of the projection of sectors of the left visual field *CG* and of the right visual field *CD* on the edges of the calcarine fissure S of the right cerebral hemisphere. The macula occupies the posterior position of the occipital lobe.

vision and a multiplication of the cells. In fact, over a hundred million rods and cones discharge on only 800,000 optic nerve fibers, which, after passage through the geniculate body, become of the order of 2 million and in the visual cortex (the thickness of which is between 1 and 2 mm) some 1400 million neurones are estimated. The multiplication is especially important in the macular area, where each square millimeter of the fovea is represented by a few square centimeters of the cortical surface, 3.7 in the baboon (Whitteridge and Daniel [26a]). According to Sholl (1954 [61a]) the impulses from a single optic nerve fiber can sometimes directly excite 5000 cortical neurones. Actually, the total surface of the human cortex devoted to vision is relatively large. Man is a creature exceptionally visual, and according to Bruesch and Arey (1942) nearly 40% of all sensory information is provided by our eyes.

Actually, the isolation of the visual cortex seems less absolute than it was formerly held to be. First, the striate area is surrounded by a ring of cortical areas referred to as *peristriate* and *parastriate areas,* which play a role in perception. Electrical excitation of these areas on the exposed cortex produces hallucinatory phosphenes, whereas the excitation of the striate area produces localized luminous but shapeless impressions. A lesion of these areas does not impair vision but abolishes the recognition of forms and colors, which attributes to these areas a role of higher *integration* necessary for the psychology of vision. Some areas of the brain even farther away intervene in this integration, and Zangwill (1948) has shown that disturbances near the language area could prevent the patient from recognizing objects although their form is perceived. Conversely, the visual cortex seems to intervene in the intellectual functions which are not essentially visual (Lashley, 1948; Parsons, 1949). This was confirmed by the studies of Levine (1952) on the comparison between subjects blind from an ocular lesion and others from a trauma to the visual cortex.

Binocular Vision

The anatomical explanation of binocular fusion of two monocular sensory messages is still unknown. Galen and Newton attributed it to the chiasma. From studies on the degeneration of the lateral geniculate body, Minkowski (1920) was inclined to think that the fusion center was located in that organ. Actually, the old hypothesis of Descartes (1677), according to whom the messages arrive independently at the striate area (of course, Descartes spoke of "animal spirits" and the pineal gland), is nowadays accepted. But the terminal unification still remains hypothetical. Is it due to simple connections among neighboring areas of the cortex where the monocular excitations terminate (Minkowski, 1922), or to a special bundle of connecting fibers which travel from one hemisphere to the other (Pfeifer, 1925), or is it due to the white "line of Gennari," which establishes a relationship between the direct and crossed endings of the optic radiations (Kleist, 1926), or even to some other mechanism?

On the other hand, pathology has provided us with a great deal of information by the clinical study of visual field defects. Any lesion along the pathway between the retina and the visual cortex produces a scotoma which is usually negative, like the blind spot (it is not a black hole but an absence of vision), but sometimes positive (as, for example, a grey area on a white background). The extraordinary variety of the observed facts of this subject only permits us to summarize the essentials and we refer the reader to the treatise of Dubois-Poulsen et al. [15] for more details.

A lesion of the optic nerve before the chiasma produces amaurosis of one eye (Fig. 82*a*). Occasionally, only part of the nerve is affected, as, for instance, when an axial neuritis destroys the papillomacular bundle and causes a central scotoma with eccentric fixation and poor visual acuity (Monbrun, 1914). On the other hand, there can be a total peripheral scotoma resulting in tunnel vision with a visual field of 1° or less; the patient can still read but is quite unable to manage otherwise. A lesion at the chiasma causes a diversity of scotomata; for example, a *bitemporal hemianopsia* (the two temporal halves of the field are lost) follows a lesion of the crossed bundles (Fig. 82*b*). A lesion along the optic tract causes a *homonymous hemianopsia,* which affects the right or left halves of the visual field (Fig. 82*c*). Actually, the separation between the blind and seeing area is not a rigid vertical line; there is often some overlap of one field on the other (particularly in the upper part of the field) and the line of separation usually spares the macula. A posterior lesion of the geniculate body is ascertained by the fact that the pupillary reflex is retained when the blind area is stimulated, because the pupillary fibers going to the nucleus of the oculomotor nerve, which controls the muscles of the iris, bifurcate in the optic tract just before the lateral geniculate body. Some lesions of the optic radiations produce quadrantanopsia (Fig. 82*d* and *e*).

The problem of the macula is difficult. It used to be held that unlike the peripheral retina, each macula was represented on both hemispheres of the visual cortex. In order to be so, Lenz (1938) assumed a bifurcation in the optic radiations, sending fibers into the other hemisphere. But this concept is contradicted by an observation of Maison et al. (1938), who made a monkey totally blind by sectioning one optic tract and the opposite radiation very near the cortex. In *lobectomies*, where part of the striate area is removed, a homonymous hemianopsia occurs with sparing of the macula in some cases and disappearance in others. It is also possible that the macula may have reduced vision but is still partially retained. It was formerly believed (Holmes) that the fovea was split into four by a horizontal and a vertical meridian intersecting through the fixation point and that these foveal quadrants were projected on the cortex in the same way as the peripheral quadrants. All these questions, the solutions to which are uncertain, are of primary importance for a better comprehension of binocular vision. For example, Wolf and Zigler (1963) noted that 10° away from the fixation point the absolute threshold of a square of 1° presents a summation except on the vertical passing through the fixation point, which suggests that summation takes place only for impulses arriving in the same cerebral hemisphere.

To end this section let us note the current concept of a point-to-point correspondence between the retina and visual cortex, supported by Meynert (1872), Henschen (1892), and Wilbrand (1913), but attacked by Monakow (1914), who supported the more plastic concept of Flourens (1824), which provides an argument heavily in favor of the ideas of Kant on the innate nature of the notion of space and opposed to the empiricism of Helmholtz. We can relate this to the conclusion that we arrived at previously in the study of the stability of corresponding points.

Cortical Electrophysiology

Even better than the use of anatomy or physiology is the electrical method of exploring the visual centers. In this way we can hope in the future to obtain important data. We will summarize briefly the results already obtained.

We refer the reader to the classical textbooks about the electrophysiology of the retina and the essential characteristics of the nerve impulse which travels along the optic nerve fibers. After the synapse in the lateral geniculate body, the impulse arriving in the visual cortex can be recorded and amplified by well-known methods. The electrodes can be placed either against the exposed cortex of the trepanated subject (or animal) or even simply on the cranium. But the latter technique is obviously less precise, because the thickness of the cranium impairs the localization of the recorded potentials. Historically these phenomena have been known since 1880. Caton and Danielewsky [in 3] noted that flashes of light produce electrical modifications in the brain of the dogs, but it has only been since 1930 that the advancement of electrical recording techniques has provided a systematic study of cerebral electrophysiology.

Single Excitation

To produce a single and well-defined excitation of the visual cortex, the best method is to remove the eye and stimulate the optic nerve directly by a brief electrical shock. Alternatively, the eye could be left intact and the retina stimulated by a flash of light. But these conditions, although more natural, are not as simple. We know in fact that the optic nerve responds by a volley of discharges when stimulated by a single flash no matter how brief it is. In both methods Bishop (1935) noted differences in electrical responses in the cortex from one experiment to the other even if the excitation was identical. There exists some sort of characteristic cycle of cortical sensitivity and we will return to that point later. Generally speaking, the response of the cortex to a single excitation consists of a brief variation of potential lasting 0.01 to 0.02 second, referred to as *primary variation*, followed by several slower secondary variations.

The primary variation was studied on several animals, particularly on the rabbit by Bartley and Bishop (1933), the dog by F. Fischer (1934), and the cat by Bishop and O'Leary (1938), Marshall et al. (1943), and Chang and Kaada (1950). We have reproduced a typical recording by these latter authors (Fig. 85). The curve is somewhat complicated. A first spike (*a*) appears after a very short latency period, sometimes

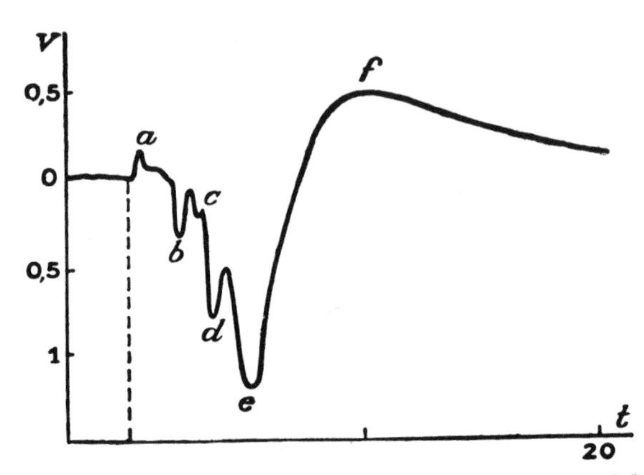

FIG. 85. Typical record of the primary cortical response of the cat (Chang and Kaada). Abscissa: The time t (in msec). The origin corresponds to the excitation of the optic nerve. Ordinate: Potential V (in mV). A displacement of the curve upward indicates that the active electrode is negative compared to the reference.

shorter than 0.5 msec, following the electrical excitation of the optic nerve, which implies that this is not a cortical phenomenon. W. A. Cobb and Morton (1952) have actually noted that the electrical propagation of noncortical phenomena through the mass of tissue often reaches the electrode before the cortical response. Then follow three waves (*b*, *c*, and *d*) that Chang (1951) interpreted as the activity of three different groups of optic nerve fibers. He inferred that these three groups could conduct the messages of red, green, and blue, according to the decreasing order of diameters of the fibers and therefore the speed of the impulse. This is in agreement with the measurements of Piéron on the difference of latency of colors. By inserting the electrode into the visual cortex, Bishop and Clare (1951) noted a modification of these spikes which, according to them, represented the action of different groups of neurones discharging in the cortex after different journeys. The next wave (*e*) would be an activity of the cortical neurones, whereas the last wave (*f*), slower and larger, would result from the diffusion of discharge from cortical neurones through the cerebral cortex. But all these interpretations are uncertain and so are the investigations of Ingvar (1959) and Lennox-Buchtal (1962) on the spectral variations of cortical responses.

Following the primary variation, several secondary variations of potential are recorded. A first wave which immediately follows the primary variation in the visual system has been described by Fields et al. (1949) and is usually interpreted as a temporary intensification of the spontaneous cortical activity which we will discuss later. A second, slower wave, with about 70 msec latency, was found in the stimulation of the sciatic nerve and was independently found in the visual system by Chang (1950) and H. Gastaut (1950). Its meaning is disputed (see Brazier, 1953). Other secondary variations have also been described. These very complex phenomena depend a great deal upon the anesthesia used to prepare the animal.

Instead of electrical excitation of the optic nerve, one can stimulate the lateral geniculate body and a similar cortical response is obtained except for a diminution of the latency of about 0.8 msec. As each *synapse* (the contact between two successive neurones in a pathway of nervous conduction) usually introduces a delay of 0.5 msec, it is concluded that there exists only one synapse between the optic nerve and the cortex, in agreement with the anatomical findings, and a total of three synapses along the whole visual pathway. The first is between the photoreceptors (cones and rods) and the bipolar cell, the second is between the bipolar and the ganglion cell, and the last is in the lateral geniculate body.

Chang (1952) noted that during the electrical excitation of the lateral geniculate body, the cortical response was more marked if the retina was stimulated simultaneously by continuous illumination. This effect is not instantaneous; it only occurs in full after about 5 seconds illumination. At the beginning of the illumination and just after extinction there is, on the contrary, a diminution of the cortical response, which is perhaps at the origin of the well-known aftereffect of glare called "blackout." These phenomena would not be of retinal origin but would be due to summations in the lateral geniculate body.

Finally, up to a certain point the direct excitation of the retina by a flash of light causes the same cortical response but with some spreading out of the variations, caused by the bursts of impulses from the nerve. Furthermore, the latencies are lengthened because of the lags within the retina. Bartley (1934–1935) has measured these latencies in the rabbit as a function of the intensity and duration of the flash as well as the area of stimulated retina. The results showed that these modifications of latencies could be attributed to the retina. In man, the primary response to a flash of light has been recorded through the cranium (Monnier, 1948; Cobb, 1950), but usually it is the slow secondary variation which is detected more easily. The simultaneous recording of the electroretinogram and of the transcranial response in man showed a delay of 5 to 10 msec for the latter (M. Monnier, 1949). According to Vaughan and Hull (1965), the latency t varies with luminance L of the stimulus following the law

$$t = t_0 + L^{-0.33}$$

When a discrete point of the retina is illuminated it is usually noted that the electrode placed in contact with the cortex yields a response over a reasonably large area, which Bartley explained by the scattering of light in the eye but which we now know is also due to lateral connections in the retina. By determining systematically the point on the retina where the response recorded with a fixed electrode is the greatest and presents the smallest latency, one can establish a geographic mapping of the retina and the cortex which enables one to verify the anatomical and pathological findings. In particular, Thompson et al. (1950) have accomplished this with accuracy on the rabbit. A very interesting but delicate inverse method consists of exciting a given point of the visual cortex by an electrical shock and localizing the phosphene which appears in the visual field. Unfortunately this is only possible on human subjects during a brain operation while the cortex is exposed, which leaves little leisure for experimentation (Penfield, 1947).

Multiple Excitations

Instead of a single stimulus, a pair of electrical stimuli is presented, with a variable time between the two. In time it is noted that the ratio of the second cortical response to the first (Fig. 86) presents an appearance of damped oscillations (Bartley, 1936). There exists a refractory period (distance *OR*) of about 20 msec during which the second excitation has no effect; then it is followed by facilitation or partial inhibition periods. That is, the second excitation produces more effect or less effect than if it were alone. This phenomenon can be compared to the cycle of cortical excitability which we have discussed briefly and to the *spontaneous cortical activity* in the absence of peripheral excitation. Berger (1930) and Adrian and B. Matthews (1934–1935) have shown that rhythmic potentials could be recorded through the bones of the skull in man. Analysis of these oscillations reveals a fundamental frequency between 8 and 13 per second and this has been called the *alpha rhythm*. This rhythm occurs when the subject is awake but his mental activity is reduced, that is, when no excitation of either internal or external origin attracts his attention. Closing the eyes facilitates the phenomenon (this cortical activity can be compared to the phenomenon described at the end of Chapter 9, that is, the spontaneous activity of the retina in the dark, and from that point of view, as in many others, the retina plays the role of a small peripheral brain). The alpha rhythm is broken up whenever a sensory message arrives in the area where the recording is made. We will not dwell any more upon these recordings,

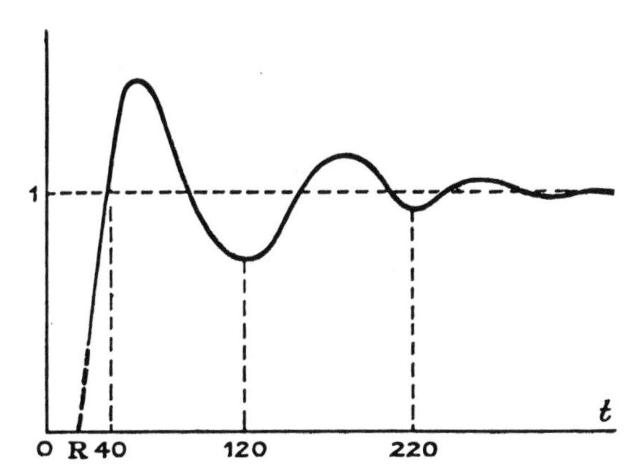

FIG. 86. Ratio of the amplitude of the second cortical response of the cat to the first, when the two successive excitations are separated by a time t (in msec) (after Gastaut et al., 1951).

or *electroencephalograms*, the clinical importance of which is considerable but which have so far provided little information on vision [21a]. We will only note, in agreement with Durup and Fessard (1938), that the latency of blocking of the alpha rhythm in the visual cortex is long (sometimes more than 0.3 second), which shows that it is a secondary effect. Following Jasper and Cruikshank (1937), Popov (1953–1955) also studied the successive blockings of the alpha rhythm after a single intense luminous excitation. These blockings could be related to after-images, the origin of which should be cortical and not retinal. Relations between spontaneous cortical activity and the cycle of cortical excitability have been studied particularly by Heinbecker and Bartley (1940), Marshall (1949), and Jarcho (1949). Let us also point out the experiments of Evans and Smith (1964) on the encephalogram during stabilization of the retinal image, and the influence of the orientation of the eyes on the alpha rhythm (Mulholland and Evans, 1965).

An interesting case of multiple excitations is that of a regular repetition. Adrian and B. Matthews (1934) noted that a light flickering at 7 to 14 cycles/second produced potentials of the same frequency in the cortex. This result has often been interpreted to mean that the alpha rhythm can be "driven" by the flickering light exposed to the eye. Actually, Grey Walter et al. (1946) have noted that there exists for each subject a "resonance" frequency which is not necessarily that of the dominant resting alpha rhythm but is often related to some components of the resting electroencephalogram. According to H. Gastaut et al. (1951), two groups of subjects can be distinguished. First, calm people, who have a slow alpha rhythm (8 to 10/second), ample and regular, with little interference of the rhythm in the presence of a luminous excitation and little or no driving by a flickering light. Second, nervous people, who have a rapid rhythm (10 to 13) in spaced-out bursts, with a blocking of the rhythm when the subject opens his eyes or even recalls a visual memory, and in whom a flickering light can drive the waves at frequencies between 1 and 30 per second, with a maximum around 4 to 6. Let us point out, by the way, that this cortical drive seems to cease earlier than that of the retina, since in the retinogram of the cat, Enroth (1952) noted a synchronization with frequencies above 70 and, in the pigeon, Dodt and Wirth (1953) sometimes went up to 140 cycles/second to obtain fusion. Recently the use of sine-wave-modulated lights in vision studies (de Lange, 1957; Levinson, 1964) has allowed a more complete study of the distortions in visual transmission. For example, in most subjects a modulation at 5 cycles/second gives a nearly pure 10 cycles/second response in the encephalogram (van der Tweel, 1964); if a visual "noise" of bandwidth 15 to 25 cycles/second is added to the signal, linearization takes place and the distortion is much reduced (Specreyse and van der Tweel, 1965).

Continuous Excitation

The best way to study the electrical response of the cortex to a continuous excitation of the retina by light consists of using *microelectrodes*. The research we have discussed so far used gross electrodes to record the potentials through the skull or an active electrode made up of a metallic wire of some tenths of a millimeter in diameter touching the exposed cortex, whereas the reference electrode was a plate fixed to the edge of the open skull. Adrian and Moruzzi (1939) have brought

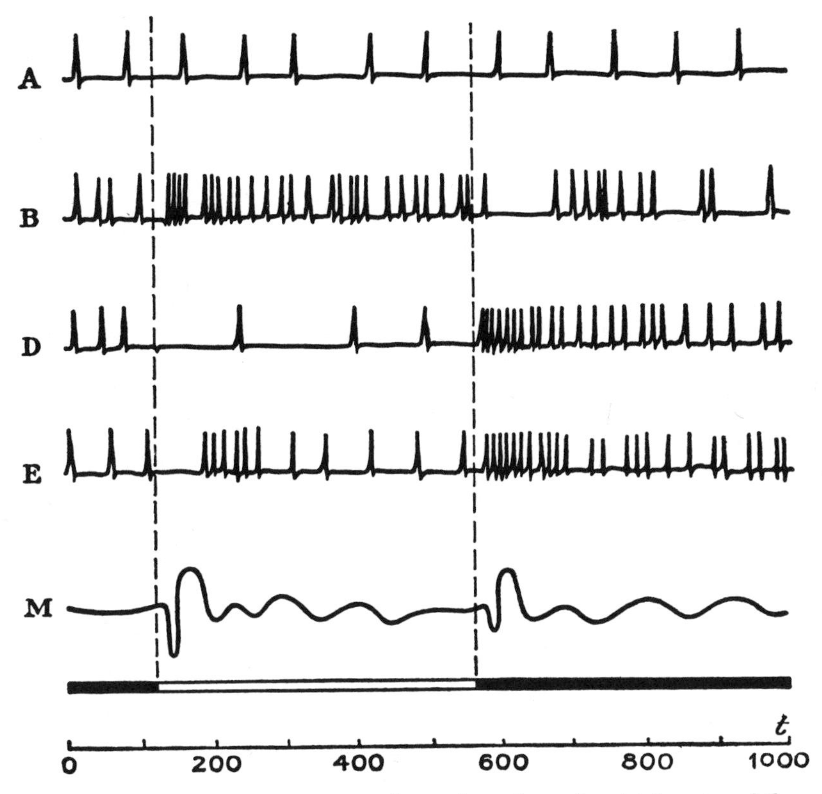

FIG. 87. Several kinds of electrical recordings from the visual cortex of the cat (after Jung). *A*, *B*, *D*, and *E*, with a microelectrode; *M*, with a gross electrode. The luminous excitation is represented by the white segment of the black line above the time axis *t*, graduated in milliseconds.

attention to the importance of using a very small electrode touching only one cortical neuron, if possible. This technique has recently been developed and the diameter of the active electrode can be below 1 μ. Under these conditions a series of short discharges can be recorded (of a duration less than 1 msec) where the active electrode is negative and has an amplitude exceeding 1 mv. As an example we have represented in Figure 87 some results obtained by Jung (1953) on the cat. Depending upon the position of the microelectrode, luminous excitation of the whole retina yields several kinds of responses. In type *A* we observe a spontaneous discharge at a frequency of 8 to 15 per second (not to be confused with the alpha rhythm, which is of the order of 4 to 8 in the cat), the illumination producing no effect. In type *B* the spontaneous frequency is of the order of 20, increases when the light is switched on, with a latency of 14 msec, reaches a maximum (100 to 400), and returns to a lower value (30 to 50); darkness produces a brief inhibition, then the rhythm continues irregularly. In type *D*, on the other hand, there is inhibition at the onset of illumination, then a burst of discharges (100 to 300) appears at extinction and gradually returns to the initial activity. Finally, type *E* gives an initial inhibition at the onset of the illumination, followed by a moderate activity (30 to 60) that returns to the spontaneous activity (10 to 15), and then produces a marked activity (60 to 100) at extinction followed by a progressive decline.

These results obviously should be compared to the electrical phenomena of the retina and the optic nerve. Types *B*, *D*, and *E* resemble the "on," "off," and on–off" fibers, respectively. But there are differences. For example, unlike the optic nerve fibers, the cortical neurones present waves of potential which may vary in amplitude (and not only in frequency) depending upon the modulation of the light. Li and Jasper (1953) have studied the distribution of the different types in the visual cortex. Type *B* is more frequent in the macular region, *D* and *E* in the peripheral region. With repeated flashes instead of a continuous luminous exitation, the discharges follow reasonably well the rhythm of the excitation up to a frequency of about 30.

To end this section we will mention the elegant experiments of Hubel (1959) on the cat. In collaboration with Wiesel (1962) he has shown that there exist cortical receptive fields elongated in shape, made up of an excitatory and an inhibitory region and very sensitive to movements on the retina; these findings appear fundamental to form and movement perception. This important problem of the receptive fields of the retina has been studied also with rectangular or striated patterns in the rabbit (van Hof, 1965) and the cat (Baumgartner et al., 1965; Rodieck, 1965). Experiments by Andrews (1965), using psychophysical methods, support the hypothesis that similar units are to be found in the human fovea.

Section B

APPLICATIONS

CHAPTER SIXTEEN

Vision Through the Atmosphere

We will end this volume with some chapters devoted to the principal practical applications of the facts and theories that have already been discussed. We will begin with the most natural case, that of a subject in the atmosphere. The vision of near objects does not offer any particular problem, but the study of distance vision through the atmosphere is difficult and has given rise to many investigations. We refer the reader to the excellent treatise of Middleton [41] for further details. We will content ourselves with a summary of the principal visual aspects of these problems.

Photometry of Scattering Media

Consider a monochromatic radiation emitted by a very distant source and therefore practically propagated by plane waves. Consider among these waves a beam of parallel rays having a section perpendicular to the direction of propagation Ox (Fig. 88), of area S. Consider F, the luminous flux passing through this area S. In a vacuum the energy would be retained in the given cylinder and F would remain constant; but this is quite different in air for the following two reasons.

Consider an element of a cylinder, of volume S dx, between two planes perpendicular to Ox, at locations x and $x + dx$. A part of the radiation enters this volume and does not emerge from it, at least in the same form, because of its transformation into heat inside the elemental volume. This is the phenomenon of *absorption*. Another part of the flux F is not found in this volume. These radiations are propagated in another direction but retain their original nature and wavelength; this is the phenomenon of *scattering*. All these effects produce a diminution of the incident flux F proportional to the thickness dx and we will write

$$dF = -\sigma F\,dx \tag{56}$$

The *extinction coefficient* σ combines the effects of absorption and scattering. Of course, if we change the wavelength λ of the monochromatic radiation, the extinction coefficient σ also changes. It is therefore a function of λ.

The equation (56) of the extinction coefficient does not present any difficulty in the case of absorption only. This is not true, however, of scattering. Let us actually isolate the cylindrical beam with an opaque screen in which there is a hole of area S. Let us look at the center M of a small scattering volume, from a distance great enough so that the dimensions of this volume are negligible and it appears as a point source (more exactly, as an element of the linear source which the beam appears to be from a great distance). This source possesses a luminous intensity in this given direction which is proportional to the volume $S\,dx$ mentioned earlier and to the illumination F/S that it receives. We will therefore write

$$dI = \beta\,F\,dx \tag{57}$$

The coefficient β is a function of the wavelength λ and of the angle Ψ which represents the direction of observation (if the incident light was polarized rectilinearly, β would depend also upon the azimuth of the direction of observation in relation to the plane of polarization); this is called the *scattering function.* The luminous flux scattered by this elementary volume between the cones of apex M and angles Ψ and $\Psi + d\Psi$ is equal to the product of the luminous intensity dI and the solid angle situated between these cones, that is, $2\pi \sin \Psi\, d\Psi$. If the absorption is neglected and only the scattering is retained (which is often the case in the atmosphere), the total flux scattered by the elementary volume is nothing but the flux dF of expression (56) reduced

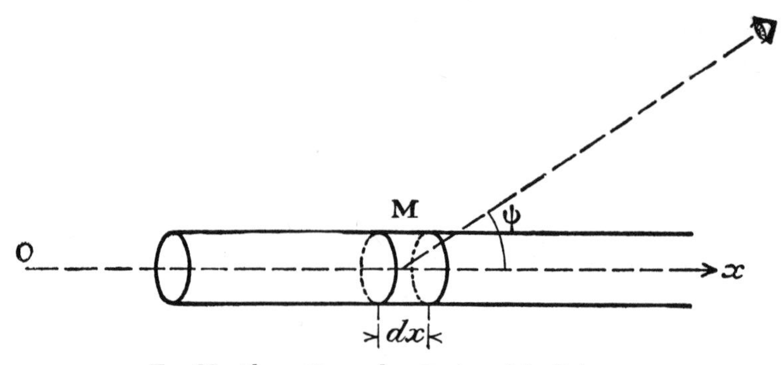

FIG. 88. Absorption and scattering of the light.

to the radiation which is propagated along the cylindrical beam, scattered in all directions. Hence, the relation

$$\sigma = 2\pi \int_0^\pi \beta(\Psi) \sin \Psi \, d\Psi \tag{58}$$

From the physical point of view the beam is not exactly a cylinder, first, because of the finite angular dimensions of the source which emits radiations, and second, because of the diffraction by the opening of area S. The result is that the lower limit of the integral (58) is not zero but possesses a value which depends upon the experimental conditions and the method of measurement. Consequently there is a practical uncertainty in the definition of σ with regard to scattering.

In a homogeneous medium where σ is independent of the coordinates, the integration of (56) is

$$F = F_0 e^{-\sigma x} \tag{59}$$

where F_0 is the flux for $x = 0$. This law of exponential decrease of the flux (as the thickness of the medium increases) was established by Bouguer (1760). It is valid only for monochromatic radiation. With complex radiation such as sunlight, it would be exact only if the coefficient σ was independent of wavelength.

Case of the Atmosphere

In the visible spectrum, the absorption of air is practically negligible, except in the red end of the spectrum (the bands of water vapor), but scattering always plays an essential role.

If the air were perfectly pure this scattering would obey Rayleigh's law (1871). The molecules of air, being of negligible dimensions compared to wavelength λ, introduce a scattering function represented by

$$\beta(\Psi) = A\lambda^{-4} (1 + \cos^2 \Psi) \tag{60}$$

where the coefficient A is a function of the density of the air. According to this law, the scattered light in the direction of the incident light ($\Psi = 0°$) or in the opposite direction ($\Psi = 180°$) is twice the amount scattered at right angles ($\Psi = 90°$) from the incident light. The polarization of the scattered light is nearly complete in the latter plane. On the other hand, the variation of the intensity of the elementary volume in any given direction is inversely proportional to the fourth power of the wavelength. This explains the predominance of the short wavelengths in the light scattered through the atmosphere and hence the blue color of the sky.

TABLE 28

Extinction Coefficient of Pure Air, by Molecular Scattering Only

λ, μ	0.3	0.4	0.5	0.6	0.7	0.8	0.9	1.0
σ, km^{-1}	0.1479	0.0436	0.0175	0.0083	0.0047	0.0026	0.0016	0.0011

The theory of Rayleigh could be perfected to take into account the asymmetry of the molecules. The values of the extinction coefficient σ are given in Table 28 for air at 0°C and pressure 1013.2 mb [41].

From definition (56) the dimensions of the extinction coefficient σ vary inversely with the path length. It is convenient to use the kilometer as a unit of length. The transparency of 1 km of pure atmosphere for $\lambda = 0.5\ \mu$, for instance, using equation (59) and Table 28, is

$$F/F_0 = e^{-0.0175} = 0.983$$

which is a reduction of less than 2% per km. The air is therefore very transparent. It is often convenient to introduce the *optical density* which is, except for the sign, equal to the common logarithm of the transparency:

$$\delta = -\log_{10} \frac{F}{F_0} = \sigma x \log_{10} e = 0.434\sigma x \tag{61}$$

Again, for $\lambda = 0.5\ \mu$ the optical density of 1 km of pure air is 0.0076.

Actually, air always contains other particles besides the molecules of air. In particular, water is condensed in liquid droplets about *nuclei* of different origins. The diameter of these droplets can be measured either optically (by the diffraction rings) or by determining the rate of fall in undisturbed air or gathering droplets on a spider's web and taking a photomicrograph on a black background. In fogs and clouds the droplets have radii between 1 and 60 μ, whereas in haze the radii are always less than 0.4 μ (Dessens, 1944–1947). Regarding this matter let us note that even in the thickest fogs the water content of the air is very low (maximum 0.4 g/m^3 of air).

As the radius a of the water droplets is of an order of magnitude comparable to or larger than the wavelength, Rayleigh's theory does not apply. First of all, the function $\beta(\Psi)$ is no longer symmetrical but presents a very marked maximum in the direction of the incident light (Ψ near 0°). These functions have often been calculated following Mie (1908), in particular by Gumprecht et al. (1952). Measurements made at night with searchlights above the sea show that with a slight haze the minimum of β is not reached for $\Psi = 90°$ as expected from equation (60), but around 120°. In the opposite direction ($\Psi = 180°$) the scattering is twice the minimum, but in the forward direction ($\Psi = 0$) it is 30 times the minimum (Chesterman and Stiles, 1948). Waldram (1945) found even higher values in the forward direction, of the order of 100 times the minimum (which can shift to $\Psi = 140°$).

Furthermore, the extinction coefficient can be represented by the following expression:

$$\sigma = KN\pi a^2 \tag{62}$$

where N is the number of droplets (they are assumed to all be of radius a) per unit of volume of air. The dimensionless factor K is a function of the refractive index of the droplets and of the variable α, where

$$\alpha = 2\pi a/\lambda \tag{63}$$

For water droplets (index 4/3) we have reproduced the curve of K as a function of α (Fig. 89) from the calculations of Houghton and Chalker (1949), utilizing the Mie theory. The law in λ^{-4} only applies for small values of α. If in a small region of the spectrum K is represented by a law in λ^n, the exponent n will be zero at about $a = \lambda$, then oscillates between positive and negative values with an average around 0. This means that a haze (or a fog) consisting of droplets of all sizes is not selective and the extinction is similar for all radiations. A haze

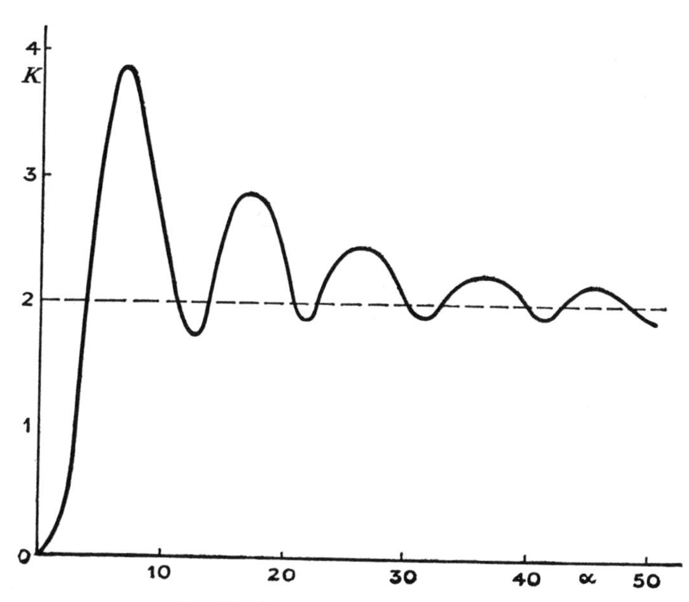

FIG. 89. Scattering of water droplets.

with droplets of equal size is more transparent for the short or long wavelengths, depending on whether n is positive or negative. According to the measurements of Arnulf et al. (1954), who made very thorough studies in the interval of wavelengths 0.4 to 10 μ, fogs are practically neutral, whereas haze is much more transparent in the far infrared than in the visible spectrum. Within the visible spectrum σ can be regarded as being always independent of λ, except for a very light haze, where the exponent n is usually negative and below 2 in absolute value. The sky is still blue but the color is less pure than the Rayleigh atmosphere.

When the droplets are very large, K tends toward 2 (a large droplet scatters twice the amount of light that would be expected from geometric consideration), which at first glance seems paradoxical. In fact, in a beam of section S and length dx, where there are $NS\ dx$ droplets, each having a cross-sectional area πa^2, the geometric effect of a screen diminishes the luminous flux in the proportion of the cross-sectional area of the drops to that of the beam, that is,

$$dF/F = -\pi a^2 NS\ dx/S \tag{64}$$

and in comparison with (56) and (62), $K = 1$. The discrepancy is explained by a reasoning which does not seem very convincing, ascribing it to Babinet's principle. Whatever the case may be, the geometric consideration is incorrect, and it is the diffraction of the light around the drops which doubles the lost flux dF. Actually, the larger the drops, the more the diffracted flux is confined to small values of the angle Ψ. This brings back the physical difficulty raised in the preceding paragraph; and in fact the measurements of K give values between 1 and 2, depending upon the angle between the incident light and the photometer (Sinclair, 1947). The case of large drops, such as in rain, raises interesting problems involving the geometric theories of the refraction and reflection of light in the drops (rainbows) to be related to the above facts. The reader could consult Bricard (1953) on this matter.

Point Signals at Night

This is the simplest case of vision through the atmosphere. If no extinction of the light occurred through the atmosphere, the illumination E produced by a source of intensity I, in the plane of the pupil of the subject looking at the source from a distance x, would be given by the classical inverse-square law

$$E = Ix^{-2}$$

In the case of monochromatic radiation or even when σ does not vary with λ, Bouguer's law (59) must also be taken into consideration and we arrive at the expression

$$E = Ix^{-2}e^{-\sigma x} \tag{65}$$

This is Allard's law (1876).

The visual range of a signal is obtained by substituting the threshold illuminance in (65); the corresponding value of x is the visual range.

In laboratory experiments and for subjects perfectly dark-adapted, the thresholds will be between 10^{-8} and 10^{-9} lux, but under the practical conditions of air and sea navigation the values usually adopted are between 1.2×10^{-7} and 1.9×10^{-7} lux [41]. Let us assume 1.5×10^{-7} lux, for example. If the distance x is expressed in kilometers, the limiting distance will be obtained by graphic solution of (65) by writing

$$0.434\sigma x = \log_{10}(I/0.15) - 2\log_{10} x \tag{66}$$

where the intensity is expressed in candelas. For example, a source of 1500 cd would theoretically be seen at 100 km if the extinction coefficient were negligible and at 60 km if the molecular scattering is taken into account (for $\lambda = 0.5\ \mu$). The term *haze* refers to atmosphere in which the optical density is less than 2 per kilometer. With this value the same source would have a limiting distance of 1.75 km. A fog of optical density equal to 10 would reduce the distance to 0.47 km. For a source of 15 cd the limiting distances in the same conditions would be 10, 9.2, 1, and 0.3 km, respectively. It should be noted that in fog very little improvement is obtained by increasing the intensity of the source.

If instead of producing constant illumination, the signal emits flashes of duration t (occulting light), according to the classical Blondel-Rey law (1911) it suffices to replace the constant 0.15 of equation (66) by the quantity $0.15\,(t + t_0)/t$. At absolute threshold the value of t_0 is usually between 0.15 and 0.3 sec, but under normal conditions of observation the measurements of Toulmin-Smith and H. Green (1933) and the analysis of Hampton (1934) arrive at an average of $t_0 = 0.11$ second.

Recognition of the color of signals poses difficult problems. If we consider the three most common colors of signal lights (red, green, yellow) the constant 0.15 of equation (66) must be increased to 0.4 according to H. Green (1935). Stiles et al. (1937) found a threshold of recognition for each color. The best choice of colors was studied statistically by MacNicholas (1936) for railway signals, then by Holmes (1941) with 256 different colors, but each was only seen three times by the observers. Finally, Hill (1947) used aviation signals with nine subjects, each of whom saw at least 20 times each of the 73 colors of point sources. His results show that the purples must be avoided as they are seen red because of the chromatic aberration of the eye. One must also be cautious of the yellowishness of white signals seen through a light-selective haze. For example, Middleton [41] calculated the effect of haze containing 3.4×10^4 drops of 0.2 μ radius per cubic centimeter; a source of color temperature 2850°K diminishes progressively in apparent color temperature as the distance increases and at 1 km is below 1500°K, that is, a rather saturated orange. The very important consequences of these findings on the efficient choice of colored signals, a choice which must also take into account the anomalies of color vision, have been gathered in the reports of the "Colour of Signals" Committee of the Commission Internationale de l'Eclairage (Stockholm, 1951; Zurich, 1955; Brussels, 1959; Vienna, 1963). It is also essential that subjects whose occupation requires a recognition of signals pass special tests, different from the usual tests which use extended colored sources (Beyne, 1954). Sloan and Habel (1955) have studied especially the problem of the recognition of red and green point sources, as a function of their intensity, by color-defective subjects.

Point Signals on a Light Background

This more complex case can occur at night when the background on which the signal is presented is illuminated, or in daytime. In principle, Allard's law (65) still applies, but the threshold E is a function of the luminance of the background (see Fig. 21). On the other hand, the fact that attention is now shared by the background and the signal (or several) modifies the psychological conditions of perception. Thus, Gerathewohl (1953) observed that a signal emitting flashes of 0.2 second duration, at a frequency of 60 per minute, was often better perceived than a constant signal of equal intensity. If the subject is not aware of the position where the source will appear, the time τ necessary to detect it is often appreciable. Langmuir and Westendorp (1931) studied the case of flashes of 0.2 second duration appearing with a period T (several seconds) anywhere in a solid angle of Ω steradians on a background of 7×10^{-2} cd/m^2 luminance. Their results have been represented as a formula by Stiles, Bennett, and Green (1937):

$$\left(\tau + \frac{T}{2}\right)(P - 1) = 50\Omega T$$

provided the intensity of the source is equal to P times its liminal value (threshold for the observer). This was done in foveal vision and the conditions probably differ in peripheral vision.

A systematic study of the recognition of sudden signal lights of cars (stop light and direction indicator light) was carried out by Devaux and his collaborators (1951). The subject was placed 10 m from a large vertical screen in the center of which five small lamps of different colors were switched on successively and regularly, one per second. The attention of the subject was concentrated on these lamps, at which he stared. At any moment a signal was switched on at some point of the screen situated between 2.7 and 17° from the center and in an arbitrary direction. The subject responded by pressing a switch and his reaction time was measured. Three levels of luminance of the background were utilized (0, 800, and 1300 cd/m^2). With the highest luminance the intensity of the signal light had to be 0.40 cd in order for the reaction time to remain less than 1 second, as long as the eccentricity was below 12°. Beyond that eccentricity the intensity increased rapidly and reached 2 cd at 17°. These experiments have also been performed for higher luminances of the background up to 7800 cd/m^2.

The effect of the color of the signal has been studied by Hill (1947), who repeated, with a light background, experiments similar to those performed on a dark background on the recognition of colors. When the luminance of the signal becomes equal to that of the background and only color difference plays a role, the phenomena become more complicated. Lucretius was already aware of the disappearance of colored points on a grey background. Also bearing on this matter are the experiments by Willmer and Wright (1945), Farnsworth and Reed (1944), and Hillman et al. (1954). According to Malone (1955), two very keen observers could still distinguish red or green points of 0.6′ apparent diameter on a grey background of equal luminance, whereas for yellow, blue, and purple the diameters needed to be much greater. These phenomena can be related to the foveal dichromatism (tritanopia; blue blindness).

Extended Sources

In the general case the theory is complicated. It can be simplified, however, with certain hypotheses. First, the case of a black object outlined against the sky just above the horizon. The observer does not see this object as black because the intervening air between him and the object sends scattered light toward him. As before, let us consider an elementary volume of thickness dx, the coordinate x measured on a line with its origin at the observer's eye, and a section, of area S, perpendicular to this line. The scattered luminous intensity toward the observer is proportional to the volume of the element

$$dI = kS\,dx$$

We will assume that k is a factor independent of x, which means that the atmosphere is homogeneous and that the conditions of illumination do not vary. This is reasonable if one gazes horizontally. In the absence of air the luminance corresponding to the element would be the quotient of the luminous intensity by the surface S, but if one takes into consideration Bouguer's law it becomes

$$dL = ke^{-\sigma x}\,dx$$

dL representing the luminance of the element as it appears to the observer. The luminance of the intervening air in front of the object which is situated at a distance r is the integral

$$L = k\int_0^r e^{-\sigma x}\,dx = \frac{k}{\sigma}(1 - e^{-\sigma r})$$

and it will be the luminance of the object itself, because, being black, it will not add any luminance itself. If the object is at infinity this expression becomes

$$L_h = k/\sigma$$

which is nothing but the luminance of the sky near the horizon in the direction of the object. Hence, finally,

$$L = L_h(1 - e^{-\sigma r}) \tag{67}$$

If instead of being black, the object had a luminance which, seen at near, was L_0, the apparent luminance of the object at a distance r would be

$$L = L_0e^{-\sigma r} + L_h(1 - e^{-\sigma r}) \tag{68}$$

This expression was established by Koschmieder (1924). Note that in the case of a black object situated at a given distance, the photometric measurement of the ratio L/L_h gives the value of σ. Arnulf et al. (1954) have verified that this gives the same values as applying Bouguer's law to a cylindrical beam reflected in a vertical mirror placed at a great distance and directed toward the observer. The deviations between these methods can be explained by the rapid variations of σ in the case of fog or by the nonhomogeneity of this fog.

The case of an object seen, not horizontally but in any direction, is much more complicated. The analysis of Duntley (1949) is oversimplified. The calculation of the luminance in a given direction within a scattering medium follows an integral obtained by King (1913). Since then this equation has been expressed in different ways but none lends itself to a simple mathematical resolution other than by approximations. Chandrasekhar (1950) indicated an original method which has the advantage of taking into account all the orders of scattering at once. Lenoble (1955) applied the latter method in certain cases of fog.

If we return to expression (68) we see that not only the luminance of the object changes but also its color. In the case of pure atmosphere σ varies greatly with the wavelength (see Table 28). The extinction coefficient varies inversely with wavelength. If $L_0 > L_h$ (exceptional case of an object lighter than the sky), the color would turn toward yellow, but toward blue in the opposite case. Everyone has encountered these effects of aerial perspective where objects in the far distance appear bluish. In the cases of fog or haze σ is practically independent of wavelength and the hue of the object does not change but the purity decreases. At large distances, practically all the colors of natural objects turn to grey or blue, and it is merely by the ratio of the luminances or *contrasts* that one can distinguish the objects. The principal factor for day vision is the alteration of these contrasts by the intervening atmosphere.

Limiting Distance of Vision

Consider an object, the luminance of which in the direction of the observer is L_0 in the absence of intervening atmosphere, and suppose L'_0 is the luminance of the background under the same conditions. The *inherent contrast* is defined by the ratio

$$C_0 = \frac{L_0 - L'_0}{L'_0}$$

If the object and the background are at about the same distance r from the observer, the apparent luminances L and L' are given by expression (68); hence

$$L - L' = (L_0 - L'_0)e^{-\sigma r}$$

and the *apparent contrast* is expressed by

$$C = \frac{L - L'}{L'} = \frac{C_0 e^{-\sigma r} L'_0}{L'} \tag{69}$$

If the sky is the background at the horizon, $L'_0 = L'$, since there is always a practically infinite thickness of air in front of the observer and it simply reduces to

$$C = C_0 e^{-\sigma r} \tag{70}$$

The contrast therefore always reduces with the distance. But at a given distance r, C does not always diminish when σ increases, because it may happen that C_0 changes at the same time if the luminance of the sky L'_0 varies. It is sometimes possible that a snow-covered mountain becomes more visible because of the darkening of the sky when σ increases (Dessens, 1944). When $C_0 = 0$ the value of C is equal to zero at all distances; an object which would have the luminance of the sky at the horizon would retain it. Darker or lighter objects tend toward the luminance of the sky as the distance increases.

When a far object subtends a sufficient angle, the only condition necessary for it to be perceived is that its apparent contrast with the background remain higher than the differential threshold. The values of this threshold measured in the laboratory (Blackwell, 1946) seem to be applicable outside, according to H. S. Coleman and Verplanck (1948). In daytime with objects subtending an apparent angle of at least 1°, a value of 0.02 for the minimum contrast perceived with certainty is commonly adopted, and replacing C by this value in (70), the limiting distance of vision of an object against the sky is

$$r = \frac{1}{\sigma} \ln | 50\, C_0 |$$

For a black object ($C_0 = -1$) it becomes simply

$$r = 3.91/\sigma \tag{71}$$

If, for example, the air is perfectly pure, from the values of Table 28 for $\lambda = 0.6\ \mu$, one arrives at a limiting distance of vision beyond 400 km. It increases greatly with λ and in the photographic infrared it can exceed 2000 km. Obviously, at these large distances (photographs taken from satellites), the curvature of the earth comes into play. Even greater distances are possible for snow-covered peaks in the sun, for which C_0 can reach 3. Through a haze of optical density 2 per kilometer, the limiting distance r for a black object would be 0.85 km and it would fall to 0.17 km through a fog of density 10. In metereology the *visibility*, that is, the limiting value of r for a black object against the sky at the horizon, constitutes the simplest measurement of the extinction coefficient of the atmosphere. Several range meters for the determination of meteorological range have been described (see, for example, Vos, 1961).

Nightfall, of course, markedly increases the contrast threshold. On the average, the limiting distance of vision of a large dark object outlined against the sky reduces to one quarter of what it is in daytime and objects less than 30′ in diameter become invisible at all distances (Seidentopf, 1948). This explains the important role of night telescopes with large exit pupils. By increasing the apparent size of the object without appreciably diminishing the luminances, the limiting distance of vision increases by reducing the minimum perceptible contrast. The effect is less pronounced in daytime because at high luminances this threshold varies less rapidly with the dimensions of the object.

The more general case of objects on a background other than the sky is very important in aerial observation. The contrast of objects on the ground seen from an aircraft flying at an altitude of 1500 m have been measured with a special telephotometer in fine weather, by Carman and Carruthers (1951). The luminances vary from about 20,000 cd/m^2 (sunlit glacier) to 33 cd/m^2 (shaded forested mountain), with an average value of 3600 cd/m^2. Between two adjacent areas of the terrain the mean contrast is of the order of 0.2. The analysis of such cases is more complicated and I refer the reader to Middleton [41].

We have admitted that the only effect of the atmosphere was an attenuation of contrast (besides the alteration of colors, which only plays a small visual role). It has sometimes been said that haze or fog could blur the contours of objects and reduce their visibility. This is purely subjective and the data of Barber (1950), in particular, have proved that through the thickest fogs no blurred edge could be discovered by photography, not even of 1′ width. On the other hand, in clear weather and near the ground on a hot day, the gradient of the refractive index of the air causes shimmer of the image which may reach an angle of several minutes of arc (see Riggs et al., 1947).

Intermediate Case

The case of objects (or sources) not points but of small dimensions is, as always, more difficult than the extreme cases. The case of sources seen at night was studied by Lash and Pridaux (1943). Sources seen in daytime through fog bring out the problem of ground lighting in the landing of airplanes. It has aroused theoretical research by de Boer (1951), Tassëel (1951), and Olivier (1954). The inherent contrast C_0 is very large (high luminance of the sources), which multiplies the limiting distance of vision by a constant ratio whatever the thickness of the fog. Therefore, ground lighting is as useful in the daytime as at night. A similar problem is that of the illumination of automobiles, but as it is further complicated by glare we will study it in Chapter 17.

Vision in Water

The problems of vision in the sea (diver, deep-sea diver, submariner, etc.) present many similarities to a hazy atmosphere. For this reason we will say a few words about it. The principal differences are:

1. Pure sea water collected far away from the coast and carefully filtered possesses the same transparency in the visible spectrum as distilled water. The maximum of transparency is at $\lambda = 0.48\ \mu$ (Jerlov, 1951) and for this value of λ the absorption can be as low as 2% per meter of water, that is, $\sigma < 0.02\ \text{m}^{-1}$. Near the red end of the visible spectrum the absorption becomes higher, for example, $\sigma = 0.5\ \text{m}^{-1}$ for $\lambda = 0.7\ \mu$. This is a true absorption, that is, a conversion of the luminous radiation into heat.

2. Molecular scattering is negligible compared to absorption, even in the region of maximum transparency (Le Grand, 1939). It follows that the blue color of the sea is caused by an utterly different mechanism than that causing the blue of the sky. The sea is blue by absorption of the long wavelengths and not by selective scattering of the short wavelengths.

3. Near the coasts, colored pigments from algae modify the transparency of the water and its color tends toward green.

4. Finally, the particles in suspension are in general of a diameter larger than a micron (Jerlov, 1951) and the scattering that they produce is consequently almost independent of wavelength.

The result is that vision in the sea is similar to vision through a dense fog. All objects shade off into a "blue wall" and disappear at a distance that is a maximum of 200 m in a horizontal direction, according to equation (71), when $\sigma = 0.02\ \text{m}^{-1}$. But it is often less and may be only 20 m in the purest coastal waters, as, for example, in the Mediterranean Sea. As the scattering is not selective, there is no hope of improving this range by colored lights or filters. The question of underwater illumination was treated by Ivanoff (1954) in particular. He also studied (1955) the polarization of daylight transmitted and scattered through the sea. For some marine animals whose eyes are sensitive to the polarization of light, this phenomenon might be an advantage in clear waters (Lythgoe and Hemmings, 1967).

CHAPTER SEVENTEEN

Visual Task and Illumination

In our modern civilization the problems of vision through the atmosphere are declining in importance in favor of the problems of vision indoors. Simultaneously, interiors are becoming more and more artificially illuminated, either entirely or in combination with daylight. It is the intention in this chapter to review the principles upon which one can attempt to create satisfactory illumination. Of course, *lighting* is both a technique and an art of which only a rapid review will be given here and the questions that we will discuss briefly must merely be considered as a sort of psychophysiological introduction to lighting. For more details the reader can consult other texts (e.g. [70a]).

Physical Factors

When the illumination is well planned the subject is able to accomplish his required visual task (e.g., reading, sewing, working at a bench, etc.) with efficiency and comfort. In general, the sources of light which produce the illumination are not intended to be looked at directly, and it is even preferable that they be placed as peripherally as possible in the visual field to avoid glare. It is only light reflected or scattered by the surfaces of objects in the visual field which assists efficiently in a visual task.

One can imagine two extreme cases, the *mirror* and the perfectly diffusing *mat* surface. From a mirror there is only regular reflection, sometimes referred to as specular, which obeys Descartes' law. The luminance L_r at a point of the surface of the mirror can be written

$$L_r = \rho_r L_s$$

where L_s is the luminance of the source that is seen by reflection from the mirror at the given point and ρ_r is a coefficient called the *regular reflection factor* of the surface of the mirror. This factor may in fact vary with the angle of incidence, the position of the plane of polarization of the incident light (if it is polarized in this plane totally or partially), and the spectral composition of the light. This factor is always less than unity.

The other extreme case is that of a perfectly diffusing surface. Its luminance L_d at a given point is then independent of the direction from which it is observed and only depends upon the illumination E which reaches the surface at that point. We then write

$$L_d = \frac{1}{\pi} \rho_d E \tag{72}$$

where the coefficient ρ_d is the *diffuse reflection factor* of the surface. This factor depends upon the spectral composition of the incident light and is always less than unity (except in the case of a surface covered with a fluorescent paint). Of course, coherent units must be employed in equation (72) and the illumination E could, for instance, be expressed in lux (lumens per square meter) and the luminance L_d in candelas per square meter.

Actually, all bodies behave both like mirrors and like mat surfaces and must be defined by two coefficients ρ_r and ρ_d, both of which can depend upon the direction from which they are observed. As an example, Table 29 contains several numerical values of different papers. These data are borrowed from the measurements of Luckiesh and Moss [34].

It is easy to see that under certain conditions of lighting a print with glossy black ink on mat paper may appear as a "negative" and if the paper is glossy it may disappear completely in the reflection of the

TABLE 29

Regular Reflection Factor ρ_r and Diffuse Reflection Factor ρ_d

	ρ_r	ρ_d
Mat black paper	0.0005	0.04
Mat white paper	0.003	0.77
Newspaper	0.004–0.007	0.68–0.73
Semiglossy paper for printing	0.007	0.75
Fairly glossy paper for printing	0.008	0.78
Semimat white photographic paper	0.014	0.83
Very glossy white photographic paper	0.048	0.83
Black printing		
ordinary black	0.009	0.027
glossy black	0.017	0.019
very glossy black	0.039	0.016

specular image. Similar considerations apply to the keys of a typewriter. In such a case the reflections are uncomfortable. On the other hand, in some cases they can be of some help; for instance, a worker using a machine tool may find them useful in appreciating the geometrical shape of objects.

Determination of Illumination

If we temporarily leave aside this question of reflections and that of color, the performance of any visual task consists of distinguishing contrasts and details. It therefore brings into play two distinct faculties, intensity discrimination and visual acuity. Actually, because intensity discrimination depends upon the dimensions of the test, and acuity upon its contrast, these two faculties are only extreme cases of the same function that we could call *visual discrimination.* It involves both the smallness of the details to be perceived and their contrast in relation to the background on which they are seen.

The only precise measurement one can make is to determine the *threshold* of visual discrimination. Observers are presented with a test of given contrast and size. The experiment consists of determining the amount of illumination necessary on the plane of the test to obtain a proportion (determined beforehand) of correct answers. Experiments of this nature have been described previously (regarding acuity, for example). But what characterizes the studies of visual discrimination, carried out for industrial purposes of determining the required illumination, is the notion of *efficiency* of the worker and consequently the achievement of a certain task in a given time. Thus the measurements of visual discrimination include limited exposure time of the test, with a forced response at the end of each presentation.

Among the research in this area, besides the studies of Ferree and Rand (1927–1928) (cited in Chapter 5), we will mention the work of Fortuin and Balder (1955), who used Landolt rings as a test, and that of Blackwell (1955) who employed a circular test of apparent diameter u, luminance L', and exposure time t, projected on a uniform background of lower luminance L. Blackwell defines the contrast of this test by the expression

$$C = \frac{L' - L}{L}$$

which assures a positive value of C as opposed to the more common expression utilized in Chapter 5. The angle u varied from 8 to 51′, the exposure time from 0.001 to 1 second. It is surprising to notice that the longer the exposure time the greater the relative contrast necessary to increase the probability of correct answers. Whereas with an exposure time of 0.001 second the minimum contrast necessary to obtain 50% correct answers must be multiplied by 1.22 to obtain 75% correct answers, by 1.43 for 90%, by 1.55 for 95%, and by 1.77 for 99%, for an exposure time of 1 second these same ratios become 1.33 for 75%, 1.63 for 90%, 1.80 for 95%, and 2.13 for 99%.

These measurements of threshold (Fig. 90) confirm what we have already said, namely, if glare is avoided and if the eye is adapted to the level of the surrounding luminance, visual discrimination increases constantly with luminance. There is no optimum, the more illumination the better one sees. The result is that the level of illumination to be provided is based only on economic considerations. All that physiological optics can do is to set *minima* below which the task is no longer possible. However, it would be unreasonable to adopt these threshold values; the task would perhaps be possible at the beginning but it would soon become very tiring. How can this difficulty be solved? Several methods have been proposed:

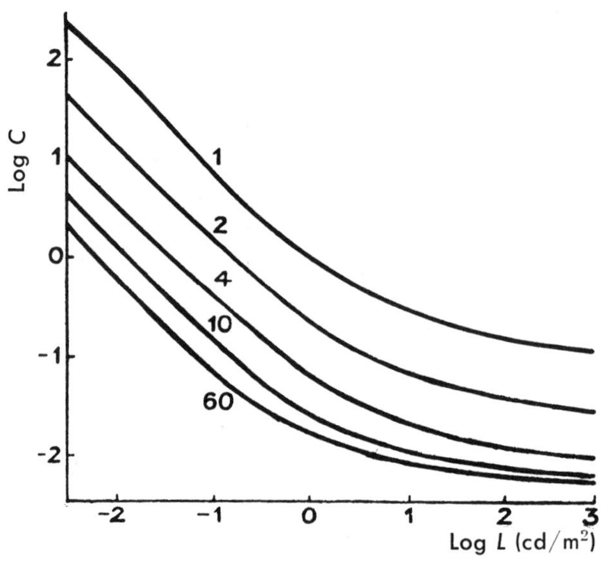

FIG. 90. Extrapolation of the measurements of Blackwell: 1-second exposure of a circular test of which the apparent diameter (in minutes of arc) is noted along each curve. The probability of correct answers is 50%. Abscissa: Decimal log of the background luminance. Ordinate: The contrast of the object compared to the background.

1. The threshold can be determined in the laboratory, then multiplied by a large "coefficient of safety," of the order of 100, for instance. This is on the whole the method utilized by Luckiesh and his collaborators [32,33,35], but the determination of the coefficient is quite arbitrary.

2. Subjects can be placed in various real conditions and they may be asked their preference. Tinker (1944) found on 30 subjects that the "best" illumination for reading varied between 100 and 850 lux. Adaptation prior to the test plays an important role in this type of experiment.

3. One can modify experimentally the level of illumination in a workshop and assess the repercussions on the work accomplished. Many studies of this nature have been carried out (D. P. Hess and Harrison, 1923; J. E. Ives, 1924; Weston and Taylor, 1935; Howell, 1948, etc.). They show that, generally, production is increased and errors are reduced as the illumination becomes greater, but it is impossible to quantify this effect because of psychological factors which play a greater role in production than illumination.

4. A variation of the preceding method was suggested by Beuttell (1933) and carried out by Weston [72]. It utilizes the notion of "*performance.*" The subject is given a sheet of paper on which is printed a number of Landolt rings of such dimensions that the gaps subtend angles between 1 and 6′ and of contrasts varying between 0.28 and 0.97. At first the illumination is 5000 lux. The examinee is instructed to cross out all the rings of a given orientation. The examiner counts the number of rings correctly crossed out after 1 minute. The test is repeated with a lower level of illumination and by definition the "relative performance" is equal to the square of the ratio of the numbers of rings correctly crossed out in the second and first experiment (of course, with the same dimension of the rings and same contrast; a correction is made to take into account the motor time necessary to cross out the rings, so that only the visual processes are taken into consideration).

When arbitrarily taking the value 90% for the "relative performance" one finds that the product LCu^3 is almost constant (L, luminance of the background in candelas per square meter; C, contrast; u, angle subtended by the gap of the Landolt ring, in minutes of arc) and possesses a value of about 200 for subjects between 30 and 40 years of age. The constant increases appreciably with age (Weston, 1949), principally due to the decrease in the speed of perception. It is from these results that the British and French codes have been calculated. They recommend the industrial values of illumination as functions of the fineness and the contrast of the visual task required by the worker. In the United States, where the cost of electricity is not so large a part of the general expenses of a factory, the values adopted are higher and correspond to approximately 95% for the relative performance.

5. Finally, along with Blackwell (1955), on the basis of the measurements of the threshold described above, one can calculate another type of "performance" by the product of the speed and the precision, the speed being inversely proportional to the exposure time and the precision directly proportional to the percentage of correct recognition. The results obtained show a much more rapid variation with luminance than in the measurements of Weston, but it is an entirely different matter. The method used by Blackwell has given rise to numerous interesting developments, described in an article by Blackwell and Smith [80] to which I refer the reader.

Visual Tasks

In general, the real tasks of workers bear very little resemblance to the operation which consists of crossing out Landolt rings or discriminating a circle on a background. It is therefore prudent to make sure that the visual functions of each worker are adequate to accomplish his required task, besides the very schematic determination of the minimum level of illumination as we have just described above. According to Kuhn [27] the essential tests are the following: near and distance visual acuity with and without correction (some caution must be applied to presbyopic workers, who often wear their near correction even to look in the distance, quite unaware of the blur of the image); far and near muscular balance tests; and color aptitude tests. A field test would also be useful, particularly with older subjects. Several systems have been devised to gather a series of visual screening tests in one instrument. One of the most common in the United States is the *Ortho-Rater*, which tests at optical distances of 8 m and 33 cm. It should be useful to examine all workers when hiring or changing posts, particularly those whose tasks demand a certain visual ability or involve danger to others.

It is essential that in this type of visual test the subject be placed in his working environment. Tiffin (1942) reported amusing errors regarding this. For example, in a textile factory, the best women workers were those who had the lowest visual acuity; they had been tested at 6 m, whereas their visual task was at 20 cm, which was much better for the myopes. It is also often noted that the workers with poor vision have production superior in quantity and inferior in quality to those with good vision because the latter correct their products themselves, which improves their work but slows it down. In the case of dangerous

TABLE 30
Minimum Visual Standards

	DISTANCE ACUITY		NEAR ACUITY		MUSCULAR BALANCE		Depth perception
	Mono.	Bino.	Mono.	Bino.	Distance	Near	
Clerical workers	0.7	0.8	0.8	0.9		normal	
Vehicle operators	0.8	0.9	0.6	0.7	normal		normal
Production inspectors	0.8	0.9	0.6	0.7		normal	
Machine operators	0.7	0.8	0.7	0.8		normal	
Carpenters, plumbers, electricians, etc.	0.7	0.8	0.8	0.9	normal	normal	normal
Laborers	0.7	0.8					

occupations the percentage of accidents increases by more than 2% in workers with poor vision, who are very numerous; 40% of the workers can improve their vision with spectacles [27]. Table 30 contains the minimum visual standards that are recommended in the U.S. according to the type of occupation. Normal muscular balance is 1.5 prism diopters as a maximum for hyperphoria and less than 3 for esophoria; exophoria must be less than 5 prism diopters at distance and 10 at near. Normal depth perception corresponds to a stereoscopic acuity of 20″ or less.

With automation there is a visual task which has become increasingly important, that is, the reading of scales. This was studied particularly by Sleight (1948) and by Grether and Kappauf [in 76]. They found that rectilinear scales (e.g., mercury thermometer) produce the maximum number of errors, whereas the minimum is obtained with a system of direct reading (e.g., mileage meter in automobiles); the maximum speed of reading is obtained with a circular dial (3.5 to 7 cm diameter) intended for near viewing; the minimum of errors is obtained for a separation of 1.5 mm between two adjacent scale marks with a longer one every 5 and a number every 10 marks. The needle must not hide any number. The precision of the interpolation between two scale marks was studied by Grether and Williams (1947) and by Leyzorek (1949). Lazet and Walraven (1959) compared the reading of linear and logarithmic scales and offered an explanation for the characteristics of these interpolations. In the case of apparatus with several dials it is preferable that at zero all the needles be aligned (horizontally or vertically; see Johnsgard, 1953).

The piloting of aircraft at night presents a special problem of illumination of the numerous control dials [8]. It is better to illuminate them in red or to use a red fluorescent paint with ultraviolet lighting, because the peripheral retina is not very sensitive to long wavelengths when dark-adapted. It is not advisable to use green fluorescent or phosphorescent paints, which appear blurred and produce a loss of adaptation. The optimum luminance is about 0.02 cd/m^2, but the ideal would be an adjustable system varying from 0.1 to 10 times this value.

The reading of scales during vibration of the aircraft is a practical case of much importance which has been studied by Crook and his collaborators [in 65]; for a constant amplitude, the impairment increases with the frequency of vibration.

Comfort of the Illumination

Well-planned illumination alone does not permit the accomplishment of a visual task; this task must also be carried out under the best conditions of well-being, to which is given the name *visual comfort*. This is a subjective concept which seems difficult to define at first glance; nevertheless many studies arrive at conclusions very much in agreement, as we shall see.

Everyone agrees that the absence of *glare* is a primary requisite for visual comfort. Unfortunately, this word glare covers a multitude of phenomena, all different, which have only one thing in common—the presence of a source in the visual field which is very intense, not in absolute value, but relative to the luminance of the surround in the visual field. First, one must distinguish between *simultaneous glare*, which is due to the coexistence at any one moment of very different luminances, and *successive glare*, which is a temporal effect due to the change in luminance. Let us put this latter type of glare aside for the moment. With simultaneous glare one can first distinguish *disability glare*. This causes a deterioration of the visual functions as a consequence of the intensity of the source. A large amount of research has been performed on the reduction in contrast discrimination caused by glare and it has been shown that it was the same as if a luminous *veil* were to be spread over the visual field, a veil of *equivalent luminance* defined by a simple mathematical expression as proportional to the illumination given by the glare source on the pupil of the observer, and inversely proportional to the square of the angle between the glare source and the test. This veil (also referred to as "veiling glare") is due partly to scattered light within the eye and partly to some nervous interaction [69a]. Visual acuity decreases, in fact, along with intensity discrimination, as Haas (1913), Sheard (1936), and many others have shown.

This measurable disability that is produced by glare plays an important role in certain cases such as night driving, but in any lighting installation worthy of this name it is not permissible. On the other hand, certain apparently satisfactory installations may in the long run produce a subjective discomfort that cannot be measured but that is a cause of disagreeableness and sometimes fatigue. This is what is called "*discomfort glare.*" Illuminating engineers have thought of several empirical methods to evaluate the comfort and discomfort of an installation, which we shall now summarize.

1. Luckiesh and Holladay (1925) were the first to attempt to establish a subjective scale of visual comfort. Then Luckiesh and Guth (1949) resumed these experiments in a systematic manner. The subjects were placed within a scattering sphere of uniform luminance L_0. At one point of the sphere was placed a circular source of solid angle Ω and luminance L. This luminance L was adjusted until the subject decided that he was at the "limit of comfort and discomfort." When the luminances were evaluated in candelas per square meter and the solid angle in steradians, 50% of the subjects decided that they reached the limit when the following relation was satisfied:

$$L = 215PL_0^{0.44}\,(\Omega^{-0.21} - 1.28) \tag{73}$$

The coefficient P is a factor of position which is equal to unity when the glare source is looked at directly by the subject. In the case where the subject gazes horizontally and the source is placed in a plane perpendicular to the line of sight at a distance x from the subject, the values of P are given in Figure 91. The source is situated at a horizontal distance y and a vertical distance z from the fixation point. Observe that for a constant angular separation from the fixation point a source produces more glare in the upper part of the visual field than laterally. On the other hand, equation (73) shows that it is not the product $L\Omega$ (that is, the illumination on the eye of the subject) which is responsible for the discomfort. Small sources with a high luminance are more disagreeable than large sources of low luminance. These two conclusions, in contradiction to the laws of disability glare, reveal the difference between these two types of glare.

2. Moon and Spencer (1945) introduced the notion of *delos*, a type of visual damage caused by glare. But there is here a combination of disability (particularly the reduction in visual acuity) and discomfort which is not a happy one.

3. In a series of researches, Logan (1939–1947) was guided by the curious idea that humanity was made to live in a tropical zone where one could go without clothes, and that the eyes must be normally

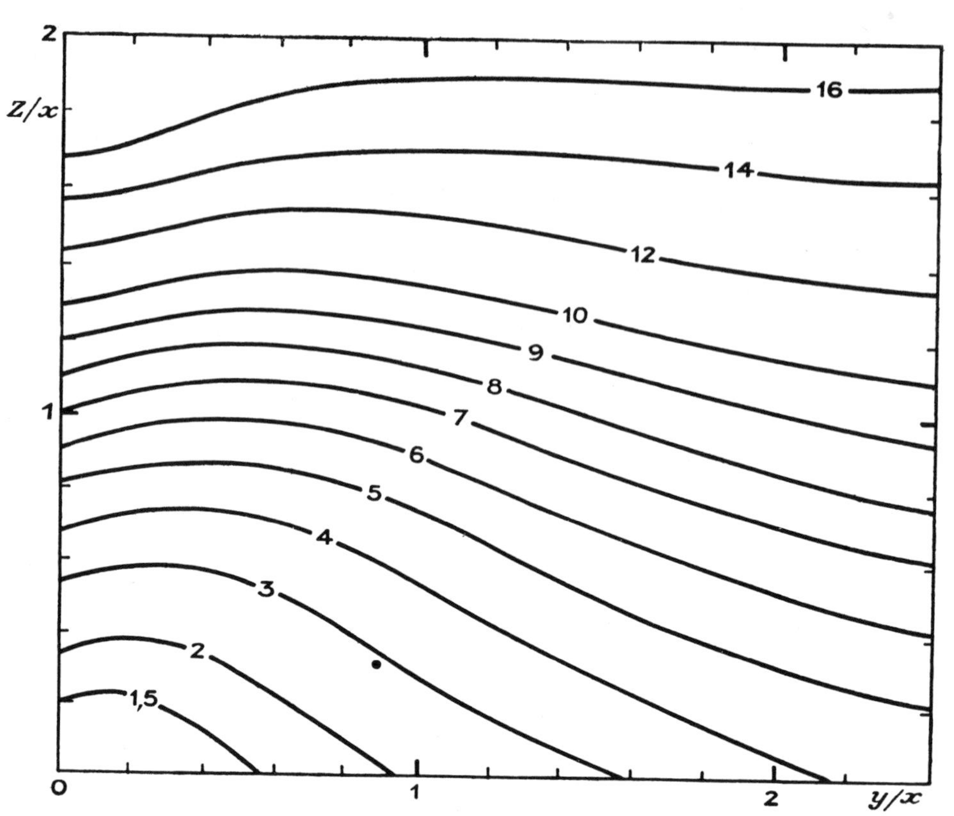

FIG. 91. Factor of position P (Luckiesh and Guth). The source is at a height z and at a lateral separation y from the point of fixation, which is situated at a horizontal distance x from the observer in a frontal plane passing through the source.

adapted to the luminances of this area of the earth. He measured the average distribution of the luminous fluxes in diverse regions of the visual field which he divided arbitrarily as central zone ($\eta < 30°$), binocular zone ($30° < \eta < 60°$), and monocular zone ($\eta > 60°$). For comfort the fluxes in these regions had to be, in relative value, between certain minima and maxima. These ideas apparently had very little success. Logan later modified them (1952) to bring them into harmony with subsequent studies.

4. Harrison (1945) introduced the notion of the *glare factor,* defined by the expression

$$G = 10^{-9} S L^2 z^{-2} p f^{-1} \tag{74}$$

where S is the surface area of the source of glare in square centimeters

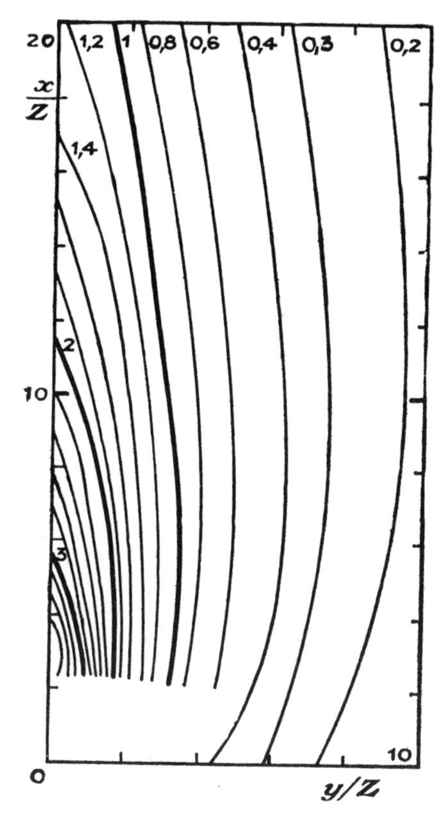

FIG. 92. Factor of position P (Harrison). The letters x, y, and z have the same meaning as in Figure 91.

(or, rather, its apparent surface area on a plane perpendicular to the line joining the observer to the source), L its luminance in candelas per square meter, z its height in meters above a horizontal plane passing through the eyes of the subject, P a factor of position (Fig. 92), and f a factor depending upon the luminance of the background near the source and the general luminance of the room (Table 31).

The advantage of the glare factors G is that they are additive if several sources are present at the same time in the visual field. If the sum of the values of G is less than 8, there will be no criticism of the installation; between 8 and 12 some subjects will complain at the end of a long task, between 12 and 20 is still satisfactory if the task does not exceed 1 hour, up to 30 is still permissible in a workshop, and values up to 80 can still be tolerated in a place where workers move about a lot (Harrison and Meaker, 1947). In spite of the difference in methods,

TABLE 31

Values of the Factor f *(Harrison)*

General luminance cd/m²	LUMINANCE OF THE BACKGROUND NEAR THE SURFACE, cd/m²						
	0.1	1	10	30	100	300	1000
0.3	0.05	0.09	0.15	0.19	0.22	0.25	0.28
1	0.07	0.15	0.26	0.33	0.39	0.46	0.53
3	0.10	0.19	0.37	0.50	0.63	0.75	0.87
10	0.17	0.27	0.51	0.74	1.0	1.2	1.5
30	0.29	0.41	0.68	1.0	1.5	2.0	2.5
100		0.67	1.0	1.3	2.0	2.9	3.9
300			1.6	2.0	2.7	4.0	6.0
1000				3.2	4.1	5.4	8.0

Harrison's formula for $G = 30$ arrives at results in good agreement with those of Luckiesh and Guth for the usual cases.

5. Vermeulen and de Boer (1948) have performed measurements similar to those of Luckiesh and Guth with diverse forms (squares, lines, etc.), on 12 subjects. However, the empirical formula that they recommend deviates markedly from the preceding results.

6. Petherbridge and Hopkinson (1950) employed small models of lighted rooms. The glare sources were kept at a fixed level, whereas the background luminance was adjustable by the subject. He had to determine four levels of luminance corresponding to four different judgments of glare; just perceptible, just acceptable, just uncomfortable, and just intolerable. If very nervous subjects are avoided, this subjective choice remains reasonably constant for each observer even during long periods of time. But from one observer to another there are systematic differences, some subects being more sensitive to glare than others, although no correlation either with age or other personal factors was found (Hopkinson, 1950–1952; Guth, 1951). The authors represented their results by a factor of additive glare of the form $\Omega^{0.8}L^{1.6}L_0^{-1}$, the letters having the same meaning as in expression (73).

Attempt at a Synthesis

A synthesis of the bulk of the preceding research (and some other research of less importance) was attempted in a report from the committee on "estimation of the comfort of illumination" at the Congress of the Commission Internationale de l'Eclairage in 1955. The glare factor which was then proposed is

$$G = L^2 L_0^{-1} \Omega P^{-2.3} f(L_1) \tag{75}$$

where L is the luminance of the source, L_0 the mean luminance of the background, L_1 the luminance very close to the source, Ω the solid angle subtended by the source, P the factor of position (from Luckiesh and Guth; Fig. 91), and f a function still not very well known (for the time being we will take it as equal to unity). The units of the factor G are those of illumination; G is additive. One will find in reference [80] an analysis of research done along this line.

A different formula was proposed by Cadiergues (1953). It involves the surface area of the subject's pupil, which from a physiological point of view seems judicious, but it is feared that the illuminating engineers may prefer an expression similar to equation (75) which will lend itself more easily to their calculations.

The relative difficulty of the formulas studied has encouraged the illuminating engineers to attempt to extract simple rules, for instance, to limit to a value between 5 and 10 the ratio of the highest luminance to the average luminance of the visual field. Unfortunately, rules of this kind are too simple to be truly useful.

It is actually uncertain whether uniformity of the luminance is ideal. With regard to this matter the experiments of P. W. Cobb and his collaborators (1913–1916) and of Lythgoe (1932) are often cited. According to them visual acuity is better when the task is surrounded by a field of uniform luminance. But Fisher (1938) noted that maximum acuity was often obtained when the background luminance was less than that of the task. What must be avoided, though, are excessive contrasts and sharp gradients of luminance; a harmonious relationship is preferable. Generally speaking it is better to replace the contrasts of luminance by color contrasts with very little difference of luminance. Of course, in industry a choice of light colors of the premises and machinery is beneficial, as it increases the level of luminance and improves the psychological environment by rendering it more cheerful (see Déribéré, 1955).

More and more, the idea of visual comfort deviates from the negative concept of discomfort to include positive elements of aesthetics; there is here for the illuminating engineer of tomorrow a large field of research where technology and art will combine (see, for example, Kalff, 1952).

Choice of Artificial Light Sources

The sources most commonly used nowadays are incandescent lights and fluorescent tubes. It is not within the scope of this book to compare these sources, but a few words should be said regarding the physiological problems that arise from them.

First, the essential difference between these sources is the color temperature. It is low for incandescent light whereas fluorescent light enables a large range of emissions more or less close to white. The choice of the color temperature as a function of the level of illumination has been considered (Kruithof, 1941; Wald, 1953). It is usually agreed that fluorescence is not advisable at low levels, less than 50 or 100 lux, because the impression which it produces is dull. The problem of the reproduction of colors by different sources has stimulated much research of which a summary is given in the reports of the Commission Internationale de l'Éclairage (1955, 1959, 1963).

At its beginning, fluorescent lighting provoked sharp criticisms. Most of them are not well founded but two are still occasionally discussed.

1. A slight proportion of ultraviolet light emitted by the mercury vapor lamp passes through the tube and may irritate the eye. This is perhaps true for tubes placed near the eye of the subject, but beyond 1 meter the energy which passes through the tube is so weak that it is very unlikely to have any effect on the conjunctiva (Latarjet, 1950). However, in all adequate installations the distance between the sources and the subject is always greater than this limit. Moreover, Wolf (1946) thought that ultraviolet light might slightly impair dark adaptation, but a systematic study by Sexton, Malone, and Farnsworth (1950) proved that reading for 1 hour, the luminance of the paper being 70 cd/m^2, affected subsequent dark adaptation in the same way whether the illumination was produced by an incandescent lamp, white fluorescent tube, or a special fluorescent tube with a filter eliminating the ultraviolet. Later studies by Wald (1952) have in fact produced some doubt about the truth of the phenomenon described by Wolf (or at least it occurs only at reasonably high levels of ultraviolet light when fluorescence of the crystalline lens comes into play; see Ogilvie and Ryan, 1955).

2. The use of alternating current of 50 or 60 cps produces a luminous emission at a frequency of 100 or 120 for both incandescent and fluorescent tubes. But the thermal inertia of the filament generally reduces the periodic fluctuations of the luminance more than for fluorescent tubes. This *flicker* has been held responsible for the discomfort felt by some subjects with fluorescent lighting. This does not seem very likely for a frequency of 100 or 120 cps which is beyond the critical fusion frequency, but some installations present an asymmetry, thereby producing a component having a frequency of 50 or 60 cps, which is perceptible. The research of Collins and Hopkinson (1954) and of Silber (1955) show that observers differ very much in this respect; a certain amount of flicker is hardly seen by some, but is intolerable for others. Therefore this flickering at 50 or 60 cps should be eliminated as much as possible by employing phosphors with long-lasting emission and eventually by mounting tubes in groups of two or three with differences of phase such that the resulting light is almost constant. However, according to Floyd (1955) the criticisms based on flicker are generally not well founded because changing the alternating current of the fluorescent tubes to direct current does not introduce any measurable physiological difference.

Visual Fatigue

In all practical studies of illumination the lack of an objective criterion for assessment of an installation is a deficiency, as the brief discussion above may indicate to the reader. An attempt has been made to remedy this difficulty in the following manner: If visual comfort is subjective, would not visual discomfort at least be converted into measurable fatigue?

In physiology the notion of fatigue cannot be reduced to simple elements. A difficult and prolonged task produces a disagreeable sensation which gives the subject the impression that if he continues, the quality and speed of his work will be impaired; but it is often the opposite which happens, and Muscio (1921) thought that the word "fatigue" should be excluded from the scientific vocabulary. Since that time, very diverse and contradictory concepts of fatigue have been presented [9]. It seems certain that the psychological factors outweigh the physiological ones. For example, Löwenstein and Friedman (1942) stimulated the pupillary reflex until it disappeared, which gives the impression of exhaustion; but if the subject is warned that when the light reappears some pain will be felt, the pupillary contraction occurs again.

Tests of visual fatigue have been sought for a long time but none is satisfactory. The speed of eye movement has often been proposed but after 6 hours of difficult reading, it is found that the speed has hardly reduced (MacFarland et al., 1942). It is slightly more marked for the reserve of convergence or the amplitude of accommodation which is linked to it, but recovery is rapid and 10 minutes after the end of a visual task that the subject thought exhausting the modifications are hardly perceptible. Luckiesh [35] proposed as tests of visual fatigue the frequency of blinking, the reduction in the heart rhythm, and the increased muscular and nervous tension, but this is now doubtful (Bitterman and Soloway, 1946; Tinker, 1946; Wood and Bitterman, 1950) or at least more elaborate methods would be necessary (Ryan et al., 1948–1950). The critical frequency of fusion—that is, the frequency beyond which a flickering light seems steady—reduces slightly with visual fatigue, according to several authors. A critical study shows that here again psychological factors frequently mask the fatigue (Ryan et al., 1953).

The idea that visual fatigue results in a reduction of the visual functions has often been proposed but without very convincing proof. For example, Ferree and Rand (1931) devised a special test of acuity with which they could evaluate the fatigue caused by various colored lights, but Peckham and Arner (1952) demonstrated that the reduction in acuity even after an arduous visual task is not significant.

According to Simonson and Brozek (1948), the only systematic variation would be that of the "performance" itself. They imposed upon their subject a tedious task consisting of copying letters 10′ in height which appeared behind a slit at the frequency of about two per second. This lasted for 2 hours without any stop and the performance was measured at the beginning, in the middle, and at the end of the task (performance was defined by the proportion of letters copied correctly during a period of 6 minutes). The frequency of blinking, the critical fusion frequency, the acuity, the speed of the eye movements, the amplitude of accommodation, and the differential threshold remained unaffected by the fatigue of the subject (except perhaps the critical fusion frequency), or on the contrary improved as the test was prolonged. Only the "performance" decreased systematically, but in reality it does not constitute a measure of visual fatigue and depends upon a considerable number of factors, most of them psychological. Incidentally, let us mention here that from their measurements Simonson and Brozek arrived at an optimum illumination of about 1000 lux but, as Moon and Spencer (1948) have pointed out, this was due simply to the non-uniformity of the luminances in their experiment.

When the light becomes very inadequate, tasks which demand keen acuity (reading, for example) rapidly produce a particular fatigue which is probably caused by the ciliary muscle. The lens oscillates constantly, searching for the best adjustment; hence, a sensation of strain such as the reported "pulling of the eyes." These are the microfluctuations of accommodation which become amplified when the light decreases. A similar fatigue occurs when a presbyopic subject fixates a point at his punctum proximum. But these particular cases where a true *ocular fatigue* takes place leave untouched and unresolved the problem of visual fatigue in general (see for example, Weston, 1953).

Automobile Lighting

To close this chapter we will say a few words on the important problem of the lighting of automobiles at night, because this is a good example of conditions of glare totally different from those that take part in the discomfort studied above. When a driver is dazzled by the glare from the headlights of oncoming cars, there occurs a first phenomenon, which is a type of visual *shock* due to the sudden appearance of the intense light in an otherwise dark visual field. This discomfort, sometimes painful, probably arises from a spasm of the pupillary muscles occurring to protect the eye because the retina itself does not seem to feel any painful sensations. This pupillary contraction has by itself a regrettable effect, because it diminishes the retinal illumination in the region already in the dark and therefore hinders the visibility of obstacles.

A second effect is the disability caused by glare that we discussed earlier, which also alters the perception of potentially dangerous objects (pedestrian, cyclist, or parked truck on the side of the road) because of the reduction of contrast discrimination of the eye (acuity plays only a small role in night driving).

Other complex phenomena may happen which are referred to by the term *successive glare* and which affect vision after the oncoming car has passed. First there appears a "black hole," a type of temporary blindness which is perhaps of central origin. It is followed by adaptation effects of which some are retinal and correspond to a slow recovery of visual functions in the region of the visual field most affected by the headlights.

The complexity of these effects makes the exact analysis of glare in automobile driving almost impossible. It is not the lack of data, however, as they are numerous. In France, the law requires that all headlights have a yellow lamp which eliminates short wavelengths from the emitted light. A diminution of the flux of the order of 8% results if one adopts the factors of photopic relative luminous efficiency, but in the opinion of the legislator this reduction is compensated for by less glare. Actually, the experiments are not very convincing; in particular, disability from glare, which is the only directly measurable effect, does not seem to be reduced by yellow light if we consider the test of intensity discrimination, for example. The acuity (which plays only a small role) is improved very slightly. Nevertheless, in France, the drivers are satisfied because the impression of the shock is less and the effects of successive glare subside more rapidly. It is probable that this apparent contradiction arises from the fact that the tests of disability involve cone vision and the other effects apply mainly to the rods, which are essential in peripheral vision at low luminances. However, for the rods the diminution of light due to the suppression of the short wavelengths is much more than 8%, and this would explain the paradox of a solution which seems good in practice to French drivers but of which they cannot convince other countries experimentally.

CHAPTER EIGHTEEN

Applications of Binocular Vision

In this last chapter we shall review some of the applications of binocular vision; here again the subject is so vast that we shall only be able to indicate the existence of the principal problems, the role played by the physiological factors, and the possibilities of practical usage. For more details we refer the reader to some excellent works that will be mentioned during the course of this chapter.

Stereoscopes

It is possible to fuse two plane images without any instrument, by placing them side by side so that the separation between corresponding points is not greater than the interpupillary distance of the subject. However, this very simple method usually requires a disassociation of the relationship between accommodation and convergence, which is easy only for myopes who remove their spectacles. It is generally advantageous to use a *stereoscope*. Discovered in 1833 [60], the original instrument was described by Wheatstone (1838). It was made up of two plane mirrors inclined at 45°, one in front of each eye. This arrangement is not very practical (it requires inverted figures) and it is nowadays only utilized in the *haploscope* of Hering, a laboratory instrument which is used to study binocular vision. Brewster (1843) designed a stereoscope with prismatic lenses which combined a prismatic effect and a magnification of the images. This instrument was widely distributed by the Parisian opticians Duboscq and Soleil; it enjoyed an extraordinary success, and over a century ago more than a half-million of them had been sold! At some later date a small stereoscope designed from an ingenious idea of Swan (1863) using total reflection (Fig. 93) became available on the market.

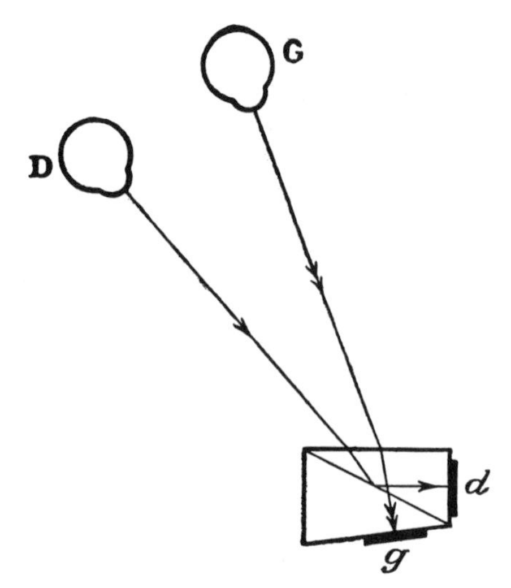

Fig. 93. Diagram of the stereoscope of Swan. The right eye *D* sees the image *d* (inverted) by total reflection on a thin layer of air separating the two parts of the prism, whereas the left eye *G* sees directly the image *g* (upright). These images are small photographic reproductions attached to the sides of the prism.

It was Helmholtz (1857) who modified the stereoscope of Brewster to its present form by incorporating achromatic lenses as eyepieces instead of prismatic lenses. He invented the *telestereoscope* (Fig. 94), which enhances the stereoscopic effect by increasing the distance between the two points of view so that it is considerably larger than the interpupillary distance. This is also the principle of the stereoscopic telemeter [40]. The *iconoscope* of Javal (1866) produces the opposite effect; it suppresses stereopsis although vision is still binocular. It enables the observer to see paintings in a more plastic manner by reducing the flattening caused by a localization at close distances. The *pseudoscope* reverses the stereopsis by presenting to the right eye the image that should be seen by the left eye, and vice versa. The original apparatus of Wheatstone (1853) consisted of two prisms with total reflection which inverted the half-view of each eye, but it is better to use mirrors (Fig. 94) as Stratton (1898) suggested. A pseudoscopic effect is obtained easily with geometric figures or medals, the impression of which becomes concave. It is more difficult with landscape and impossible with a familiar object such as a human face. The subject feels that the impression is strange but does not sense the reversal in the stereoscopic effect, which shows the importance of perceptive factors in this instance (Eaton, 1919). With an ordinary stereoscope one can easily

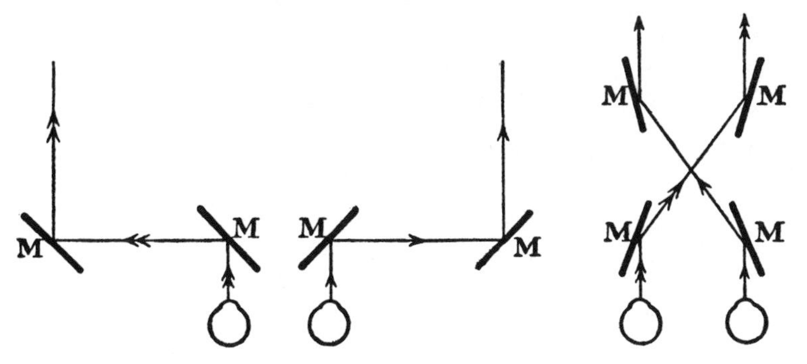

FIG. 94. Principle of the telestereoscope (left diagram) and of the pseudoscope (right). *M*, mirrors.

experience the effects of the iconoscope (two identical images) and of the pseudoscope (interchanged images). Edridge-Green (1920) showed in this way how one could test with certainty for of stereopsis with two stereoscopic half-images. To obtain stereoscopic images, the most usual way is by means of photography (Quételet, 1841), but one can enjoy constructing drawings by the method of perspective geometry (see, for example, Sauer, 1937).

The finesse of stereoscopic perception due to very small disparities enables one to assess certain differences between two given patterns that would not be perceived otherwise. For instance, two editions of the same text, or a genuine bill and a false one can be differentiated in a stereoscope; some of the letters or patterns seem to appear in front or behind the document. Comas Solà (1915) utilized this same principle to reveal the proper motions of the stars or small planets on two photographs taken within a certain interval of time; one sees in front or behind the plane of the fixed stars any body which has changed position in the interval of time. This method is open to many errors, analyzed especially by J. Rösch (1943). Rösch studied particularly the binocular vision of photographic spots other than points and showed that they were fused on the "barycentres optiques," that is, on the centers of gravity of the logarithms of the luminances, conforming to the classical Fechner's law. This same principle is applicable to certain problems of heterochromatic photometry (A. Monnier and J. Rösch, 1946).

Anaglyphs

Instead of separating geometrically the images to be viewed by each eye, the stereoscopic effect can be obtained by the use of color or polarization. For example, if on a white paper two stereoscopic drawings, one in red and the other in blue, are superimposed and the observer puts on red and blue goggles, the eye with the red glass sees only the blue drawing (which appears black on a red background) and the other sees the red drawing (which appears black on a blue background). This method was conceived by Rollman (1853) but the first realization was by d'Almeida (1858), who projected the pictures onto a screen by means of two projectors provided with color filters. In this case the effect is the opposite; each eye sees the image of the same color as the filter it is wearing. This principle, called *anaglyphs* by Ducos du Hauron (1891), can be used to present stereoscopic pictures to a large audience. It was tried in the early days of the cinema [16] but without much success because of the effects of binocular rivalry due to the differences of color and luminance of the two fields. To reduce this disadvantage, L. Lumière (1935) designed blue and orange filters equated for luminance and used Rollman's method, printing on the same film the superimposed images in orange and blue.

Instead of a difference of colors the lights corresponding to the two images can be polarized at right angles. The screen, however, must not appreciably depolarize the images by diffusion. This method was utilized by Anderton (1891) with stacked glass plates as polarizers, but became more practicable recently with the arrival on the market of polarized sheets, with which viewing spectacles can be made at a very low cost. Obviously the instrument must be modified so that two polarized images are projected simultaneously on the screen. The images of each eye could also be projected alternately, but to avoid flicker the rate of speed of the film, and therefore its length, must be doubled.

Parallax Stereogram

Claudet (1857) observed that an image projected onto a ground-glass plate of a camera obscura gave a stereoscopic effect when it was viewed with both eyes. This phenomenon is due to the fact that each eye sees best the ray that falls on the ground glass in the direction of its own visual axis and therefore produces a slight difference in points of view. Some 40 years later this method was repeated independently by Berthier and F. Ives [in 16]. They used a large objective in front of which there is a diaphragm with two apertures separated by a distance equal to the interpupillary distance $\tilde{\omega}$. A grid consisting of alternately transparent and opaque parallel lines of equal width d is placed at a small distance x in front of the photographic plate (Fig. 95*a*). In this fashion the paral-

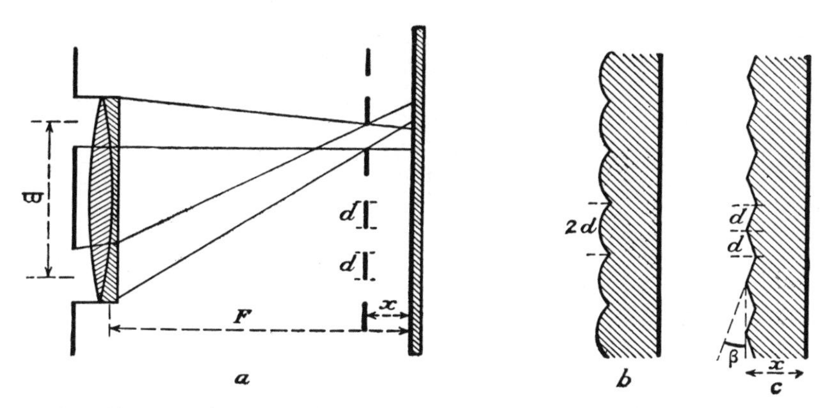

Fig. 95. Principle of parallax stereogram; *a*, sectional plan of the camera suitable for the production of a parallax stereogram; *b*, cylindrical optical light-condensing elements; *c*, prismatic optical light-condensing elements. (The ribs are actually much more numerous than is shown in this figure.)

lel lines of the grid produce adjacent rectangles on the photographic plate on which images are projected alternately from one and the other of the apertures of the objective. A positive transparency is made from the negative parallax stereogram and it is then viewed by an observer from a distance F (which was the focal length of the objective), with a grid having the same spacing as that of the grid used in the camera situated at the same distance x from the transparency. Each eye sees the part of the stereogram which is not shielded by the grid. This distance F is such that both eyes see the whole picture in stereopsis. If the observer moves laterally, a distance equal to $\tilde{\omega}$, he will experience a pseudoscopic effect and if he moves farther, a distance $2\tilde{\omega}$, a true stereoscopic effect will reappear, and so on. The positions where the stereopsis is correct are sharply defined, and to minimize this disadvantage a grid can be employed with a ratio of opaque width to transparent width of about 4 or 5 to 1. Another fault of this process is the "striped" appearance of the image. It may be bearable if the angle d/F is about 1 minute of arc.

This method has been improved by substituting for the grid and the plate a number of optical light-condensing elements of cylindrical (Fig. 95*b*) or prismatic (Fig. 95*c*) shape. The ribs or lenticulations are embossed on the base of the actual film carrying the photographic emulsion. Furthermore, the stereogram is more easily produced by employing a camera with two objectives separated by the distance $\tilde{\omega}$. This principle has often been applied in projection and especially in cinematography, but the obvious disadvantage is that only a limited number of spectators can appreciate the stereoscopic effect. Two more elaborate techniques, one devised by Lipmann (1908) and known as "integral photography" and the other by Kanolt (1915) known as "parallax panoramagram," may partially minimize this difficulty. However, the complexity of these techniques has so far hindered their practical use in cinematography [16], but recently stereoscopic photographs using this method have appeared in some periodicals.

Orthostereoscopy

A stereoscopic effect is perceived when each eye sees the image which corresponds to it, but this is not sufficient to produce a correct stereoscopic effect. Equal dimensions of the reproduced object, whether in a frontal or a sagittal plane, must appear equal to an observer. This implies that the *magnification* g of the stereo image is the same in all directions (it refers to linear magnification and not to angular magnification). If this condition is fulfilled it may be said that there is *orthostereoscopy.*

The first condition of orthostereoscopy is that the angular parallax between the two images of a point must vary as a function of g in the same ratio as the visual angles subtended by the frontal lengths of the object. The frontal lengths vary as $1/d$, where d is the distance of

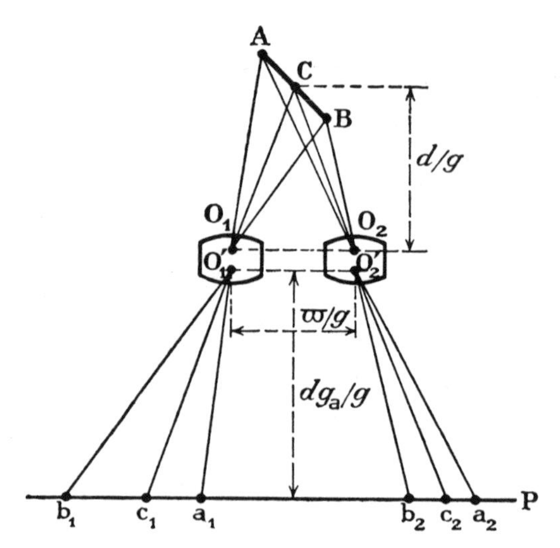

FIG. 96. Diagram of the photographic system used to take a stereoscopic picture.

the "center of interest" C of the (reconstituted) object from the observer, whereas the stereoscopic parallaxes vary as $\tilde{\omega}/d^2$. This ratio will remain constant if $\tilde{\omega}$ varies proportionately to d. Hence, the following rule: to obtain orthostereoscopy with magnification g, the object must be placed at a distance d/g from the camera and the two objectives must be separated by a distance $\tilde{\omega}/g$.

Let us consider the camera (Fig. 96). O_1 and O_2 are the first nodal points and O'_1 and O'_2 the second nodal points of its identical objectives of power D_a. The magnification of the camera is

$$g_a = \frac{O'_1c_1}{O_1C} = \frac{O'_2c_2}{O_2C} = \frac{g}{g + dD_a} \tag{76}$$

Let us now consider the stereoscope (Fig. 97). The two eyepieces have a power D_s. We shall assume that they are thin, in order to neglect the distance between the first and second nodal points. The image of the stereogram P is projected to P' at a distance d. The separation of the eyepieces is adjusted so that the center of interest C' of the image is in the plane P'. If δ is the distance from the eyes to the eyepiece, the magnification of the stereoscope is

$$g_s = 1 + (d - \delta)\, D_s \tag{77}$$

and the total magnification is the product

$$g = g_a g_s \tag{78}$$

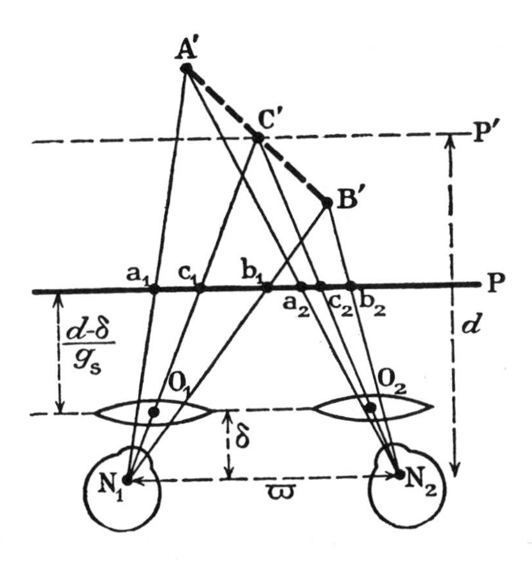

FIG. 97. Observation through a stereoscope.

Expressions (76) to (78) summarize the whole problem of orthostereoscopy. We have assumed that the separation of the centers of interest, c_1c_2, on the plate P and that of the eyepieces were adjusted so that the binocular convergence is on C'. It thus assures a better synergy with accommodation. Hence we have the following additional relations:

$$c_1c_2 = \frac{\tilde{\omega}(d - \delta)(g_s - 1)}{dg_s} \tag{79}$$

$$O_1O_2 = \frac{\tilde{\omega}(d - \delta)}{d} \tag{80}$$

In case the value c_1c_2 does not satisfy equation (79), the distance O_1O_2 can be modified to utilize the prismatic effect of the eyepieces to reestablish the correct synergy. The axes of the objectives could also be inclined slightly when taking the photograph so that they converge at C, if this will improve the quality of the image. The eyepieces of the stereoscope could also be inclined. The two pictures cannot, however, be taken on plates that are similarly inclined so as to be perpendicular to the mean rays, unless it is also possible to incline the two halves of the stereogram P in the stereoscope in order to recover the orthostereoscopy. Nevertheless, this technique would be complicated and would introduce some deformations [12a].

If we consider the static perspective (eyes immobile), the points N_1 and N_2 (Fig. 97) are the first nodal points of the eyes situated about 7 mm behind the cornea. It is, however, more customary to consider the dynamic perspective where the eyes scan the image, and the points N_1 and N_2 are then the centers of rotation situated, on the average, 13.5 mm behind the cornea.

Special Cases

Along with Kurtz (1937) we shall consider some special cases of orthostereoscopy.

1. $g = 1$, d is large. This is, for example, the case of a landscape seen in a stereoscope. The background must appear at infinity. δ is therefore negligible and the only solution of equations (76) to (78) is $D_a = D_s$. The camera and the stereoscope have equal focal lengths of any value and the objectives are separated by a distance $\tilde{\omega}$.

2. $g = 1$, d is small (less than 1 m). We can assume $g_a = g_s = 1$, hence $D_a = D_s = 0$. The photographs, which are natural size, are

viewed in a mirror stereoscope. This is the case of stereoscopic radiography.

3. $g < 1$, d is small. This could be the case in aerial photography. The subject looks at a picture of the terrain under very small magnification. Usually, however, in photogrammetry the stereopsis is exaggerated to enhance the precision of the drawing of contour maps.

4. $g > 1$. This is the case of "stereomicrophotography." A microscope with only one objective can be used, provided that between the exposures of the two pictures of the stereoscopic pair the camera is displaced by a distance $\tilde{\omega}/g$ and the photographic plate by a distance $\tilde{\omega}(1 + g_a)/g$. In general, the microscope must be used with a diaphragm to improve the depth of field. The best position of the diaphragm is the point whose images in the objective and the eyepiece are the nodal points of the complete microscopic system (Martin and Wilkins, 1937). This method is useful for viewing in depth the tracks of high-energy atomic particles in photographic emulsion.

5. Projections (especially motion pictures). This is a little more complicated because the first image is taken with an objective of focal length $f = 1/D_a$. This image is then projected on a screen with a magnification g_p (the ratio of the size of the image on the screen to the image on the film). To obtain a correct perspective the objects seen by the spectators at a distance d from the screen must subtend a visual angle equal to the angle subtended by the object when the picture was taken. Hence we have the following condition:

$$d = fg_p \tag{81}$$

In this case the separation of the objectives of the camera is $\tilde{\omega}$ (since $g = 1$). The projection should be adjusted in such a way that the objects situated at a distance d from the camera have their two images merge on the screen, but objects situated at a distance less than $d/2$ will produce marked visual discomfort for the spectator. Also, when doing a close-up, the separation between the objectives of the camera must be diminished so that the center of interest falls on the screen. This introduces a magnification g and the distance (81) must be multiplied by g. Under these conditions the stereopsis of a human head will be correct but it will appear as a giant's head, g times as large as in reality. Consequently movies in three dimensions can be orthostereoscopic only at one distance d from the screen, and furthermore the value d changes during the projection of the same film if the focal length, f, varies or if the producer does a close-up. Rule (1941) indicated a simple graphical construction to evaluate the distortions of stereopsis, which are considerable. The recent vogue in three-dimensional movies has produced many studies (see for example, Aschenbrenner, 1952 and MacAdam, 1954), which, however, is not new, since Clerc around 1920 [12a] had already described the essentials of the subject. Obviously, the screens known as "panoramic" do not constitute the solution to three-dimensional cinema, but the widening of the field helps the images to acquire depth.

Visual Instruments

Just a few words on visual instruments to end this chapter. The theory is the same, the intermediate image plays the role of the photographic plate P (see [60] and Günther, 1955). Binocular telescopes have been in use since the 17th century. Let us assume the instrument is afocal and therefore adjusted so that the focal points of the objectives and eyepieces coincide. An object situated at a distance d_a in front of the objectives gives an image which is practically in the second focal plane of the objectives, the focal lengths of which are f_a. The magnification is thus

$$g_a = \frac{f_a}{d_a}$$

More exactly, this image is at a distance $f_a{}^2 d_a$ from the coincident focal points and consequently the final image is at the distance d' from the second focal point of the eyepiece,

$$d' = d_a (f_s/f_a)^2$$

The magnification of the eyepiece is

$$g_s = d'/f_s$$

Finally, the total magnification is

$$g = g_a g_s = f_s/f_a = 1/G$$

where G is the angular magnification. Orthostereoscopy will only be obtained if the separation between the two objectives is $\bar{\varphi}G$, which rarely occurs, and therefore there is evident a certain loss of depth.

In microscopes the possibility of stereoscopic vision was envisaged by Wheatstone (1853), who showed the necessity of an erecting system. The theory of orthostereoscopy was established in this instance by Czapski and Gebhardt (1897). At low magnifications a pair of objectives and eyepieces can be used with erecting prisms. At high magnifications the beam which enters one objective is split into two, each traveling toward one eye (Riddell, 1853).

Conclusion

I shall end this book with one recommendation to my readers. They must not by any means believe that this text is complete or that it represents everything that is known—or thought to be known—about vision. The subject is immense and I have had to make choices, inevitably arbitrary. I ask that my colleagues whose important work I may have omitted to cite will excuse me. And especially that young men do not think that all is known in physiological optics and that all has been said in a world that is too old. As it progresses, science arouses more questions than it resolves and an immense field of research is open to all those who are attracted by the mysteries of this act, so simple and yet so incomprehensible: to see.

References

1. Alpern (M.), "Movements of the Eyes," in Davson, *The Eye*, Vol. 3, Academic Press, New York, 1962, pp. 3–229.
1a. Arnulf (A.), "Etude de la limite de séparation visuelle," *Réunions de l'Institut d'Optique*, Paris, 1937.
1b. Arnulf (A.), "Le depistage des insuffisances visuelles," in *Mieux voir*, Conferences, Paris, 1962, pp. 65–100.
2. Aubert (H.), *Physiologie der Netzhaut*, Morgenstern, Breslau, 1865.
2a. Ballard (S. S.), Ed., *Visual Search Technique*, National Academy of Sciences, Washington, D.C., 1960.
3. Bartley (S. H.), *Vision*, Van Nostrand, Princeton, N.J., 1941.
4. Bonaventura (E.), *La parallasse binoculare*, Bandetini, Florence, 1931.
5. Bose (J. C.), *Réactions de la matière vivante*, Masson, Paris, 1926.
5a. Bouman (M. A.), and Vos (J. J.), Eds. *Performance of the Eye at Low Luminances*, Excerpta Medica Foundation, Amsterdam, 1966.
6. Bourdon (B.), *La perception visuelle de l'espace*, Schleicher, Paris, 1902.
7. Brandt (H. F.), *The Psychology of Seeing*, The Philosophical Library, New York, 1945.
7a. Brindley (G. S.), *Physiology of the Retina and Visual Pathway*, Arnold, London, 1960.
8. Calvert (E. S.), *Cockpit Lighting*, Royal Aircraft Establishment Monograph 2.3.01, 1950.
9. Carmichael (L.) and Dearborn (W. F.), *Reading and Visual Fatigue*, Harrap, London, 1948.
10. Carr (H. A.), *An Introduction to Space Perception*, Longmans, Green, London, 1935.
11. Chrétien (H.), *Cours de calcul des combinaisons optiques*, 3rd ed., Revue d'Optique, Paris, 1938.
12. Clemmesen (V.), "Central and Indirect Vision of the Light-Adapted Eye," *Acta Physiol. Scand., Suppl.*, Vol. 2, 1944.
12a. Clerc (L. P.), *Applications de la photographie aérienne*, Doin, Paris, 1920.

13. Danjon (A.), and Couder (A.), *Lunettes et télescopes*, Revue d'Optique, Paris, 1935.
14. Davson (H.), *The Physiology of the Eye*, Churchill, London, 1949.
14a. Deutsch (J. A.), *The Structural Basis of Behaviour*, University of Chicago Press, Chicago, Ill., 1960.
15. Dubois-Poulsen (A.), François (P.), Tibi (A.), and Magis (C.), *Le champ visuel*, Masson, Paris, 1952.
16. Dudley (L. P.), *Stereoptics*, MacDonald, London, 1951.
17. Duke-Elder (W. S.), *Textbook of Ophthalmology*, Kimpton, London, 1932, and Mosby, St. Louis, 1933.
17a. Fischer (G. L.), Pollock (D. K.), Radack (B.), and Stevens (M. E.), Eds., *Optical Character Recognition*, Spartan Books, Washington, D.C., 1962.
18. Fortuin (G. F.), "Visual Power and Visibility," thesis, Groningen, 1951.
18a. Fry (G. A.), *Blur of the Retinal Image*, Ohio State University Press, Columbus, 1955.
19. Gibson (J. J.), *The Perception of the Visual World*, Houghton, Boston, 1951.
19a. Graham (C. H.), "Visual Perception," in Stevens, *Handbook of Experimental Psychology*, Wiley, New York, 1951, pp. 868–920.
19à. Graham (C. H.), Ed. *Vision and Visual Perception*, Wiley, New York, 1965.
19ä. Hake (H. W.), *Contributions of Psychology to the Study of Pattern Vision*, USAF, WADC Tech. Rept., Wright-Patterson Air Force Base, Columbus, Ohio, 1957.
19b. Hardy (L. H.), Rand (G.), Rittler (M. C.), Blank (A. A.), and Boeder (P.), *The Geometry of Binocular Space Perception*, Columbia University Press, New York, 1953.
19c. Harrison (V. G. W.), *Definition and Measurement of Gloss*, Printing and Allied Trades Research Association, Cambridge, England, 1945.
19d. Hartridge (H.), *Recent Advances in the Physiology of Vision*, Churchill, London, 1950.
20. Helmholtz (H. von), *Physiological Optics* (transl. by Javal and Klein), Masson, Paris, 1867; 3rd ed., 1909 (transl. by Southall), with an Appendix by Gullstrand, Optical Society of America, Washington, D.C., 1924–1925.
21. Hering (E.), "Der Raumsinn und die Bewegungen der Auges," in Hermann's *Handbuch der Physiologie*, Vol. 3, Vogel, Leipzig, 1879.
21a. Hill (D.), and Parr (G.), Eds., *Electroencephalography*, Macmillan, New York, 1963.
22. Hillebrand (F.), *Lehre von den Gesichtsempfindungen*, Springer, Vienna, 1929.
23. Hofmann (F. B.), "Physiologie Optik," in *Graefe-Saemisch Handbuch Ges. Augenheilk.*, Vol. 3, 2nd ed., 1925.

23a. Hopkins (H. H.), "The Application of Frequency Response Technique in Optics," in *Proceedings of the Conference on Optical Instrument Techniques*, Chapman and Hall, London, 1961, pp. 480–514.

24. Ittleson (W. H.), *The Ames Demonstrations in Perception*, Princeton University Press, Princeton, N. J., 1952.

24a. Ittelson (W. H.), *Visual Space Perception*, Springer, Berlin, 1960.

25. Ivanoff (A.), *Les abérrations de l'oeil*, Revue d'Optique, Paris, 1953.

26. Javal (E.), *Manuel théorique et pratique du strabisme*, Masson, Paris, 1896.

26a. Jung (R.), and Kornhuber (H.), Eds., *Neurophysiologie und Psychophysik des visuellen System*, Springer, Berlin, 1961.

26b. Kronfeld (P. C.), McHugh (G.), and Polyak (S. L.), *The Human Eye in Anatomical Transparencies*, Bausch & Lomb, Rochester, New York, 1943.

27. Kuhn (H. S.), *Eyes and Industry*, 2nd ed., Mosby, St. Louis, 1950.

28. Lapicque (C.), "L'optique de l'oeil et la vision des contours," thesis, Paris, 1938; see also *Rev. Opt.*, Vol. 15, 1936, p. 121.

29. Le Grand (Y.), "Essai sur la diffusion de la lumière dans l'oeil," thesis, Paris, 1937; see also *Rev. Opt.*, Vol. 16, 1937, pp. 201, 241.

30. L'Hermitte (J.), "Anatomie des voies optiques," in Baillairt, *Traite d'Ophtalmologie*, Vol. 1, Masson, Paris, 1939, pp. 598–660.

30a. Livingstone (R. B.), "Central Control of Receptors and Sensory Transmission System," in *Handbook of Physiology*, Vol. I, Williams & Wilkins, Baltimore, 1959, Sec. 1, pp. 741–760.

31. Lord (M. P.), and Wright (W. D.), "The Investigation of Eye Movements," *Rept. Progr. Phys.*, Vol. 13, 1950, pp. 1–23.

32. Luckiesh (M.), *Light and Work*, Van Nostrand, Princeton, N.J., 1924.

33. Luckiesh (M.), and Moss (F. K.), *Seeing*, Ballière, Tindall, and Cox, London, 1931.

34. Luckiesh (M.), and Moss (F. K.), *Reading as a Visual Task*, Van Nostrand, Princeton, N.J., 1942.

35. Luckiesh (M.), *Light, Vision and Seeing*, Van Nostrand, Princeton, N.J., 1945.

36. Luneburg (R. K.), *Mathematical Analysis of Binocular Vision*, Princeton University Press, Princeton, N.J., 1947.

37. Luneburg (R. K.), "Metric Methods in Binocular Visual Perception," in *Courant Anniversary Volume*, Interscience (Wiley), New York, 1948.

38. Mach (E.), *Bieträge zur Analyse der Empfindungen*, Fischer, Jena, 1886.

38a. Malbran (J.), *Strabisme et paralysies*, Heraly, Charleroi, Belgium, 1953.

38b. Maréchal (A.), and Françon (M.), *Diffraction, structure des images*, Revue d'Optique, Paris, 1960.

39. Marshall (W. H.), and Talbot (S. A.), in "Visual Mechanisms," *H. Klüver Biological Symposia*, Vol. 7, Cattell, Lancaster, Pa., 1942, pp. 117–164.

40. Mazuir (P.), *Traité de télémétrie*, Revue d'Optique, Paris, 1931.
41. Middleton (W. E. K.), *Vision through the Atmosphere*, Toronto University Press, Toronto, Can., 1952.
42. Moon (P.), *The Scientific Basis of Illuminating Engineering*, McGraw-Hill, New York, 1936.
43. Müller (J.), *Beiträge zur vergleichenden Physiologie des Gesichtsinnes*, Cnobloch, Leipzig, 1826.
43a. Muller-Limmroth (W.), *Elecktrophysiologie des Gesichtsinnes*, Springer, Berlin, 1959.
44. Nagel (A.), *Das Sehen mit zwei Augen*, Winter, Leipzig, 1861.
44a. Neff (W. D.), Ed., *Contributions to Sensory Physiology*, Academic Press, New York, 1965.
45. Nordmann (J.), "Physiologie de la motilité oculaire extrinsèque," in Baillairt, *Traité d'Ophtalmologie*, Vol. 2, Masson, Paris, 1939, pp. 191–259.
46. Ogle (K. N.), *Researches in Binocular Vision*, Saunders, Philadelphia, 1950.
46a. Ogle (K. N.), "The Optical Space Sense," in Davson, *The Eye*, Vol. 4, Academic Press, New York, 1962, pp. 209–417.
47. Opin (L.), *Vision binoculaire et spatiale*, in Bailliart, *Traité d'Ophtalmologie*, Vol. 2, Masson, Paris, 1939, pp. 769–844.
48. Osterberg (G.), *Topography of the Layer of Rods and Cones in the Human Retina*, Arnold Busch, Copenhagen, 1935.
49. Panum (P. L.), *Physiologische Untersuchungen über das Sehen mit zwei Augen*, Kiel, 1858.
50. Parinaud (H.), *Stéréoscopie et projection visuelle*, Paris, 1904.
51. Piéron (H.), "Physiologie de la vision," in Bailliart, *Traité d'Ophtalmologie*, Vol. 2, Masson, Paris, 1939, pp. 497–768.
52. Piéron (H.), *La sensation, guide de vie*, 2nd ed., Gallimand, Paris, 1955.
52a. Piéron (H.), "Le maniement de la perception," in *Traité de Psychologie appliquée*, Vol. 5, 1955, pp. 959–1091.
53. Pirenne (M. H.), *Vision and the Eye*, Chapman and Hall, London, 1948.
53a. Pirenne (M. H.), "Quantum Physics of Vision," *Progr. Biophys.*, Vol. 2, 1951, p. 193.
53b. Pirenne (M. H.), Marriott (F. H. C.), and O'Doherty (E. F.), *Individual Differences in Night Vision Efficiency*, Med. Res. Council Spec. Rept. Ser., H.M. Stationery Office, London, 1957.
53c. Pirenne (M. H.), "Visual Functions in Man," in Davson, *The Eye*, Vol. 2, Part I, Secs. 6–9, Academic Press, New York, 1962, pp. 123–195.
54. Polack (A.), "Le chromatisme de l'oeil," in *Bull. Soc. Ophtalmol. France*, No. 9 bis, 1923.
55. Polyak (S.), *The Retina*, University of Chicago Press, Chicago, Ill., 1941.
55a. Polyak (S)., *The Vertebrate Visual System*, University of Chicago Press, Chicago, Ill., 1957.

56. Pulfrich (C.), *Die Stereoskopie im Dienste der Photometrie und Pyrometrie*, Springer, Berlin, 1923.
57. Purkinje (J.), *Beobachtungen und Versuche zur Physiologie der Sinne*, Reimer, Prague and Berlin, 1819–1825.
58. Purkinje (J.), *Opera Omnia*, Prague, 1918.
59. Rochon-Duvigneaud (A.), *Les yeux et la vision des vertébrés*, Masson, Paris, 1943.
60. Rohr (M. von), *Die binokularen Instrumente*, Springer, Berlin, 1907.
60a. Rosenblatt (F.), *Principles of Neurodynamics*, Spartan Books, Washington, D.C., 1962.
60b. Rosenblith (W. A.), Ed., *Sensory Communication*, Massachusetts Institute of Technology Press, Cambridge, Mass., 1961.
61. Senden (M. V.), *Raum- und Gestaltauffassung bei operierten Blindgeboren*, Barth, Liepzig, 1932.
61a. Sholl (D. A.), *The Organization of the Cerebral Cortex*, Methuen, London, 1956.
61b. Spigel (I. M.), Ed., *Readings in the Study of Visually Perceived Movement*, Harper & Row, New York, 1965.
62. Stern (W.), *Die Wahrnehmung von Bewegungen*, Berlin, 1894.
63. Takala (M.), "Asymmetries of Visual Space," *Ann. Acad. Scienti. Fennicae, Ser. B*, Vol. 72, 1951, p. 175.
63a. Tolansky (S.), *Optical Illusions*, Pergamon, Oxford, 1964.
64. Traquair (H. M.), *An Introduction to Clinical Perimetry*, 5th ed., Kimpton, London, 1948.
64a. Trumbull (R.), Ed., *Visual Problems of the Armed Forces*, National Academy of Sciences, Washington, D.C., 1962.
65. Tschermak (A. von), "Raumsinn und Augenbewegungen," *Bethe Handb. norm. u. Pathol. Physiol.*, Vol. 12, 1950, pp. 883–1094.
65a. Tschermak (A. von), *Einfuhrung in die Physiologische Optik*, Bergmann, Munich, and Springer, Berlin, 1942.
66. Tscherning (M.), *Optique physiologique*, Carré et Naud, Paris, 1898.
67. Tscherning (M.), "Vision," in d'Arsonval, *Traité de physique biologique*, Vol. 2, Masson, Paris, 1903.
68. Vernon (M. D.), *A Further Study of Visual Perception*, Cambridge University Press, Cambridge, England, 1952.
69. Volkmann (A. W.), *Physiologische Untersuchungen im Gebiete der Optik*, Breitkopf & Hartel, Leipzig, 1863, 1864.
69a. Vos (J. J.), *On Mechanisms of Glare*, Institute for Perception, Soesterberg, 1963.
70. Walls (G.), *The Vertebrate Eye and its Adaptive Radiations*, Cranbrook Press, Bloomfield Hills, Mich., 1942.
70a. Walsh (J. W. T.), *Planned Artificial Lighting*, Odhams, London, 1956.
71. Weber (E. H.), in *Wagner Handworterb. d. Physiol.*, Vol. 3, 1846.
72. Weston (H. C.), *Sight, Light and Efficiency*, Lewis, London, 1949.

73. Weyl (H.), *Philosophy of Mathematics and Natural Science*, Princeton University Press, Princeton, N.J., 1949.
73a. Whiteside (T. C. D.), *The Problems of Vision in Flight at High Altitudes*, Butterworth, London, 1957.
74. Wright (W. D.), *Photometry and the Eye*, Hatton, London, 1949.
75. Wundt (W.), *Beitrage zur Theorie der Sinneswahrnehmung*, Engelmann, Leipzig, 1862.
76. *Handbook of Human Engineering Data for Design Engineers*, 2nd ed., Tech. Rept. SDC 199–1–2a, U.S. Office of Naval Research, Port Washington, N.Y., 1951.
77. *The Illuminating Engineering Society Lighting Handbook*, New York, 1952.
78. *Studies in Visual Acuity*, Bull. 743, U.S. Govt. Printing Office, Washington, D.C., 1948.
79. *Binocular Visual Acuity of Adults*, National Center for Health Statistics, U.S. Govt. Printing Office, Washington, D.C., 1964.
80. *Compte-Rendus de la 15ᵉ Session de la C.I.E.*, 4 vols., Revue d'Optique, Paris, 1963.

General References

Adler and Fliegelman, *AMA Arch. Ophthalmol.*, **12**, 75 (1934).
Adler and Meyer, *Trans. Am. Ophthalmol. Soc.*, 71st meeting, 1935.
Adrian and B. Matthews, *Brain*, **57**, 355 (1934); **58**, 323 (1935).
Adrian and R. Matthews, *J. Physiol. (London)*, **65**, 273 (1928).
Adrian and Moruzzi, *J. Physiol. (London)*, **97**, 153 (1939).
Aeby, Z. *Rat. Med.*, 1861.
Aguilar, *Ann. Opt. Oculaire (Paris)*, **3**, 99 (1955).
Airy, *Cambridge Phil. Soc. Trans.*, **1834**, 283.
Albe-Fessard, *J. Physiol. (Paris)*, **49**, 521 (1957).
Albertotti, *Arch. Ital. Biol.*, **6**, 341 (1884).
Alexander, *Am. J. Ophthalmol.*, **34**, 895 (1951).
Allard, Mémoire sur l'intensité et la portée des phares, Paris, 1876.
Allen, *Am. J. Optom.*, **30**, 78, 293 (1953).
Alpern, *JOSA*, **43**, 648 (1953).
Alpern and David, *Indust. Med. Surg.*, **27**, 551 (1958).
Alpern and Larson, *Am. J. Ophthalmol.*, **49**, 1140 (1960).
Ames, *Am. J. Ophthalmol.*, **28**, 248 (1945).
Ames and Proctor, *JOSA*, **5**, 22 (1921).
Ames, *Psychol. Monogr.*, **65**, No. 7 (1951).
Ames, Ogle, and Gliddon, *JOSA*, **22**, 538, 575 (1932); *AMA Arch. Ophthalmol.*, **7**, 576 (1932).
Amigo, *JOSA*, **53**, 630 (1963).
Ancona, *Contrib. Lab. Psichiat. Univ. Sac. Cuore*, **14**, 55, 81 (1950); **15**, 37 (1952).
Anderton, Brit. Pat. 11520 (1891).
André, *Ann Ecole Norm. Suppl.*, **5**, 275 (1876).
Andrews, *Nature*, **205**, 1250 (1965).
Angelucci and Aubert, *Pflüger's Arch. Ges. Physiol.*, **22**, 69 (1880).
Ansbacher, *J. Exptl. Psychol.*, **34**, 1 (1944).
Anstis and Atkinson, *Nature*, **212**, 1614 (1966).
Apter, *J. Neurophysiol.*, **8**, 123 (1945); **9**, 73 (1946).

Aquilonius, *Opticorum Libri VI*, Anvers, 1613.
Arden and Liu, *Acta Physiol. Scand.*, **48**, 36, 49 (1960).
Arden and Söderberg, *Experientia*, **15**, 163 (1959).
Arden and Weale, *J. Physiol. (London)*, **125**, 417 (1954).
Armington, *JOSA*, **45**, 1058 (1955).
Arndt, *Z. Tech. Physik*, **15**, 296 (1934).
Arnulf and Dupuy, *Compt. Rend.*, **250**, 2757 (1960).
Arnulf, Richard, and Verét, *Rev. Opt.*, **33**, 658 (1954).
Arnulf, Dupuy, and Flamant, *Compt. Rend.*, **228**, 1057 (1949); **230**, 1791 (1950); **232**, 351, 439 (1951); Florence Meeting, 1954; *Ann. Opt. Oculaire (Paris)*, **3**, 109 (1955).
Arnulf, Flamant, and Françon, *Rev. Opt.*, **27**, 741 (1948).
Arrer, *Phil. Studies*, **13**, 116, 222 (1896).
Asch and Witkin, *J. Exptl. Psychol.*, **38**, 325, 455, 603, 672 (1948).
Aschenbrenner, *Photogramm. Eng.*, **18**, 818 (1952).
Aserinsky and Kleitman, *J. Appl. Physiol.*, **8**, 1 (1955).
Asher, *J. Physiol. (London)*, **112**, 40 (1951).
Aubert, *Physiologie der Netzhaut*, Breslau, 1865; *Pflüger's Arch. Ges. Physiol.*, **39**, 347 (1886).
Aubert and Förster, *Graefe's Arch. Ophthalmol.*, **3**, 1 (1857).
Auerbach and von Kries, *Arch. Anat. Physiol.* Physiol. Abt., 1877, p. 297.
Avetisov, *Vestn. Oftal.*, **78**, 6 (1965).

Bach-y-Rita, *Acta Neurol. Latinoam.*, **5**, 17 (1959).
Baker (H)., *JOSA*, **39**, 172 (1949).
Baker (K. E.), *JOSA*, **39**, 567 (1949).
Ball, *Trans. Intern. Opt. Congr.*, 1951, p. 92.
Banister, *Rept. Joint Disc. Vision*, Phys. Soc., London, 1932, p. 227.
Banister and Pollock, *Brit. J. Psych.*, **19**, 394 (1928).
Banister, Hartridge, and Lythgoe,, *Proc. Opt. Conv.*, 1926, p. 551.
Bappert, *Z. Sinnesphysiol.*, **90**, 167 (1922).
Barany, *Acta Ophthalmol.*, **24**, 63, 93 (1946); *Acta Physiol. Scand.*, **14**, 296 (1947).
Barany and Hallden, *J. Neurophysiol.*, **11**, 25 (1948).
Barber, *Proc. Phys. Soc. (London)*, **B63**, 364 (1950).
Barlow, *J. Physiol. (London)*, **116**, 290 (1952); **119**, 58, 69 (1953).
Barlow, Brindley, and Hill, *Nature*, **200**, 1345, 1347 (1963).
Barnard, *Popular Astron.*, **5**, 285 (1897).
Barnes and Czerny, *Z. Physik*, **79**, 436 (1932).
Bartlett, *Brit. J. Psych.*, **8**, 222 (1916).
Bartlett and Gagné, *J. Exptl. Psychol.*, **25**, 91 (1939).
Bartley, *Am. J. Physiol.*, **108**, 397 (1934); **110**, 666 (1935); *J. Cellular Comp. Physiol.*, **8**, 41 (1936).
Bartley and Bishop, *Am. J. Physiol.*, **103**, 159 (1933).

Bartley and Fry, *JOSA*, **24**, 342 (1934).
Basler, *Pflüger's Arch. Ges. Physiol.*, **115**, 82 (1906).
Baudoin, Fischgold, Caussé, and Lerique, *Bull. Acad. Med. Paris*, **121**, 688 (1939).
Baumgardt, *Arch. Sci. Physiol.*, **1**, 257 (1947); *Compt. Rend. Soc. Biol.*, **143**, 786 (1949); *Année Psychol.*, **53**, 431 (1953).
Baumgardt and Segal, *Année Psychol.*, **48**, 54 (1947).
Baumgartner, Brown (J. L.), and Schulz, *J. Neurophysiol.*, **28**, 1 (1965).
Bedford, *Proc. IRE*, **38**, 1003 (1950).
Bedford and Wyzecki, *JOSA*, **47**, 564 (1957).
Bein, *Veroeffentl. Inst. Meeresk.*, **9**, 28 (1935).
Bell (C.), *Anatomy and Physiology of the Human Body*, London, 1829.
Bell (L.), *Elec. World*, **57**, 1136, 1569 (1911).
Benari, *JOSA*, **51**, 371 (1961).
Bender and Teuber, *Arch. Neurol. Psychiat.*, **55**, 627 (1946); **56**, 300 (1946).
Benham, *Nature*, **51**, 113 (1894).
Bennett-Clark, *Opt. Acta*, **11**, 301 (1964).
Benussi, *Pflüger's Arch. Ges. Physiol.*, **32**, 396 (1914).
Berger (C.), *Skand. Arch. Physiol.*, **74**, 27 (1936); **83**, 39 (1939); *Acta Ophthalmol.*, **26**, 517 (1948); **28**, 243 (1950); *Acta Physiol. Scand.*, **30**, 161 (1954).
Berger (C.), and Buchtal, *Skand. Arch. Physiol.*, **79**, 15 (1938).
Berger (H.), *J. Psychol. Neurol.*, **40**, 160 (1930).
Berger (P.), and Segal, *Compt. Rend.*, **234**, 1308 (1952).
Berny, thesis, Paris, 1965.
Bernyer, Durup, and Piéron, *Année Psychol.*, **40**, 15 (1939).
Berry, *J. Exptl. Psychol.*, **38**, 708 (1948).
Berry, Riggs, and Duncan, *J. Exptl. Psychol.*, **40**, 349 (1950).
Best, *Graefe's Arch. Ophthalmol.*, **51**, 458 (1900); **149**, 413 (1949).
Bethe, *Pflüger's Arch. Ges. Physiol.*, **121**, 1 (1908).
Beutell (1933), see *Nature*, **176**, 717 (1955).
Beyne, *Med. Aeron. (Paris)*, No. 1 (1954).
Beyne and Monnier (A.), *Med. Aeron. (Paris)*, No. 3 (1947).
Beyne and Worms, *Reunion Inst. Opt.*, **2**, 107 (1931).
Bhatia and Verghese, *JOSA*, **53**, 283 (1963); **54**, 948 (1964).
Bidwell, *Proc. Roy. Soc. (London)*, **40**, 368 (1896).
Biedermann, *Z. Ophthal. Opt.*, **15**, 1 (1927); **16**, 34 (1927).
Biot, *Ann. Soc. Sci. Bruxelles*, **60**, 138 (1946); *Congr. Natl. Sci. Bruxelles, 3rd Congr.*, 1950, p. 196.
Bishop, *Am. J. Physiol.*, **103**, 213 (1935).
Bishop and Clare, *J. Neurophysiol.*, **14**, 497 (1951).
Bishop and O'Leary, *J. Neurophysiol.*, **1**, 391 (1938).
Bishop, Burke, and Davis, *Science*, **130**, 506 (1959).
Bitterman and Soloway, *J. Exptl. Psychol.*, **36**, 134 (1946).

Blackburn, *Am. J. Optom.*, **14**, 365 (1937).
Blackwell, *JOSA*, **36**, 624 (1946); **38**, 1097 (1948); **42**, 606 (1952); **43**, 456, 815 (1953); *CIE*, 1955.
Blank, *JOSA*, **43**, 717 (1953); **51**, 335 (1961).
Blondel and Rey, *Compt. Rend.*, **153**, 54 (1911).
Bloom and Garten, *Pflüger's Arch. Ges. Physiol.*, **72**, 372 (1898).
Blumenfeld, *Z. Sinnesphysiol.*, **65**, 241 (1913).
Boehm, *Acta Ophthalmol.*, **18**, 109, 143 (1940).
Boerma and Walther, *Graefe's Arch. Ophthalmol.*, **39**, 71 (1893).
Boll. *Arch. Anat. Physiol.*, 1877.
Bordier, *De l'Acuité visuelle*, Paris, 1893.
Boring, *Am. J. Psychiat.*, **53**, 293 (1940); *Am. J. Phys.*, **11**, 55 (1943).
Bornschein, *Z. Biol.*, **110**, 210 (1958).
Borries, *Fixation und Nystagmus*, Copenhagen, 1926.
Bouguer, *Traité d'optique sur la gradation de la lumière*, Paris, 1760.
Bouma, *Philips Tech. Rev.*, **1**, 215 (1936).
Bouman, *JOSA*, **42**, 820 (1952); **43**, 209, 895 (1953); **45**, 36 (1955); *Opt. Acta*, **1**, 177 (1955).
Bouman and Blokhuis, *JOSA*, **42**, 525 (1952).
Bouman and van den Brink, *Ophthalmologica*, **123**, 100 (1952); *JOSA*, **43**, 895 (1953).
Bouman and van der Velden, *JOSA*, **37**, 908 (1947); **38**, 231, 570 (1948).
Bouman and Walraven, *JOSA*, **47**, 834 (1957).
Bouman, ten Doesschate, and du Marchie Sarvaas, *Ophthalmologica*, **122**, 368 (1951).
Bourdon, *Rev. Phil.*, **46**, 124 (1898); **49**, 74 (1900).
Bourdy, *Rev. Opt.*, **36**, 449, 570 (1957); **39**, 64 (1960); *Ann. Oculist. (Paris)*, **194**, 1048 (1961).
Boutry, Billard, and Le Blan, *Onde Elect.*, *1954*, p. 824.
Bowditch and Hall, *J. Physiol. (London)*, **3**, 297 (1880).
Boynton, *JOSA*, **43**, 442 (1953); **44**, 351 (1954).
Boynton and Clarke, *JOSA*, **54**, 110 (1964).
Boynton, Bush, and Enoch, *JOSA*, **44**, 56, 879 (1954).
Braunstein, *JOSA*, **56**, 835 (1966).
Brazier, *Electroencephalog. Clin. Neurophysiol. Suppl.*, **4**, 93 (1953).
Brecher, *Pflüger's Arch. Ges. Physiol.*, **246**, 315 (1942).
Breese, *Psychol. Rev.*, **16**, 410 (1909).
Brewster, *Ann. Physik*, **29**, 339 (1835); *Trans. Roy. Soc. Edinburgh*, **15**, 349 (1843).
Bricard, *Physique des nuages*, Paris, 1953.
Bridges and Bitterman, *Am. J. Psychol.*, **67**, 525 (1954).
Brindley, *J. Physiol. (London)*, **121**, 332 (1953); *Proc. Phys. Soc. (London)*, **B67**, 673 (1954).
Broca, *J. Physiol. Pathol. Gen.*, **3**, 384 (1901).
Brown (J. F.), *Psych. Forsch.*, **14**, 199, 233, 249 (1931).
Brown (J. F.), and Mize, *Psych. Forsch.*, **16**, 355 (1932).

Brown (J. L.), *JOSA*, **44**, 48 (1954); **52**, 580 (1962).
Brown (J. L.), Graham, Leibowitz, and Ranken, *JOSA*, **43**, 197 (1953).
Brown (K. T.), *JOSA*, **43**, 464 (1953); **45**, 301 (1955).
Brown (R. H.), *JOSA*, **45**, 189 (1955).
Brücke, *Arch. Anat. Physiol.,1841*, p. 459.
Brückner and von Brücke, *Pflüger's Arch. Ges. Physiol.*, **107**, 263 (1905).
Bruesch and Arey, *J. Comp. Neurol.*, **77** 631 (1942).
Brumberg and Vavilov, *Bull. Acad. Sci. USSR*, **1**, 919 (1933).
Bryngdahl, *JOSA*, **54**, 1152 (1964).
Buddenbrock, *Grundriss der vergleichenden Physiologie*, 2nd ed., Berlin, 1937.
Buisson, *J. Physique*, **7**, 68 (1917); *Réunion Inst. Opt.*, **3**, 4 (1932).
Bull, *Bull. Soc. Franc. Ophtalmol., 1891*, p. 208; *Périmétrie*, Bonn, 1895.
Burchard, *Internationale Sehproben*, Cassel, 1882.
Burian, *Arch. Verg. Ophthalmol.*, **136**, 172 (1936); *AMA Arch. Ophthalmol.*, **21**, 486 (1939); **37**, 336, 504, 618 (1947).
Burnham, *Am. J. Psychol.*, **67**, 492 (1954).
Burnham and Newhall, *JOSA*, **43**, 899 (1953).
Burt, Cooper, and Martin, *Brit. J. Stat. Psychol.*, **8**, 29 (1955).
Buswell, *Suppl. Educ. Monogr., No. 21*, 1922.
Buxton, *Monthly Notices Roy. Astron. Soc.*, **81**, 547 (1921).
Buys and Coppez, *Bull. Soc. Franc. Ophtalmol.*, **30**, 9 (1913).
Byford, *Opt. Acta*, **9**, 223 (1962).
Byram, *JOSA*, **34**, 571, 718 (1944).

Cabello, *Anal. Fis. Quim.*, **41**, 439, 449 (1945).
Cabello and Stiles, *Anal. Fis. Quim.*, **A46**, 251 (1950).
Cadiergues, *Ann. Opt. Oculaire (Paris)*, **2**, 41, 115 (1953).
Cajal, *Histologie du système nerveux*, Paris, 1909–1911.
Campbell, *J. Physiol. (London)*, **123**, 357 (1954); **125**, 11P, 29P (1954); *JOSA*, **50**, 738 (1960).
Campbell and Green, *Nature*, **208**, 190 (1965).
Campbell and Gregory, *JOSA*, **50**, 831 (1960).
Campbell and Primrose, *Trans. Ophthalmol. Soc.*, **73**, 353 (1953).
Campbell and Westheimer, *JOSA*, **49**, 568 (1959); *J. Physiol. (London)*, **151**, 285 (1960).
Campbell, Robson, and Westheimer, *J. Physiol. (London)*, **145**, 579 (1959).
Caneja, *Arch. Oftalmol. Hispan.*, **A27**, 1 (1927); *Ann. Oculist. (Paris)*, **165**, 721 (1928).
Carman and Carruthers, *JOSA*, **41**, 305 (1951).
Carreras, *Arch. Soc. Oftalmol. Hispanam.*, **11**, 1443 (1952).
Casperson, *J. Exptl. Psychol*, **40**, 668 (1950).
Cattell, *Brain*, **8**, 295 (1885).
Cavonius and Schumacher, *Science*, **27**, 1276 (1966).
Chandrasekhar, *Radiative Transfer*, Oxford, 1950.

Chang, *J. Neurophysiol.*, **13**, 235 (1950); *Année Psychol.*, **50**, 135 (1951); *J. Neurophysiol.*, **15**, 5 (1952).
Chang and Kaada, *J. Neurophysiol.*, **13**, 305 (1950).
Chapanis, *Human Biol.*, **22**, No. 1 (1950).
Chariton and Lea, *Proc. Roy. Soc. (London)*, **A122**, 304 (1929).
Charnwood, *Trans. Intern. Opt. Congr.*, *1951*, p. 165.
Charpentier, *Compt. Rend.*, **102**, 1155, 1462 (1886); *Compt. Rend. Soc. Biol.*, **40**, 469 (1888).
Chatterjee, *Indian J. Psychol.*, **29**, 155 (1954).
Chesterman and Stiles, *Symposium on Searchlights*, London, 1948, p. 75.
Chevreul, *De la loi du contraste simultané des couleurs*, Paris, 1849.
Chiba, *Pflüger's Arch. Ges. Physiol.*, **121**, 150 (1926).
Chiewitz, *Intern. Monatsch. Anat. Physiol.*, *1883*.
Chin and Horn, *JOSA*, **46**, 60 (1956).
Chodin, *AMA Arch. Ophthalmol.*, **23**, 92 (1877).
Cibis and Haber, *JOSA*, **41**, 676 (1951).
Clark, *Am. J. Psychol.*, **46**, 325 (1934); **48**, 82 (1936).
Clarke, *Opt. Acta*, **8**, 121 (1961).
Clarke and Belcher, *Vis. Res.*, **2**, 53 (1962).
Claudet, *Proc. Roy. Soc. (London)*, **8**, 569 (1857).
Clowes, *Opt. Acta*, **9**, 65 (1962).
Clowes and Ditchburn, *Opt. Acta*, **6**, 252 (1959).
Coates, *Dioptric Rev.*, **1**, 127 (1940); **2**, 102 (1941).
Cobb (P. W.), *Psychol. Rev.*, **21**, 23 (1914); *Am. J. Physiol.*, **36**, 335 (1915); *J. Exptl. Psychol.*, **1**, 419 (1916); **6**, 138 (1923).
Cobb (P. W.), and Giessler, *Psychol. Rev.*, **20**, 425 (1913).
Cobb (P. W.), and Moss, *J. Franklin Inst.*, **200**, 239 (1925); **205**, 831 (1928).
Cobb (W. A.), *Electroencephal. Clin. Neurophysiol.*, **2**, 104 (1950).
Cobb (W. A.), and Morton, *Electroencephal. Clin. Neurophysiol.*, **4**, 547 (1952).
Cohn, *Arch. Augenheilk.*, **31**, 197 (1895).
Coleman (H. S.), and Coleman (M. F.), *JOSA*, **37**, 572 (1947).
Coleman (H. S.), and Verplanck, *JOSA*, **38**, 250 (1948).
Coleman (H. S.), Fridge, and Harding, *JOSA*, **39**, 766 (1949).
Collier, *J. Exptl. Psychol.*, **47**, 47 (1954).
Collins and Hopkinson, *Trans. IES*, **19**, 135 (1954).
Comas Solà, *Compt. Rend.*, **161**, 121 (1915).
Commission Internationale d'Optique, *JOSA*, **40**, 881 (1935).
Connor and Ganoung, *JOSA*, **25**, 287 (1935).
Conrady, *Monthly Notices Roy. Astron. Soc.*, **79**, 575 (1919).
Cords, *Graefe's Arch. Ophthalmol.*, **118**, 771 (1927).
Cornsweet, *JOSA*, **48**, 808 (1958).
Craik, *Proc. Phys. Soc. (London)*, **56**, 351 (1944).
Craik and Vernon, *Brit. J. Psychol.*, **32**, 206 (1942).
Crawford, *Proc. Roy. Soc. (London)*, **B128**, 552 (1940); **B129**, 94 (1942).
Crawford and Pirenne, *J. Physiol. (London)*, **126**, 404 (1954).

Creed and Granit, *J. Physiol. (London)*, **66**, 281 (1928).
Crescitelli and Dartnall, *Nature*, **174**, 216 (1954).
Crook, *J. Psychol.*, **3**, 541 (1937).
Crouch, *Illum. Eng.*, **40**, 747 (1945).
Crouzy, *Compt. Rend.*, **253**, 559 (1961); **260**, 1773 (1965); thesis, Paris, 1963.
Cruikshank, *J. Gen. Physiol.*, **58**, 327 (1941).
Crutchfield and Edwards, *J. Exptl. Psychol.*, **39**, 561 (1949).
Curtis, *Lick Obs. Bull.*, **2**, 67 (1901).
Czapski and Gebhardt, *Z. Wiss. Mikroskopie*, **14**, 289 (1897).
Czermak, *Wien. Sitzber.*, **12**, 322 (1854).

Dale, cited by Clemmesen 12.
d'Alembert, *Mem. Acad. Paris*, 1767, p. 81.
d'Almeida, *Compt. Rend.*, **47**, 61, 337 (1858).
Danjon, *Rev. Opt.*, **7**, 205 (1928); *Réunion Inst. Opt.*, **3**, 16 (1932).
Davidson, *Am. J. Ophthalmol.*, **18**, 350 (1935).
Daza de Valdès, *Uso de los antojos*, Madrid, 1623.
de Boer, *Philips Res. Rept., No. 6224*, 1951.
Dechales, *Cursus seu Mundus mathematicus*, Lyon, 1690.
de Haan, dissertation inaug., Utrecht, 1862.
de la Hire, *Acad. Roy. Sci.*, 1660.
de Lange, thesis, Delft, 1957.
Dement and Kleitman, *J. Exp. Psychol.*, **53**, 339 (1957).
de Mott, *JOSA*, **49**, 6 (1959).
Dennis, *J. Gen. Psychol.*, **46**, 340 (1934).
Denton, *J. Physiol. (London)*, **124**, 16P (1954).
Denton and Pirenne, *J. Physiol. (London)*, **123**, 417 (1954).
de Palma and Lowry, *JOSA*, **52**, 328 (1962).
Déribéré, *La couleur dans les activités humaines*, Paris, 1955.
de Silva, *Am. J. Psychol.*, **37**, 469 (1926); *Brit. J. Psychol.*, **19**, 268 (1929).
de Silva and Bartley, *Brit. J. Psychol.*, **20**, 241 (1930).
Desaguiliers, *Cours de physique experimentale*, London, 1717.
Descartes, *Tractatus de Homine*, Amsterdam, 1677.
Dessens, *Ann. Geophys.*, **1**, 161 (1944); **2**, 68, 243 (1946); **3**, 68 (1947).
de Toni, *Zentralbl. Ges. Ophthalmol.*, **30**, 229 (1934).
DeValois, Abramov, and Jacobs, *JOSA*, **56**, 966 (1966).
Devaux, Fleury, Nicolle, and Pagès, *Perception d'un signal lumineux*, Paris, 1951.
de Vries, *Physica*, **10**, 553 (1943); **14**, 319, 367 (1948).
de Vries, Sielof, and Spoor, *Nature*, **166**, 938 (1950).
Devries and Washburn, *Am. J. Psychol.*, **20**, 131 (1909).
Dichman, Preston, and Mull, *Am. J. Psychol.*, **57**, 83 (1944).
Diefendorf and Dodge, *Brain*, **31**, 451 (1908).
Dietzel, *Z. Biol.*, **80**, 289 (1924).
Dimmick and Karl, *J. Exptl. Psychol.*, **13**, 365 (1930).

Ditchburn, *Opt. Acta*, **1**, 171 (1955); **10**, 325 (1963); *Nature*, **198**, 630 (1963).
Ditchburn and Fender, *Opt. Acta*, **2**, 128 (1955).
Ditchburn and Ginsborg, *Nature*, **170**, 36 (1952); *J. Physiol. (London)*, **119**, 1 (1953).
Ditchburn, *Proc. Roy. Irish Acad.*, **A139**, 58 (1930).
Dobrowolsky and Gaine, *Pflüger's Arch. Ges. Physiol.*, **12**, 411 (1876).
Dodge, *Am. J. Physiol.*, **8**, 307 (1903); *Psychol. Rev.*, **11**, 1 (1904); *Psychol. Bull.*, **2**, 193 (1905); *Psychol. Monogr.*, **8**, No. 4 (1907); *J. Exptl. Psychol.*, **4**, 165 (1921).
Dodge and Cline, *Psychol. Rev.*, **8**, 145 (1901).
Dodt and Wirth, *Acta Physiol. Scand.*, **30**, 80 (1953).
Dodwell and Sutherland, *Nature*, **191**, 578 (1961).
Dohlman, *Acta Oto-Laryngol. Suppl.*, **5**, 78 (1925); **23**, 50 (1935).
Doleck and de Launay, *JOSA*, **35**, 676 (1945).
Dollond, *Phil. Trans. Roy. Soc. (London)*, **79**, 256 (1789).
Donaldson, *Quart. J. Exptl. Physiol.*, **45**, 25 (1960).
Donders, *Holl. Beitr. Anat. Physiol. Wiss.*, **1**, 104, 384 (1847); *Onderz. Physiol. Lab Utrechtse Hoogeschool*, **6**, 134 (1852); *Arch. Holl. Beitr.*, **3**, 260, 560 (1862); *On the Anomalies of Accommodation and Refraction of the Eye*, London, 1864; *Graefe's Arch. Ophthalmol.*, **17**, 27 (1871).
Dor, *Graefe's Arch. Ophthalmol.*, **19**, 316 (1873).
Dorsey, *Properties of Ordinary Water Substance*, New York, 1940.
Dove, *Ann. Physik, 1841*, p. 251; **83**, 169 (1850).
Dratz, *Notes Centre Rech. Sci. (Marseille)*, 1945; *J. Physique (Paris)*, **8**, 27S (1947).
Druault, *Arch. Ophtalmol. (Paris)*, **18**, 685 (1898); **20**, 21 (1900); *J. Physiol. Pathol. Gen.*, **16**, 649 (1914); *Arch. Ophtalmol. (Paris)*, **40**, 458, 536 (1923).
Ducos du Hauron, French Pat. 216,465 (1891).
Duffieux, *L'Intégrale de Fourier et ses applications à l'optique*, Rennes, 1946.
Duncker, *Psychol. Forsch.*, **12**, 180 (1929).
Dunnewold, thesis, University of Utrecht, 1964.
Duntley, *JOSA*, **38**, 179, 237 (1949).
Dupuy-Dutemps, *Bull. Soc. Franc. Ophtalmol., 1926*, p. 60.
Duran, *Anal. Fis. Quim.*, **39**, 567 (1943).
Durup, *Année Psychol.*, **32**, 151 (1931).
Durup and Fessard, *Année Psychol.*, **30**, 73 (1929); **39**, 227 (1938).
Dvorak, *Wein. Sitzber.*, **61**, 1864.

Eaton, *Brit. J. Ophthalmol.*, **3**, 63 (1919).
Ebbinghaus, *Sitzber. Preuss. Akad. Wiss., 1887*, p. 994.
Ebenholtz and Walchli, *Vis. Res.*, **5**, 455 (1965).
Edridge-Green, *Colour Blindness and Colour Perception*, London, 1909; *The Physiololgy of Vision*, 2nd ed., London, 1920; *Nature*, **106**, 375 (1920).
Edwards, *Am. J. Psychol.*, **66**, 449 (1953).

Eguchi, *CIE* (1931).
Einthoven, *Graefe's Arch. Ophthalmol.*, **31**, 211 (1885).
Emmert, *Klin. Monatsbl. Augenheilk.*, **19**, 443 (1881).
Emsley, *Proc. Phys. Soc. (London)*, **56**, 293 (1944).
Engleking and Poos, *Graefe's Arch. Ophthalmol.*, **116**, 196 (1924).
Enoch and Stiles, *Opt. Acta*, **8**, 329 (1961).
Enroth, *Acta Physiol. Scand. Suppl.*, **27**, 100 (1952).
Erdmann and Dodge, *Physiologische Untersuchungen über das Lesen*, 1898.
Erggelet, *Z. Ophthal. Opt.*, **3**, 170 (1916).
Erickson, *JOSA*, **56**, 491 (1966).
Escher-Desrivières and Jonnard, *Réunion Inst. Opt.*, **3**, 6 (1932).
Euler, *Mem. Acad. Berlin*, *1747*, p. 285.
Evans, *An Introduction to Clinical Scotometry*, Yale University Press, 1938.
Evans and Smith, *Nature*, **204**, 303 (1964).
Ewald, *Pflüger's Arch. Ges. Physiol.*, **115**, 514 (1906).
Exner, *Wien. Ber.*, **72**, 162 (1875).

Fabry, *Réunion Inst. Opt.*, **1**, 8 (1930); *Mem. Sci. Phys.*, **24**, 48 pp. (1934).
Farnsworth and Reed, *U.S. Naval Med. Res. Lab. Rept.*, 39 (1944).
Fazakas, *Graefe's Arch. Ophthalmol.*, **120**, 555 (1928).
Fechner, *Ann. Physik.*, **45**, 227 (1838); *Abhandl. Sachs. Ges. Wiss.*, **7**, 416 (1860); *Elemente des Psychophysik*, Breitkopf u. Härtel, Leipzig, 1860.
Feilchenfeld, *Z. Sinnesphysiol.*, **35**, 1 (1904).
Fender, *Brit. J. Ophthalmol.*, **39**, 65, 294 (1955).
Fender and Nye, *Kibernetik*, **1**, 81 (1961).
Ferree and Rand, *J. Exptl. Psychol.*, **11**, 295 (1917); *Trans. Am. Ophthalmol. Soc.*, 162 (1918); *Trans. IES*, **22**, 79 (1927); **23**, 507 (1928); *Am. J. Ophthalmol.*, **14**, 1018 (1931); *Person. J.*, **9**, 475 (1931); **10**, 108 (1931).
Ferree, Rand, and Hardy, *AMA Arch. Ophthalmol.*, **5**, 717 (1931).
Fick, *Arch. Augenheilk.*, **18**, 279 (1888); *Pflüger's Arch. Ges. Physiol.*, **43**, 445 (1888); *Graefe's Arch. Ophthalmol.*, **45**, 336 (1898).
Fick and du Bois-Reymond, *Arch. Anat. Physiol.*, *1853*, p. 396.
Fields, King, and O'Leary, *J. Neurophysiol.*, **12**, 117 (1949).
Fincham, *Proc. Phys. Soc. (London)*, **49**, 444 (1937); *Brit. J. Ophthalmol., Mon. Suppl.*, **8**, 33 pp. (1937); *Brit. J. Ophthalmol.*, **35**, 381 (1951); *Brit. Med. Bull.*, **9**, 18 (1953); *Vis. Res.*, **1**, 425 (1962).
Fiorentini, *Atti Fond. Giorgio Ronchi*, **5**, 195 (1950); **9**, 470 (1954); **10**, 54 (1955).
Fiorentini and Bittini, *Opt. Acta*, **10**, 55 (1963).
Fiorentini and Radici, *Vis. Res.*, **1**, 244 (1961).
Fiorentini, Jeanne, and Toraldo di Francia, *Atti Fond. Giorgio Ronchi*, **10**, 371 (1955).
Fischer (F.), *Pflüger's Arch. Ges. Physiol.*, **233**, 738 (1934).
Fischer (F. P.), *Pflüger's Arch. Ges. Physiol.*, **204**, 203, 234, 247 (1924).
Fischer (M. H.), *Pflüger's Arch. Ges. Physiol.*, **188**, 161 (1921).
Fischer (M. H.), and Kornmuller, *J. Psychol. Neurol.*, **41**, 273, 383 (1931).

Fischer (R.), *Graefe's Arch. Ophthalmol.*, **37** (1891).
Fisher, *J. Exptl. Psychol.*, **23**, 215 (1938).
Flamant, *Compt. Rend.*, **230**, 1977 (1950); *Ann. Opt. Oculaire (Paris)*, **3**, 85 (1955); *Rev. Opt.*, **34**, 433 (1955).
Fleischer, *Z. Sinnesphysiol.*, **141**, 383 (1937); **147**, 65 (1939).
Fleischl, *Wien. Sitzber.*, **86**, 17 (1882); **87**, 246 (1883).
Flom, Weymouth, and Kahneman, *JOSA*, **53**, 1027 (1963).
Flourens, *Recherches expérimentales sur le système nerveux*, Paris, 1824.
Floyd, *Nature*, **176**, 718 (1955).
Foley, *JOSA*, **54**, 684 (1964).
Foley, *JOSA*, **56**, 822 (1966).
Ford, White, and Lichtenstein, *JOSA*, **49**, 287 (1959).
Fortin, *Bull. Soc. Biol. Buenos Aires* (1927).
Fortuin and Balder, *CIE*, 1955.
Foucault, *Ann. Obs. Paris*, **5**, 197 (1859).
Foucault and Regnault, *Compt. Rend.*, **28**, 78 (1849).
Foxell and Stevens, *Nature*, **176**, 717 (1955).
Françon, *Rev. Opt.*, **28**, 292 (1949).
Frank, *Pflüger's Arch. Ges. Physiol.*, **109**, 63 (1905).
Fraunhofer, *Gilberts Ann.*, 304 (1814); **56**, 264, 297 (1817).
Freeman, *JOSA*, **22**, 285 (1932).
French, *Trans. Opt. Soc.*, **21**, 172 (1920).
Friedman, *AMA Arch. Ophthalmol.*, **6**, 663 (1931).
Fröhlich, *Z. Sinnesphysiol.*, **52**, 89 (1921); **54**, 58 (1923); *Die Empfindungzeit*, Jena, 1929.
Fruböse and Jaensch, *Z. Biol.*, **78**, 119 (1923).
Fry, *Am. J. Physiol.*, **108**, 701 (1934); *JOSA*, **37**, 166 (1947); *Am. J. Optom.*, **27**, 126 (1950); *JOSA*, **43**, 814 (1953); **51**, 560 (1961); **53**, 94 (1963); **55**, 108 (1965).
Fry and Bartley, *Am. J. Ophthalmol.*, **16**, 687 (1933); *Am. J. Physiol.*, **111**, 335 (1935); **112**, 414 (1935).
Fry and Cobb, *Am. Acad. Ophthalmol.*, 1935, p. 1.
Fugate, *JOSA*, **44**, 771 (1954).

Gaarder, *Nature*, **212**, 321 (1966).
Galifret, *Année Psychol.*, **41**, 168 (1945).
Galton, *J. Anthropol. Inst.*, **14**, 205 (1885).
Gardner, *JOSA*, **31**, 94 (1941).
Gastaut (H.), *Electroencephal. Clin. Neurol. Suppl.*, **2**, 69 (1950).
Gastaut (H.), Gastaut (Y.), Roger, Corriol, and Naquet, *Electroencephal. Clin. Neurol.*, **3**, 401 (1951).
Gehrke, *Ann. Physiol.*, **2**, 345 (1948).
Geldard, *J. Gen. Psychol.*, **1**, 123, 578 (1928).
Gemelli, *J. Psychol.*, **25**, 97 (1928).
Gemelli, Colombi, and Schupfer, *Contrib. Lab. Psichiat. Univ. Sac. Cuor.*, **15**, 26 (1952).

Gerathewohl, *JOSA*, **43**, 567 (1953).
Gertz, *Z. Physiol.*, **19**, 229 (1905); *Skand. Arch. Physiol.*, **20**, 357 (1908); *Z. Sinnesphysiol.*, **49**, 29 (1914); *Acta Ophthalmol.*, **13**, 192 (1935).
Gibbins, *JOSA*, **51**, 457, 805 (1961).
Gibson and Glaser, *Motion Picture Testing and Research*, 1947, Chap. 9.
Gibson (E. J.), Gibson (J. J.), Smith (O. W.), and Flock, *J. Exptl. Psychol.*, **58**, 40 (1959).
Gibson (J. J.), *J. Exptl. Psychol.*, **20**, 553 (1937).
Gibson (J. J.), Olum, and Rosenblatt, *Perceptual and Motor Skills*, 1953.
Giese, *J. Appl. Psychol.*, **30**, 91 (1946).
Gilbert and Hopkinson, *Brit. J. Ophthalmol.*, **33**, 305 (1949).
Ginsborg, *Nature*, **169**, 412 (1952).
Giraud-Teulon, *Ann. Oculist. (Paris)*, **81**, 215 (1879).
Glanville, *Am. J. Physiol.*, **45**, 592 (1933).
Goethe, *Zur Farbenlehre*, 1810.
Gogle, *Vis. Res.*, **3**, 101 (1963); *JOSA*, **54**, 411 (1964).
Goldmann, *Ophthalmologica*, **105**, 240 (1943); **114**, 147 (1947).
Gordon, *Am. J. Psychol.*, **60**, 202 (1947); *Brit. J. Ophthalmol.*, **35**, 339 (1951).
Göthlin, *Skand. Arch. Physiol.*, **55**, 271 (1929).
Grabke, *Pflüger's Arch. Ges. Physiol.*, **47**, 327 (1924).
Graham, *JOSA*, **53**, 1019 (1963).
Graham and Cook, *Am. J. Psychol.*, **49**, 654 (1937).
Graham and Granit, *Am. J. Physiol.*, **98**, 664 (1931).
Graham and Kemp, *J. Gen. Physiol.*, **21**, 635 (1938).
Graham, Baker, Hecht, and Lloyd, *J. Exptl. Psychol.*, **38**, 205 (1948).
Graham, Brown, and Mote, *J. Exptl. Psychol.*, **24**, 555 (1939).
Granit, *Z. Sinnesphysiol.*, **58**, 95 (1927); *Rept. Joint Disc. Vision, 1932*, p. 235; *Receptors and Sensory Perception*, Yale University Press, 1955.
Granit and Therman, *J. Physiol. (London)*, **83**, 859 (1935).
Grant, *J. Exptl. Psychol.*, **31**, 89 (1942).
Graybiel and Clark, *J. Aviation Med.*, **16**, 111 (1945).
Graybiel, Clark, Macorquodale, and Hupp, *Am. J. Psychol.*, **59**, 259 (1946).
Greeff, in *Graefe-Saemisch Handb. Augenheilk.*, Engelmann, Leipzig, 1900.
Green (H.), *Trans. IES*, **28**, 146 (1935).
Green (J)., *Trans. Am. Ophthalmol. Soc.*, 1889, p. 449.
Gregory, *Nature*, **207**, 16 (1965).
Gregory and Cane, *Nature*, **176**, 1272 (1955).
Grether and Williams, in Fitts, Ed., *Psychological Research on Equipment Design*, Washington, 1947.
Grey Walter, Dovey, and Shipton, *Nature*, **158**, 540 (1946).
Grindley, *Med. Res. Council Rept.*, 163 (1931).
Groenouw, *Arch. Augenheilk.*, **26**, 85 (1893); *Graefe's Arch. Ophthalmol.*, **35**, 29 (1899).
Gross, *Z. Sinnesphysiol.*, **62**, 38 (1931).
Guastalla, *JOSA*, **56**, 960 (1966).
Güggenbühl, *Ophthalmologica*, **115**, 193 (1948).

Guild, *Proc. Phys. Soc. (London)*, **29**, 311 (1917); **56**, 352 (1944); *Trans. Opt. Soc.*, **27**, 106 (1930).
Guilford and Dallenbach, *Am. J. Psychol.*, **40**, 83, 401 (1928).
Guilford and Helson, *Am. J. Psychol.*, **41**, 595 (1929).
Guillery, *Pflüger's Arch. Ges. Physiol.*, **75**, 466 (1899).
Gumprecht, Sung, Chin, and Slipcevich, *JOSA*, **42**, 226 (1952).
Günther, *Die Struktur des Sehraumes*, Stuttgart, 1955.
Guratzsch, *Arch. Ges. Psychol.*, **70**, 257 (1929).
Guth, *CIE*, 1951; *Illum. Eng.*, **46**, 65 (1951).

Haas, *Compt. Rend.*, **192**, 281 (1931).
Haggard and Rose, *J. Exptl. Psychol.*, **34**, 45 (1944).
Haidinger, *Ann. Physik*, **63**, 29 (1844).
Hamburger, *Das Sehen in der Dämmerung*, Vienna, 1949.
Hamdi and Whitteridge, *J. Physiol. (London)*, **121**, 44 (1953).
Hammer, *Am. J. Psychol.*, **62**, 337 (1949).
Hampton, *Trans. IES*, **27**, 46 (1934).
Hannover, *Das Auge*, Leipzig, 1852.
Hardy, Rand, and Rittler, *AMA Arch. Ophthalmol.*, **42**, 551 (1949); **45**, 53 (1951).
Harms and Aulhorn, *Graefe's Arch. Ophthalmol.*, **157**, 3 (1955).
Harrison, *Illum. Eng.*, **40**, 525 (1945).
Harrison and Meaker, *Illum. Eng.*, **40**, 525 (1945); **42**, 1953 (1947).
Hartridge (H.), *J. Physiol. (London)*, **57**, 52 (1922).
Hartridge, *J. Physiol. (London)*, **52**, 175 (1918); *Phil. Trans. Roy. Soc. London*, **B232**, 519 (1947).
Hartridge and Thomson, *Brit. J. Ophthalmol.*, **32**, 581 (1948).
Hassenfratz, *Ann. Chim. (Paris)*, **72**, 5 (1809).
Hayhow, *J. Comp. Neurol.*, **110**, 1 (1958).
Hazelhoff and Wiersma, *Z. Psychol.*, **96**, 171 (1924); **97**, 174 (1925); **98**, 366 (1926).
Hebbard, *JOSA*, **52**, 75 (1962).
Hecht, *J. Gen. Physiol.*, **11**, 255 (1928); *Proc. Natl. Acad. Sci. U.S.*, **14**, 237 (1928).
Hecht and Mintz, *J. Gen. Physiol.*, **22**, 593 (1939).
Hecht, Haig, and Chase, *J. Gen. Physiol.*, **20**, 831 (1937).
Hecht, Ross, and Mueller, *JOSA*, **37**, 500 (1947).
Hecht, Shlaer, and Pirenne, *J. Gen. Physiol.*, **25**, 819 (1942).
Heijirmans, *Arch. Neerl. Physiol.*, **19**, 384 (1934).
Heinbecker and Bartley, *J. Neurophysiol.*, **3**, 219 (1940).
Heine, *Graefe's Arch. Ophthalmol.*, **51**, 563 (1900).
Heinsius, *Graefe's Arch. Ophthalmol.*, **147**, 1 (1944).
Helmholtz, *Ann. Physik*, **101**, 494 (1857); *Graefe's Arch. Ophthalmol.*, **10**, 1 (1864).
Helson and Guilford, *J. Gen. Psychol.*, **9**, 58 (1933).
Hendley, *J. Gen. Physiol.*, **31**, 433 (1948).

Henschen, *Pathologie des Gehirns*, 1892.
Hering, *Beitr. Physiol.*, **5**, 297 (1862); *Arch. Anat. Physiol.*, *1864*, p. 27; 1865, p. 153; *Wein. Z.*, **79**, 137 (1879); *Pflüger's Arch. Ges. Physiol.*, **60**, 519 (1895); *Ber. Ges. Leipzig*, **51**, 16 (1899); *Grundzüge der Lehre vom Lichtsinn*, Springer, Berlin, 1905.
Hermann, *Pflüger's Arch. Ges. Physiol.*, **27**, 291 (1882).
Hermans, *J. Exptl. Psychol.*, **32**, 307 (1943).
Herzau, *Graefe's Arch. Ophthalmol.*, **121**, 756 (1929).
Herzau and Ogle, *Graefe's Arch. Ophthalmol.*, **137**, 327 (1937).
Hess (C.), *Graefe's Arch. Ophthalmol.*, **45**, 157 (1899).
Hess (C.), and Pretori, *Graefe's Arch. Ophthalmol.*, **40** (1894).
Hess (D. P.), and Harrison, *Trans. IES, Nov. 1923.*
Hick, *Quart. J. Exptl. Psychol.*, **2**, 33 (1950).
Hidano, cited by F. P. Fischer, *Pflüger's Arch. Ges. Physiol.*, **204**, 247 (1924).
Higgins and Jones, *J. Soc. Motion Picture Engr.*, **58**, 277 (1952).
Higgins and Stultz, *JOSA*, **38**, 756 (1948); **40**, 135 (1950); **43**, 1136 (1953).
Hill, *Proc. Phys. Soc. (London)*, **59**, 560, 574 (1947).
Hillebrand, *Z. Psychol.*, **16**, 71 (1898); *Wien. Sitzber.*, **72**, 255 (1902); *Jahrb. Psychiat. Neurol.*, **40**, 213 (1920).
Hillman, Connolly, and Farnsworth, *U.S. Naval Med. Res. Lab. Rept.*, No. 257 (1954).
Hirsch and Weymouth, *J. Aviation Med.*, **19**, 56 (1948).
Hirschberg, *Arch. Anat. Physiol.*, *1878*, p. 324.
Hofe, *Graefe's Arch. Ophthalmol.*, **117**, 40 (1926).
Hoffman, *JOSA*, **52**, 75 (1962).
Hoffmann, *Pflüger's Arch. Ges. Physiol.*, *1913*, p. 22.
Hofstetter, *Industrial Vision*, Chilton Co., Philadelphia, 1956.
Hofstetter, *Intern. Opt. Congr. Dublin, 1965.*
Holland, *The Spiral After-Effect*, Pergamon, Oxford, 1965.
Holm, *Acta Ophthalmol.*, **1**, 49 (1923).
Holmes, *Brit. J. Ophthalmol.*, **2**, 352, 449, 506 (1918); *Trans. IES*, **6**, 71 (1941); *Proc. Roy. Soc. (London)*, **B132**, 348 (1945).
Holmgren, *Congr. Intern. Sci. Med. Copenhagen*, **1**, 93 (1884).
Holt, *Psychol. Monogr.*, **4**, 3 (1903).
Holth, *Norsk Mag. Laegev.*, **11**, 148 (1896).
Holtz, *Z. Biol.*, **95**, 102 (1934).
Holway and Boring, *Am. J. Psychol.*, **54**, 21 (1941).
Hooke, *Oeuvres posthumes*, 1705.
Hopkinson, *Quart. J. Exptl. Psychol.*, **2**, 124 (1950); *Nature*, **169**, 40 (1952).
Horner and Purslow, *Nature*, **160**, 23 (1947).
Horsten and Winkelman, *Ann. Oculist. (Paris)*, **187**, 961 (1954).
Houghton and Chalker, *JOSA*, **39**, 155 (1949).
Houston and Shearer, *Phil. Mag.*, **10**, 433 (1930).
Howard, *Am. J. Ophthalmol.*, **2**, 656 (1919); *Trans. Am. Ophthalmol. Soc.*, **17**, 401 (1919); *Optician*, **58**, 285 (1919–1920).
Howard and Evans, *Vis. Res.*, **3**, 447 (1963).

Howard and Templeton, *Vis. Res.*, **4**, 433 (1964).
Howell, *J. Textile Inst.*, **39**, 487 (1948).
Hubel, *J. Physiol. (London)*, **147**, 226, 574 (1959); **150**, 91 (1960).
Hubel and Wiesel, *J. Physiol. (London)*, **155**, 385 (1961); **160**, 106 (1962).
Hueck, *Arch. Anat. Physiol.*, *1840*, p. 82.
Hummelsheim, *Graefe's Arch. Ophthalmol.*, **45**, 357 (1898).
Huschke, Z. *Ophthalmol.*, **4**, 272 (1835).
Hyde, *Am. J. Ophthalmol.*, **48**, 85 (1959).
Hyde and Eason, *J. Neurophysiol.*, **22**, 666 (1959).

Ingvar, *Acta Physiol. Scand.*, **46**, S159 (1959).
Irvine and Ludvigh, *AMA Arch. Ophthalmol.*, **15**, 1037 (1936).
Issel, Diss. Freiburg, 1907.
Ittleston, *Am. J. Psychol.*, **64**, 54 (1951).
Ivanoff, *Compt. Rend.*, **223**, 1185 (1946); **224**, 1183, 1453 (1947); *Rev. Opt.*, **26**, 479 (1947); *Compt. Rend.*, **227**, 234 (1948); *Ann. Opt. Oculaire (Paris)*, **2**, 97 (1953); *Rev. Opt.*, **33**, 369 (1954); *Compt. Rend.*, **241**, 1809 (1955); *JOSA*, **46**, 901 (1956).
Ivanoff and Bourdy, *Ann. Opt. Oculaire (Paris)*, **3**, 70 (1954).
Ives (J. E.), *U.S. Public Health Bull.*, **140** (1924).
Ives (R. L.), *JOSA*, **32**, 119 (1942).

Jackson (E.), *Am J. Ophthalmol.*, **20**, 16 (1937).
Jackson (R.), *Trans. Am. Ophthalmol. Soc.*, **5**, 141 (1888).
Jacobs, *J. Comp. Physiol. Psychol.*, **56**, 116 (1963).
Jacobs and DeValois, *Nature*, **206**, 487 (1965).
Jacobson, *Am. J. Physiol.*, **91**, 567 (1930).
Jaensch, Z. *Psychol.*, **89**, 116 (1922).
Jaensch and Reich, Z. *Sinnesphysiol.*, **86**, 278 (1921).
Jahn, *JOSA*, **36**, 76, 83, 595, 659 (1946).
Jainski, *CIE* (1955).
Jarcho, *J. Neurophysiol.*, **12**, 449 (1949).
Jasper and Cruikshank, *J. Gen. Psychol.*, **17**, 29 (1937).
Jastrow, *Am. J. Psychol.*, **5**, 220 (1893).
Javal, *Compt. Rend.*, **63**, 927 (1866); *Ann. Oculist. (Paris)*, **79**, 97, 240 (1878); **80**, 135 (1879); **82**, 242 (1879).
Jean and O'Brien, *JOSA*, **39**, 12, 1057 (1949).
Jerlov, *Rept. Swed. Deep-Sea Exped.*, **3**, Nos. 1 and 3 (1951).
Jiménez-Landi and Cabello, *Anal. Fis. Quim.*, **39**, 597 (1943).
Johannsen, *J. Gen. Psychol.*, **4**, 282 (1930).
Johns and Sumner, *J. Psychol.*, **26**, 25 (1948).
Johnsgard, *J. Appl. Psychol.*, **37**, 407 (1953).
Jones and Higgins, *JOSA*, **37**, 217 (1947); **38**, 398 (1948).
Judd (C. H.), McAllister, and Steele, *Psychol. Mon.*, **7**, No. 1 (1907).
Judd (D. B.), *J. Res. Natl. Bur. Std. U.S.*, **2**, 441 (1929); *Ann. Opt. Oculaire (Paris)*, **2**, 90 (1953).

Julesz, *JOSA*, **53**, 996 (1963).
Jung, *Electroencephal. Clin. Neurol. Suppl.*, **4**, 57 (1953).
Jurin, in Smith's *Optics*, pp. 96 and 156 (1738).

Kahn, *Pflüger's Arch. Ges. Physiol.*, **227**, 213 (1931).
Kaïla, *Psychol. Forsch.*, **3**, 60 (1923).
Kalff, *Trans. IES*, **17**, 227 (1952).
Kanolt, U. S. Pat. 1,260,682 (1915).
Karpinska, *Z. Psychol.*, **57**, 1 (1910).
Karslake, *J. Appl. Psychol.*, **24**, 417 (1940).
Katz (D.), *The World of Colour*, London, 1935; *Klin. Monatsbl. Augenheilk.*, **111**, 219 (1945).
Katz (M. S.), and Schwartz, *JOSA*, **45**, 523 (1955).
Keesy, *JOSA*, **50**, 769 (1960).
Keller, *J. Exptl. Psychol.*, **28**, 407 (1941).
Kelly, *Psychol. Bull.*, **32**, 569 (1935); *Appl. Opt.*, **4**, 435 (1965).
Kenkel, in Koffka, *Beitr. Psychol. Gestalt, 1919.*
Kepler, *Paralipomena ad Vitelionem*, Frankfurt, 1604.
Kestenbaum, *Graefe's Arch. Ophthalmol.*, **105**, 799 (1921); *Z. Augenheilk.*, **57**, 557 (1925).
Kincaid and Blackwell, *JOSA*, **42**, 873 (1952).
King, *Phil. Trans. Roy. Soc. London*, A212, 375 (1913).
Kirchhof, *Z. Biol.*, **100** (1940).
Kirschberg, *AMA Arch. Ophthalmol.*, **36**, 155 (1946).
Kirschmann, *Phil. Studies*, **6**, 417 (1890).
Kishto, *Vis. Res.*, **5**, 313 (1965).
Klein, *Licht*, **5**, 31 (1934); *Arch. Psychol.*, **39**, 71 (1942).
Kleint, *Z. Psychol.*, **149** (1940).
Kleist, *Klin. Wochschr.*, **5**, 3 (1926).
Knoll, *Am. J. Optom.*, **29**, 69 (1952).
Knoll, Tousey, and Hulburt, *JOSA*, **36**, 480 (1946).
Koffka, *The Growth of the Mind*, London, 1924; *Brit. J. Psychol.*, **14**, 269 (1924); *Principles of Gestalt Psychology*, London, 1935.
Koffka and Harrower, *Psychol. Forsch.*, **15**, 193 (1931).
Kohler (I.), *Wien. Sitzber.*, **227**, 118 (1951).
Köhler (W.), and Emery, *Am. J. Psychol.*, **60**, 159 (1947).
Köhler (W.), and Wallach, *Proc. Am. Phil. Soc.*, **88**, 269 (1944).
Kolers, *Vis. Res.*, **3**, 191 (1963).
Kölliker, *Mikroskopische Anatomie oder Gewebelehre des Menschen*, Leipzig, 1854.
Köllner, *Die Storüngen der Farbensinns*, Berlin, 1912.
Kolmer in Möllendorf, *Handb. Mikroskop. Anat.*, **3**, 295 (1936).
Kompaneisky, *Probl. Fiziol. Opt.*, **2**, 183 (1944).
König, *Sitzber. Preuss. Akad. Wiss., 1897*, p. 559.
Königshöfer, thesis, Erlangen, 1876.

Koomen, Tousey, and Scolnik, *JOSA*, **39**, 370 (1949); **41**, 80 (1951); **43**, 27 (1953).
Kopfermann, *Psychol. Forsch.*, **13**, 293 (1930).
Koschmeider, *Beitr. Physik Freien Atm.*, **12**, 33, 171 (1924).
Krause, *Graefe's Arch. Ophthalmol.*, **26**, 102 (1880).
Krauskopf, *JOSA*, **53**, 741 (1963); **54**, 715 (1964).
Krauskopf, Cornsweet, and Riggs, *JOSA*, **50**, 572 (1960).
Kreiker, *Graefe's Arch. Ophthalmol.*, **111**, 128 (1928).
Kröncke, *Z. Sinnesphysiol.*, **52**, 217 (1921).
Kruithof, *Philips Tech. Rev.*, **6** (1941).
Kuffler, *J. Neurophysiol.*, **16**, 37 (1953).
Kühl, *Physik-Z.*, **1** (1928); *Fragen der Technik*, Munich, 1950.
Kundt, *Ann. Physik*, **120**, 118 (1863).
Kunst, Dissertation, Freiburg, 1895.
Kurtz, *JOSA*, **27**, 323 (1937).

Ladd-Franklin, see Appendix to Helmholtz, *Physiological Optics* (English Transl.), Vol. 2, Optical Society of America, Washington D.C., 1924.
Lagrange, *Bull. Soc. Franc. Ophtalmol.*, 1934, p. 87.
Lamansky, *Pflüger's Arch. Ges. Physiol.*, **2**, 418 (1869).
Lamar, Hecht, Shlaer, and Hendley, *JOSA*, **37**, 531 (1947); **38**, 741 (1948).
Lamare, *Bull. Soc. Franc. Ophtalmol.*, **10**, 354 (1892).
Landolt, *Graefe-Saemisch Handb. Ges. Augenheilk.*, **3** (1874); *Arch. Ophtalmol. (Paris)*, **1**, 385 (1881); **11**, 385 (1891).
Landolt and Nuel, *Graefe's Arch. Ophthalmol.*, **19**, 301 (1872).
Langlands, *Trans. Opt. Soc. (London)*, **28**, 45 (1926–1927).
Langmuir and Westendorp, *Physics*, **1**, 273 (1931).
Langley, *Phil. Mag.*, **27**, 1 (1889).
Langsroth, Johns, Wolfson, and Batho, *Can. J. Res.*, **A25**, 58 (1947).
Lapicque, *Réunion Inst. Opt.*, **7**, 12 (1936).
Lash and Pridaux, *Illum. Eng.*, **38**, 481 (1943).
Lashley, *Gen. Psychol. Mon.*, **37**, 107 (1948).
Latarjet, *Rev. Opt.*, **29**, 65 (1950).
Latour, *Vis. Res.*, **2**, 261 (1962).
Lau, *Z. Sinnesphysiol.*, **53**, 1 (1921).
Lau, Mütze, and Weber, *Graefe's Arch. Ophthalmol.*, **157**, 92 (1955).
Lauenstein, *Psychol. Forsch.*, **22**, 267 (1938).
Laurens, *Z. Sinnesphys.*, **48**, 233 (1914).
Law and DeValois, *Papers Mich. Acad. Sci.*, **43**, 171 (1957).
Lawson, *Nature*, **161**, 154 (1948); **162**, 531 (1948); **165**, 81 (1950).
Lazet and Walraven, *Inst. Percept., Soesterberg Rept.* **IZF 1959–16** (1959).
Lefèvre, *Ann. Opt. Oculaire (Paris)*, **3**, 147 (1955).
Le Grand, *Rev. Opt.*, **11**, 313 (1932); *Compt. Rend.*, **200**, 490 (1935); **202**, 592, 939 (1936); *Ann. Inst. Oceanog.*, **19**, 393 (1939); *Rev. Opt.*, **21**, 71 (1942); *Compt. Rend.*, **215**, 547 (1942); **229**, 1089 (1949); *Onde*

Elect., **31**, 173 (1951); *Ann. Opt. Oculaire (Paris)*, **3**, 154 (1955); *Trans. Intern. Ophthal. Opt. Congr.*, 1961, p. 67; *Vis. Res.*, **3**, 281 (1963); AGARD Avion. Panel Symp., Athens, 1963.
Le Grand and Geblewicz, *Compt. Rend.*, **208**, 1845 (1939).
Le Gros Clark, *J. Anat.*, **75**, 225, 295 (1941); *Brit. J. Ophthalmol.*, **56**, 264 (1942).
Leibowitz and Moore, *JOSA*, **56**, 1120 (1966).
Leibowitz, *JOSA*, **42**, 416 (1952); **43**, 902 (1953); **45**, 829 (1955).
Lena, *Graefe's Arch. Ophthalmol.*, **72**, 1 (1909).
Lennox-Buchtal, *Vis. Res.*, **2**, 1 (1962).
Lenoble, *Compt. Rend.*, **241**, 567 (1955).
Lenz, *Klin. Monatsbl. Augenheilk.*, **80**, 398 (1938).
Levelt, *On Binocular Rivalry*, Inst. Perception, Soesterberg, 1965.
Levine, *Arch. Neurol. Psychiat.*, **67**, 310 (1952).
Levinson, *Doc. Ophthalmol.*, **18**, 36 (1964).
Leyzorek, *J. Exptl. Psychol.*, **39**, 270 (1949).
Li and Jasper, *J. Physiol. (London)*, **121**, 117 (1953).
Liang and Piéron, *Année Psychol.*, **48**, 1 (1947).
Liebermann, *Z. Sinnesphysiol.*, **44**, 428 (1910).
Liebmann, *Psychol. Forsch.*, **9**, 300 (1927).
Lindemann, *Psychol. Forsch.*, **2**, 5 (1922).
Lipkin, *JOSA*, **52**, 1287 (1962).
Lipmann, *Compt. Rend.*, **146**, 446 (1908).
Lipps, *Z. Psychol.*, **1**, 60 (1890).
Lister (1842), in *J. Roy. Microscop. Soc.*, **33**, 34 (1913).
Listing, *Z. Rat. Med.*, **4**, 801 (1854).
Lit, *Am. J. Psychol.*, **62**, 159 (1949); *JOSA*, **49**, 746 (1959); **50**, 321, 970 (1960); **54**, 83 (1964).
Livingstone, *Am. J. Ophthalmol.*, **27**, 349, 428 (1944).
Livshitz, *Dokl. Akad. Nauk SSSR*, **28**, 429 (1940).
Loewe, *Ann. Physik*, **88**, 451 (1852).
Logan, *Illum. Eng.* (Sept. 1939; Dec. 1941; March 1947); *Trans. IES*, **17**, 265 (1952).
Löhle, *Z. Sinnesphysiol.*, 54, 137 (1929).
London, *Arch. Electrol. Méd.* (Feb. 1904).
Lord, *Proc. Phys. Soc.*, **61**, 489 (1948); *Nature*, **170**, 670 (1952); *Ann. Opt. Oculaire (Paris)*, **1**, 13 (1952); *Brit. J. Phys. Opt.*, 1953, p. 85.
Lord and Wright, *Nature*, **162**, 25, 803 (1948).
Lotze, *Wagners Handworterb. Physiol.*, **3**, 183 (1846).
Low, *Am. J. Phys.*, **146**, 573 (1946); **148**, 124 (1947).
Löwenstein and Friedman, *AMA Arch. Ophthalmol.*, **27**, 969 (1942).
Luckiesh and Guth, *Illum. Eng.*, **44**, 650 (1949).
Luckiesh and Holladay, *Illum. Eng.*, **20**, 221 (1925).
Luckiesh and Moss, *J. Franklin Inst.*, **215**, 401 (1933); **230**, 431 (1935); *Illum. Eng.*, **35**, 19 (1940); *AMA Arch. Ophthalmol.*, **23**, 941 (1940); *JOSA*, **31**, 401 (1941).

Ludvigh, *AMA Arch. Ophthalmol.*, **15**, 1037 (1936); *Science*, **105**, 176 (1947); **108**, 63 (1948); *Naval Res. Rept.*, **142023** (1953).
Ludvigh and Miller, U.S. Naval School Aviation Med., 1953.
Luhr and Eckel, *AMA Arch. Ophthalmol.*, **9**, 625 (1933); *Zentralbl. Ges. Ophthalmol.*, **30**, 229 (1934).
Lumière, *Compt. Rend.*, **200**, 701 (1935).
Lythgoe (J. N.), and Hemmings, *Nature*, **213**, 893 (1967).
Lythgoe, *Med. Res. Council. Rept.*, **134** (1929); **173** (1932); *Proc. Phys. Soc. (London)*, **50**, 231 (1938); *Nature*, **141**, 474 (1938).
Lythgoe and Quilliam, *J. Physiol. (London)*, **93**, 24 (1938); **94**, 399 (1938).

MacAdam, *Nature*, **160**, 664 (1947); *J. Soc. Motion Picture Engr.*, **62**, 271 (1954).
McColgin, *JOSA*, **50**, 774 (1960).
MacCulloch and Crush, *AMA Arch. Ophthalmol.*, **36**, 171 (1946).
MacCollough, *J. Exptl. Psychol.*, **49**, 141 (1955).
McCready, *Vis. Res.*, **5**, 189 (1965).
MacDonald, *JOSA*, **43**, 290 (1953).
MacDougall, *Mind*, **12**, 473 (1903); *Brit. J. Psychol.*, **1**, 78, 151 (1904).
Macé de Lépinay and Nicati, *Ann. Chim. Phys.*, **24**, 334 (1881); **30**, 145 (1883).
MacFadden, thesis, Ohio State University, 1940.
MacFarland and Halpern, *J. Gen. Physiol.*, **23**, 613 (1940).
MacFarland, Holway, and Hurvich, *Studies on Visual Fatigue*, Harvard, 1942.
MacFie, Piercy, and Zangwill, *Brain*, **73**, 167 (1950).
Mach, *Wien. Sitzber.*, **52**, 303 (1865); *Analyse der Empfindungen*, Jena, 1886.
MacKeon and Wright, *Proc. Phys. Soc. (London)*, **52**, 464 (1940).
MacLaughlin, *Proc. Armed Forces NRC Vision Comm.*, 35th meeting, 1954.
MacNicholas, *J. Res. Natl. Bur. Std., U.S.*, **17**, 955 (1936).
MacQuarrie, *Optom. Ext. Progr.*, **19**, 23 (1955).
Maison, Settlage, and Grether, *Arch. Med.*, **40**, 981 (1938).
Malin, *Am. J. Optom.*, **32**, 30 (1955).
Malone, cited by Farnsworth, *Colloq. Intern. Heidelberg*, 1955.
Mandelbaum and Sloan, *Am. J. Ophthalmol.*, **30**, 581 (1947).
Marg, *Am. J. Optom.*, **30**, 417 (1953).
Mariotte, *Phil. Trans. Roy. Soc. London*, **3**, 668 (1668).
Mariott, *JOSA*, **49**, 1022 (1959).
Marshall, *Brit. J. Ophthalmol.*, **19**, 177 (1935); *J. Neurophysiol.*, **12**, 277 (1949).
Marshall and Day, *Australian J. Psychol.*, **3**, 1 (1951).
Marshall, Talbot, and Ades, *J. Neurophysiol.*, **6**, 1 (1943).
Martens, *Z. Angew. Psychol.*, **15**, 374 (1919).
Martin, Day, and Kaniowski, *Brit. J. Ophthalmol.*, **34**, 89 (1950).

Martin and Pearse, *Brit. J. Ophthalmol.*, **31**, 129 (1947).
Martin and Wilkins, *JOSA*, **27**, 340 (1937).
Martius, *Phil. Studies*, **5**, 601 (1889).
Marx and Trendelenburg, *Z. Sinnesphysiol.*, **45**, 87 (1911).
Maskelyne, *Phil. Trans. Roy. Soc. London*, **79**, 258 (1789).
Matin, *JOSA*, **52**, 1276 (1962); **54**, 1008 (1964).
Matthews (M. L.), *JOSA*, **56**, 1401 (1966).
Matthiessen, *Compt. Rend.*, **24**, 875 (1847); *Graefe's Arch. Ophthalmol.*, **32**, 97 (1883); *Die neuren Fortschritte in unserer Kenntniss von dem optischen Baue des Auges der Wirbeltiere*, Hamburg, 1891.
Maxwell, *Rept. Brit. Assoc.*, **2**, 12 (1856).
Mayer, *Comment. Gotting.*, **4**, 97, 135 (1754).
Merkel, *Phil. Studies*, **9**, 400 (1894).
Metcalf, *JOSA*, **55**, 72 (1965).
Metzger, *Psychol. Forsch.*, **8**, 114 (1926).
Meur, *Vis. Res.*, **5**, 1965.
Meyer (H.), in *Roser und Wunderlichs Archiv*, 1842, Bd. I.
Meyer (R.), *Arch. Ges. Psychol.*, **96** (1936).
Meyer (T.), *Ann. Physik*, **89**, 429 (1853).
Meyers, *Arch. Neurol. Psychol.*, **21**, 901 (1929).
Meynert, in *Handb. Lehre Geweben Menschen*, **2**, 694 (1872).
Mibai, *Psychol. Monogr.*, **42**, 1931, 91 pp.
Michon and Kirk, *Inst. Percept. Soesterberg, Rept.* **IZF 1962–17** (1962).
Michotte, *La perception de la causalité*, Paris, 1946; *Bull. Acad. Roy. Belg. Classe Lettres*, **34**, 268 (1948).
Middleton and Holmes, *JOSA*, **39**, 582 (1949).
Mie, *Ann. Physik*, **25**, 377 (1908).
Miles, *Am. J. Phys.*, **72**, 239 (1925); *J. Exptl. Psychol.*, **14**, 311 (1931); *Science*, **90**, 404 (1939); **91**, 456 (1940); *Federation Proc.*, **2**, 109 (1943).
Miller and Ludvigh, *JOSA*, **51**, 57 (1961).
Millodot, *Brit. J. Physiol. Opt.*, **23**, 75 (1966); *Atti Fond. Giorgio Ronchi*, **22**, 86 (1967).
Minkowski, *Schweiz. Arch. Neurol. Psychiat.*, **6**, 201 (1920); **7**, 268 (1920); *L'Encéphale*, **17**, 65 (1922).
Mitchell and Ellerbrock, *Am. J. Optom.*, **32**, 520 (1955).
Mitchell (R. T.), and Liaudansky (L. H.), *JOSA*, **45**, 831 (1955).
Mollweide, *Gilberts Ann.*, **17**, 328 (1805).
Monakow, *Die Lokalisation in Grosshirn*, Wiesbaden, 1914.
Monbrun, Thèse Méd., Paris, 1914.
Monjé, *Z. Sinnesphysiol.*, **65**, 239 (1934); *Arch. Ges. Psychol.*, **91**, 475 (1934).
Monjé and Schober, *Klin. Monatsbl. Augenheilk.*, **117**, 561 (1950).
Monnier (A.), *J. Physiol. Pathol. Gen.*, **38**, 135 (1943); *J. Physiol. (Paris)*, **38**, 238 (1945).

Monnier (A.), and Rösch (J.), *Arch. Intern. Physiol.*, **54**, 225 (1946).
Monnier (M)., *Helv. Phys. Acta*, **6**, 61 (1948); **7**, C52 (1949); *Electroencephal. Clin. Neurophysiol.*, **1**, 87 (1949).
Moon and Spencer, *JOSA*, **34**, 319, 605, 744 (1944); **35**, 233 (1945); **38**, 650 (1948).
Moreland, *Luzern Intern. Colour Meeting*, 1965.
Morino, *Arch. Oftalmol. Hispanoam.* (April 1933).
Morris, *JOSA*, **43**, 815 (1953).
Morris and Dimmick, *JOSA*, **40**, 795 (1950).
Morrison, *Trans. Intern. Opt. Congr.*, 1951, p. 55; 1961, p. 95.
Mote, *JOSA*, **45**, 7 (1955).
Motokawa, *J. Neurophysiol.*, **12**, 291, 475 (1949); **13**, 413 (1950).
Mowrer, Ruch, and Miller, *Am. J. Physiol.*, **114**, 423 (1936).
Mueller, *Am. J. Psychol.*, **63**, 92 (1950); *J. Gen. Physiol.*, **34**, 463 (1951).
Mueller and Lloyd, *Proc. Natl. Acad. Sci. U.S.*, **34**, 223 (1948).
Mulholland and Evans, *Nature*, **207**, 36 (1965).
Müller (G. E.), *Z. Psychol.*, Erg. Bd., **9**, 11 (1917).
Müller (H.), *Verhandl. Physik. Med. Ges. Würtzburg*, **4**, 96 (1853).
Müller (H. K.), *Z. Sinnesphysiol.*, **59**, 157 (1928); *Graefe's Arch. Ophthalmol.*, **125**, 614 (1931).
Müller (J.), *Zur Physiologie des Gesichtsinnes*, Leipzig, 1826; *Ueber die Phantastaschen Gesichtserscheinungen*, Coblentz, 1826.
Müller-Lyer, *Z. Psychol.*, **9**, 1 (1896); **10**, 421 (1896).
Münster, *Z. Sinnesphysiol.*, **69**, 245 (1941).
Münsterberg, *Beitr. Exptl. Psychol.*, **2**, 125, Mohr, Freiburg (1889).
Musatti, *Arch. Ital. Psichiat.*, **3**, 105 (1924).
Muscio, *Brit. J. Psychol.*, **12**, 31 (1921).
Musylev, *Acta Ophthalmol.*, **15**, 216 (1937).

Nachmias, *JOSA*, **48**, 726 (1958); **49**, 901 (1959); **50**, 569 (1960); **51**, 761 (1961).
Necker, *Phil. Mag.*, **3**, 329 (1832).
Nelson, Stark, and Young, *Mass. Inst. Technol. Res. Lab. Electron. Quart. Progr. Rept.*, **67**, 214 (1962).
Netusil, *Elektrotech. Cesk.*, **1941**, 104.
Neuberger, *JOSA*, **30**, 258 (1940).
Neuhaus, *Z. Sinnesphysiol.*, **137**, 87 (1936).
Newhall, *Am. J. Psychol.*, **40**, 628 (1928); *JOSA*, **25**, 63 (1935); **27**, 165 (1937).
Newton, *Optics*, London, 1704.
Niven and Brown, *JOSA*, **34**, 738 (1944).
Noji, *Graefe's Arch. Ophthalmol.*, **122**, 571 (1929).
Nordmann and Lieou, *Rev. Neurol. Ophthalmol.*, **6**, 81, 373 (1928).
Nussbaum, *Arch. Augenheilk.*, **87**, 142 (1920).
Nutting, *Proc. Roy. Soc. (London)*, **A90**, 440 (1914); *Trans. Am. IES*, **9**, 633 (1914).

O'Brien, *JOSA*, **40**, 796 (1950); **41**, 882 (1951); **42**, 74 (1952); *Air Res. Tech. Rept.*, **53.206**, 1953.
O'Brien and Dickerman, *JOSA*, **38**, 1096 (1948).
O'Brien and Miller, *JOSA*, **42**, 289 (1952).
O'Brien and O'Brien, *JOSA*, **41**, 882 (1951).
O'Brien and Perrin, *JOSA*, **27**, 63 (1937).
Ogden and Brown, *J. Neurophysiol.*, **27**, 682 (1964).
Ogilvie and Ryan, *JOSA*, **45**, 206 (1955).
Ogilvie, Ryan, Cowan, and Querengesser, *J. Appl. Psychol.*, **7**, 519 (1955).
Ogle, *JOSA*, **30**, 145 (1940); **41**, 517 (1951); **43**, 907 (1953); **50**, 307 (1960); **52**, 1035 (1962); **53**, 1296 (1963).
Ogle and Martens, *AMA Arch. Ophthalmol.*, **57**, 702 (1957).
Ogle and Reiher, *Vis. Res.*, **2**, 439 (1962).
Ogle and Weil, *AMA Arch. Ophthalmol.*, **59**, 4 (1958).
Ogle, Massey, and Prangen, *Am. J. Ophthalmol.*, **32**, 1069 (1949).
Ohm, *Z. Augenheilk.*, **32**, 4 (1914); *Graefe's Arch. Ophthalmol.*, **120**, 235 (1928).
Ohrwall, *Upsala Lakareforenings Forh.*, **17**, 441 (1912); *Skand. Arch. Physiol.*, **27**, 65, 304 (1912).
Oliva and Aguilar, *Opt. Acta*, **3**, 36 (1956).
Olivier, *Météorologie*, No. 36, 19 (1954).
Oppel, *Jahresber. Frank. Vereins*, **1854**, 52.
Oppenheimer, *Psychol. Forsch.*, **20**, 1 (1934).
Orbison, *Am. J. Psychol.*, **52**, 31, 309 (1939).
Osterberg, *Acta Ophthalmol.*, **11**, 204 (1933); **13** (Suppl. 6), 103 pp. (1935).
Otero, *Sobre las causas de las ametropias naturales de la vision nocturna*, Madrid, 1949.
Otero and Aguilar, *Anal. Fis. Quim.*, **A46** (1950);*JOSA*, **41**, 1061 (1951).
Otero and Duran, *Anal. Fis. Quim.*, **37**, 459 (1941).
Otero, Plaza, and Rios, *Anal. Fis. Quim.*, **A44**, 293 (1948).
Otero, Plaza, and Salaverri, *JOSA*, **39**, 167 (1949).
Otero, Vigon, and Galvez, *Anal. Fis. Quim.*, **A46**, 73 (1950).

Pages, *Journèes de l'Eclairage*, Monaco, 1954, p. 32.
Palacios, *Portugaliae Physicae*, 1944, **1**, 47 (1944).
Palmer, *Opt. Acta*, **9**, 311 (1962).
Panum, *Untersuchungen über das Sehen mit zwei Augen*, Kiel, 1858; *Arch. Anat. Physiol.*, *1861*, 63, 178.
Parent, *Diagnostic et détermination de l'astigmatisme*, Paris, 1881.
Park, *Am. J. Ophthalmol.*, **19**, 967 (1936).
Park and Park, *Am. J. Physiol.*, **104**, 545 (1933); *AMA Arch. Ophthalmol.*, **23**, 1216 (1940).
Parsons, *Lancet*, **256**, 1120 (1949).
Patel, *JOSA*, **56**, 689 (1966).

Paterson (A.), and Zangwill, *Brain*, **67**, 331 (1944).
Paterson (D. G.), and Tinker, *J. Appl. Psychol.*, **13**, 120, 205 (1929); **14**, 211 (1930); **15**, 72, 241, 471 (1931); **16**, 388, 525, 605 (1932); **20**, 128, 132 (1936); *How to Make Type Readable*, New York, 1940; *J. Appl. Psychol.*, **26**, 227 (1942); **30**, 161 (1946).
Peckham, *AMA Arch. Ophthalmol.*, **12**, 562 (1934); *Am. J. Psychol.*, **48**, 43 (1936).
Peckham and Arner, *JOSA*, **42**, 621 (1952).
Penfield, *Proc. Roy. Soc. (London)*, **B134**, 329 (1947).
Perez-Cirera, *Arch. Augenheilk.*, **105**, 453 (1932).
Pergens, *Ann. Oculist. (Paris)*, **135**, 11 (1906).
Perrin and Altman, *JOSA*, **43**, 780 (1953).
Peterman, *Arch. Ges. Psychol.*, **41**, 351 (1924).
Petherbridge and Hopkinson, *Trans. IES*, **15**, 39 (1950).
Peyrou and Piatier, *Compt. Rend.*, **223**, 589 (1946).
Pfaffman, *Am. J. Psychol.*, **61**, 383 (1948).
Pfeifer, *Myelogenetisch anat. Unters. über den zentralen Abschnitt der Sehbahn*, Berlin, 1925.
Pfeiffer, *Arch. Ges. Psychol.*, **98** (1937).
Pheiffer, *Am. J. Optom.*, **32**, 540 (1955).
Pi, *Trans. Ophthalmol. Soc. U.K.*, **45**, 393 (1925).
Piaget, *La naissance de l'intelligence chez l'enfant*, Nuechatel, 1936; *La construction du réel chez l'enfant*, Neuchatel, 1937; *La formation du symbole chez l'enfant*, Neuchatel, 1945; *Les notions de mouvement et de vitesse chez l'enfant*, Paris, 1946.
Pickard, *Brit. J. Ophthalmol.*, **19**, 481 (1935).
Pickford, *Nature*, 159, 268 (1947).
Pierce, *Science*, **10**, 425 (1899).
Piéron, *Année Psychol.*, 1927, 592; 1929, 87, 223; *J. Psych. Paris*, **31**, 1 (1939); *Année Psychol.*, 1951, 161.
Pieron, *Bull. Inst. Psychol. Intern.*, **1**, 202 (1901).
Pinegin, *Compt. Rend. Acad. Sci. URSS*, **30**, 206 (1941); also in "Visual Problems of Colour," *Natl. Phys. Lab. (Gt. Brit.) Symp.*, **2**, 727 (1958).
Pirenne, *Nature*, **154**, 741 (1944); *Proc. Cambridge Phil. Soc.*, **42**, 78 (1945); *J. Physiol. (London)*, **107** (1948); *Brit. J. Phil. Sci.*, **3**, 169 (1952); *Brit. Med. Bull.*, **9**, 61 (1953).
Pirenne and Denton, *JOSA*, **41**, 426 (1951); *Nature*, **170**, 1039 (1952).
Pirenne and Marriott, *Opt. Acta*, **1**, 161 (1954); *JOSA*, **45**, 909 (1955).
Pitts and MacCulloch, *Bull. Math. Biophys.*, **9**, 127 (1947).
Plateau, *Mem. Acad. Bruxelles*, **3**, 7, 364 (1836); **11** (1838).
Pokrowski, Z. *Physik*, **35**, 776 (1926).
Polack, thesis, Paris, 1900.
Polliot, *Arch. Ophtalmol. (Paris)*, **41**, 98, 547 (1921).
Ponder and Kennedy, *Quart. J. Exptl. Psychol.*, **18**, 89 (1927).
Popov, *Compt. Rend.*, **236**, 744 (1953); **237**, 930, 1439 (1953); **238**, 2026 (1954); **239**, 1859 (1954); **240**, 1268 (1955).

Poppelreuter, *Die psychischen Schädigungen durch Kopfschuss,* **1**, 149; *Z. Physiol. Sinnerorg., Abt. 1,* **58**, 200 (1911).
Porterfield, *On the Eye,* **2**, 285 (1759).
Pouillard, *J. Psychol.,* **30**, 887 (1933).
Pourfour du Petit, *Mem. Acad. Sci. Paris, 1728,* 408.
Prentice, *J. Exptl. Psychol.,* **38**, 284 (1948).
Pritchard, *Quart. J. Exptl. Psychol.,* **10**, 77 (1958).

Quételet, *Bull. Acad. Bruxelles,* **8**, 160 (1841).

Rady and Ishak, *JOSA,* **45**, 530 (1955).
Raman, *Phil. Mag.,* **38**, 568 (1919).
Ranke, *Arbeitsphysiol.,* **15**, 427 (1954).
Rashbass, *Nature,* **183**, 897 (1959); *JOSA,* **50**, 642 (1960).
Rashbass and Westheimer, *J. Physiol. (London),* **159**, 149 (1961).
Ratliff, *J. Exptl. Psychol.,* **43**, 163 (1952).
Ratliff, *Mach Bands,* Holden-Day, San Francisco, 1965.
Ratoosh, *Proc. Natl. Acad. Sci. U.S.,* **35**, 257 (1949).
Rayleigh, *Phil. Mag.,* **41**, 107, 274 (1871); *Nature,* **25**, 64 (1881); *Proc. Cambridge Phil. Soc.,* **4**, 324 (1883).
Reese, *JOSA,* **29**, 519 (1939); *Psychophys. Res. Rept.,* **131**, 194 pp. (1953).
Reeves, *Astrophys. J.,* **46**, 167 (1917); **47**, 141 (1918); **48**, 76, 79 (1918); *Phil. Mag.,* **35**, 174 (1918).
Richards, *JOSA,* **43**, 331 (1953).
Richardson, *Psychol. Rev.,* **37**, 218 (1930).
Riddell, *Silliman J.,* **15** (1853).
Riggs and Niehls, *JOSA,* **50**, 913 (1960).
Riggs and Ratliff, *JOSA,* **39**, 630 (1949); *J. Exptl. Psychol.,* **40**, 687 (1950).
Riggs, Armington, and Ratliff, *JOSA,* **44**, 315 (1954).
Riggs, Mueller, Graham, and Mote, *JOSA,* **37**, 415 (1947).
Riggs, Ratliff, and Cornsweet, *JOSA,* **43**, 495 (1953).
Ripps and Weale, *JOSA,* **27**, 350 (1937).
Ritter, *Psychol. Mon.,* **23**, 1 (1917).
Roaf, *Proc. Roy. Soc. (London),* **B106**, 276 (1930).
Robertson, *AMA Arch. Ophthalmol.,* **15**, 423 (1936).
Rochat, *Arch. Neerl. Physiol.,* **7**, 263 (1922).
Rodieck, *Vision Res.,* **5**, 583 (1965).
Rodin and Newell, *AMA Arch. Ophthalmol.,* **12**, 525 (1934).
Roelofs, *Arch. Neerl. Physiol.,* **2**, 199 (1918).
Roelofs and van den Bend, *Arch. Augenheilk.,* **102**, 551 (1930).
Roelofs and Zeeman, *Arch. Neerl. Physiol.,* **3**, 562 (1919).
Rogers, *Edinburgh J.,* **3**, 210 (1855); *Silliman J.,* **20**, 204, 318 (1855); **21**, 80, 173, 439 (1855); **30**, 387, 404 (1860).
Roget, *Phil. Trans. Roy. Soc. London,* **1**, 131 (1825).
Rollmann, *Ann. Physik,* **90**, 186 (1853).
Ronchi (L.), *Atti Fond. Giorgio Ronchi,* **4** (1949); **5**, 71, 200 (1950); **6**,

14, 54 (1951); **7**, 255 (1952); **10**, 285 (1955); *Opt. Acta*, **2**, 47 (1955).
Ronchi (L.), and Zoli, *Opt. Acta*, **2**, 30, 112 ,215, 380 (1955).
Ronchi (V.), *Ottica*, **8**, 153 (1943).
Rönne, *Klin. Monatsbl. Augenheilk.*, **45**, 455 (1910); *Arch. Augenheilk.*, **78**, 284 (1915).
Rood, *Silliman J.*, **30**, 264, 385 (1860).
Rösch (J.), Mesures stéréoscopiques appliquées à l'astronomie et recherches connexes d'optique physiologique, thesis, Paris, 1943.
Rösch (S.), *Sond. Praxis Phys. Chem. Photogr.*, **3**, 331 (1954).
Roscoe, *J. Appl. Phys.*, **32**, 649 (1948).
Rose, *JOSA*, **38**, 196 (1948).
Rosenberg, Flax, Brodsky, and Abelman, *Am. J. Optom.*, **30**, 244 (1953).
Rosenbloom, *Am. J. Optom.*, **30**, 563 (1953).
Rosenbrück, *Optik*, **16**, 135 (1959).
Rouleau, *Publ. Sci. Tech. Min. Air France*, **48** (1934).
Rubin, *Visuell Wahrgennomene Figuren*, Copenhagen, 1921.
Ruffer, *Ind. Psychotech.*, **5**, 161 (1928).
Rule, *JOSA*, **31**, 124, 325 (1941).
Ruppert, *Z. Sinnesphysiol.*, **42**, 409 (1908).
Rushton, *Nature*, **164**, 473 (1949).
Russel, *Astrophys. J.*, **45**, 60 (1917).
Ryan, Cottrell, and Bitterman, *Am. J. Psychol.*, **63**, 317 (1950); *Illum. Eng.*, **43**, 1074 (1948); **48**, 385 (1953).

Samsonova, *Graefe's Arch. Ophthalmol.*, **135**, 30 (1936).
Sauer, *JOSA*, **27**, 350 (1937).
Scheerer, *Klin. Monatsbl. Augenheilk.*, **74**, 688 (1925).
Schlag, *L'activité spontanée des cellules du système nerveux central*, Brussels, 1959.
Schmidt, *Kolloid-Z.*, **85**, 137 (1938).
Schmidt-Rimpler, *Jahresbuch Virchow und Hirsch*, **2**, 476 (1879).
Schober, *Graefe's Arch. Ophthalmol.*, **148**, 171 (1947); *Optik.*, **11**, 282 (1954).
Schön, *Graefe's Arch. Ophthalmol.*, **22**, 31 (1816); **24**, 27 (1878).
Schön, *Graefe's Arch. Ophthalmol.*, **31** (1885); *Arch. Augenheilk.*, **27**, 268 (1893).
Schön and Mosso, *Graefe's Arch. Ophthalmol.*, **20**, 289 (1874).
Schott, *Deut. Arch. Klin. Med.*, **140**, 79 (1922).
Schröder, *Ann. Physik*, **105**, 298 (1858).
Schultze, *Arch. Mikroskop. Anat.*, **2**, 165, 175 (1866).
Schupfer, *Ottica*, **8**, 196 (1943); *Boll. Assoc. Ottalmol. Ital.*, **18** No. 113 (1944).
Schwartz and Sandberg, *U.S. Naval Med. Res. Lab. Rept.*, **253** (1954).
Schweizer (1858), cited by Exner, *Z. Psychol.*, **12**, 313 (1896).
Scott and Sumner, *J. Psychol.*, **27**, 479 (1949).

Seashore, *Ann. Psychol. Lab. Yale*, **1**, 56 (1893).
Seliger and MacElroy, *Light, Physical and Biological Action*, Academic Press, New York, 1965, p. 302.
Selwin, *Proc. Phys. Soc. (London)*, **55**, 286 (1943).
Senden, *Raum- und Gestaltauffassung, bei operierten Blindgeboren*, Leipzig, 1932.
Sewall, *Am. J. Physiol.*, **108**, 409 (1934).
Sexton, Malone, and Farnsworth, *U.S. Naval Med. Res. Lab. Rept.*, **9**, 301 (1950).
Shaad, *J. Exptl. Psychol.*, **18**, 79 (1935).
Shade, *Natl. Bur. Std. U.S. Circ.*, **526**, 1954, p. 248.
Shaxby, *Nature*, **160**, 24 (1947).
Sheard, *JOSA*, **12**, 79 (1926); *Am. J. Optom.*, **13**, 281 (1936).
Sherif, *The Psychology of Social Norms*, New York, 1936.
Sherrington, *Brit. J. Psychol.*, **1**, 26 (1904).
Shlaer, *J. Gen. Physiol.*, **21**, 165 (1937).
Shlaer, Smith, and Chase, *J. Gen. Physiol.*, **25**, 553 (1942).
Sholl, *J. Physiol. (London)*, **124**, 23 (1954).
Shortess, *JOSA*, **53**, 1423 (1963).
Shortess and Krauskopf, *JOSA*, **51**, 555 (1961).
Shurcliff, *JOSA*, **45**, 399 (1955).
Siedentopf, *Z. Meteorol.*, **2**, 110 (1948).
Siedentopf, Meyer, and Wempe, *Z. Instrumentenk.*, **61**, 372 (1941).
Silber, *Hygiène et Santé* (in Russian), Feb. 1955, p. 20.
Silva, *J. Comp. Anat.*, **106**, 463 (1956).
Simonson and Brozek, *JOSA*, **38**, 384 (1948).
Sinclair, *JOSA*, **37**, 475 (1947).
Singer, *JOSA*, **51**, 61 (1961).
Sinsteden, *Ann. Physik*, **111**, 336 (1860).
Skubich, *Z. Psychol. Physiol. Sinn.*, **96**, 353 (1925).
Sleight, *J. Appl. Psychol.*, **32**, 170 (1948).
Sleight and Mowbray, *J. Psychol.*, **31**, 121 (1951).
Sleight, Mowbray, and Austin, *J. Psychol.*, **33**, 279 (1952).
Sloan, *Am. J. Ophthalmol.*, **30**, 705 (1947); **33**, 1077 (1950); *AMA Arch. Ophthalmol.*, **45**, 704 (1951); *JOSA*, **45**, 402 (1955).
Sloan and Altman, *JOSA*, **43**, 473 (1953).
Sloan and Habel, *JOSA*, **45**, 599 (1955).
Slotopolsky, *Zentralbl. Ges. Ophthalmol.*, **24**, 513 (1931).
Smirnov, *Biofizika*, **6**, 52 (1961).
Smith (A. G.), *JOSA*, **45**, 482 (1955).
Smith (B. B.), *Nature*, **189**, 776 (1961); **191**, 732 (1961).
Smith (J. R.), *J. Gen. Psychol.*, **14**, 318 (1936).
Smith (R.), *A Complete System of Optics*, Cambridge, 1738.
Smith (S.), *J. Exptl. Psychol.*, **36**, 518 (1946).
Smith (W. M.), and Warter, *JOSA*, **50**, 245 (1960).
Snellen, *Letterproeven ter bepaling der Gezigtsscherpte*, Utrecht, 1862; *Congr. Ophthalmol. Paris, 1862.*

Solis, Aguilar, and Plaza, *Anal. Fis. Quim.*, **A49**, 274 (1953).
Sparrow, *Astrophys. J.*, **44**, 76 (1916).
Spekreyse and van der Tweel, *Nature*, **205**, 913 (1965).
Sperling, *JOSA*, **44**, 351 (1954).
Sperling and Farnsworth, *U.S. Naval Med. Res. Lab. Rept.*, **9**, 128 (1950).
Spring and Stiles, *Brit. J. Ophthalmol.*, **32**, 347 (1948).
Stanworth and Naylor, *Brit. J. Ophthalmol.*, **34**, 282 (1950).
Stavrianos, *Arch. Psychol.*, **296** (1945).
Stein. *JOSA*, **37**, 944 (1947).
Stevens (G. T.), *Graefe's Arch. Ophthalmol.*, **26**, 181 (1897); *Norris-Olivers Handb.*, 1897.
Stevens (H. C.), *Psychol. Rev.*, **15**, 69 (1908).
St. George, *J. Gen. Phys.*, **35**, 495 (1952).
Stigler, *Arch. Ges. Physiol.*, **134**, 365 (1910).
Stiles, *Proc. Roy. Soc. (London)*, **B104**, 322 (1929); **B105**, 131 (1929); *Proc. Phys. Soc. (London)*, **58**, 41 (1946); *Doc. Ophthalmol.*, **3**, 138 (1949).
Stiles, Bennett, and H. Green, *Brit. Aeron. Res. Commun. Tech. Rept.*, **1793** (1937).
Stine, *Am. J. Ophthalmol.*, **13**, 101 (1930).
Stoddard and Morgan, *Am. J. Optom.*, **19**, 460 (1942).
Stratton, *Psychol. Rev.*, **3**, 611 (1896); **4**, 341, 463 (1897); **5**, 632 (1898); **7**, 429 (1900); **9**, 433 (1902); **13**, 82 (1906).
Sundberg, *Skand. Arch. Physiol.*, **34** (1917); **36** (1918).
Sutherland, *Nature*, **197**, 118 (1963).
Swan, *Brit. J. Photogr.*, **10**, 367 (1863).
Sweet and Bartlett, *Am. J. Psychol.*, **51**, 400 (1948).
Swenson, cited by Piéron, *La sensation, guide de vie*, Paris, 1932.
Szily, *Pflüger's Arch. Ges. Physiol.*, **105**, 964 (1921).

Tagawa, *Arch. Augenheilk.*, **99**, 587 (1928).
Tani, Ogle, Weaver, and Martens, *AMA Arch. Ophthalmol.*, **55**, 174 (1956).
Tassëel, *J. Sci. Meteor.*, **3**, 9 (1951).
Taylor (D. W.), and Boring, *Am. J. Psychol.*, **55**, 102 (1942).
Taylor (E. A.), *Controlled Reading*, Chicago, 1937.
Teichner, Kobrick, and Dusek, *JOSA*, **45**, 913 (1955); **46**, 122 (1956).
ten Doesschate, *Ophthalmologica*, **112**, 1 (1946).
Teuber and Bender, *J. Gen. Psychol.*, **40**, 37 (1949).
Thelin, *J. Exptl. Psychol.*, **10**, 321 (1927).
Thomas, Dimmick, and Luria, *Vis. Res.*, **1**, 108 (1961).
Thomas (J. P.), *JOSA*, **55**, 521 (1965).
Thomas (P. L.), and Strong, *Nature*, **212**, 51 (1966).
Thompson, Woolsey, and Talbot, *J. Neurophysiol.*, **13**, 277 (1950).
Thomson, *J. Physiol. (London)*, **106**, 59, 368 (1947); *Brit. Med. Bull.*, **9**, 50 (1953).

Thomson and Wright, *J. Physiol. (London)*, **105**, 316 (1947).
Thorner, *Arch. Ges. Psychol.*, **71**, 127 (1929).
Thouless, *Brit. J. Psychol.*, **21**, 339 (1931); **22**, 1 (1931); *Nature*, **142**, 418 (1938).
Tibi, *Ann. Oculist. (Paris)*, **184**, 606 (1951).
Tiffin, *Industrial Psychology*, Prentice-Hall, Englewood Cliffs, N.J., 1942.
Tinker, *Am. J. Psychol.*, **43**, 115 (1931); *Industrial Psychology*, New York, 1942; *Am. J. Optom.*, **21**, 213 (1944); *J. Exptl. Psychol.*, **36**, 453 (1946).
Tonner, *Pflüger's Arch. Ges. Physiol.*, **247**, 169 (1943).
Toraldo di Francia, *Nuovo Cimento*, **5**, 589 (1948); *JOSA*, **39**, 342 (1949).
Toraldo di Francia, Meeting on Contemporary Optics, Firenze, 1954.
Toraldo di Francia and Ronchi, *JOSA*, **42**, 782 (1952); *Ann. Opt. Oculaire (Paris)*, **2**, 33 (1953).
Toulmin-Smith and Green, *Illum. Eng.*, **26**, 304 (1933).
Tousey and Hulburt, *JOSA*, **38**, 886 (1948).
Tousey and Koomen, *JOSA*, **43**, 117, 927 (1953).
Travis (1949), cited in reference [76].
Travis and Martin, *J. Exptl. Psychol.*, **17**, 773 (1934).
Trendelenburg, *Z. Sinnesphysiol.*, **48**, 199 (1913); *Der Gesichtsinn*, Berlin, 1943.
Trincker, *Psychol. Forsch.*, **24**, 513 (1954).
Troelstra, Zuber, Miller, and Stark, *Vis. Res.*, **4**, 585 (1964).
Troxler, *Oph. Bib. Iena*, **2**, 7 (1804).
Tschermak, *Graefe's Arch. Ophthalmol.*, **47**, 508 (1899); *Pflüger's Arch. Ges. Physiol.*, **204**, 177 (1924); **241**, 455 (1939).
Tschermak and Höfer, *Pflüger's Arch Ges. Physiol.*, **98**, 299 (1903).
Tschermak and Kiribuchi, *Pflüger's Arch. Ges. Physiol.*, **81**, 328 (1900).

Uhthoff, *Graefe's Arch. Ophthalmol.*, **32**, 171 (1886); **36**, 33 (1890).

Van de Geer and Moraal, *Inst. Perception Soesterberg Rept.* **IZF 1963–5** (1963).
van den Brink, *Acta Electron.*, **2**, 44 (1957); *Opt. Acta*, **5**, Spec. No. (1958); *Vis. Res.*, **2**, 233 (1962).
van den Brink and Bouman, *JOSA*, **47**, 612 (1957).
van den Horst and van Essen, *Arch. Ges. Psychol.*, **87**, 287 (1933).
van der Meulen, *Graefe's Arch. Ophthalmol.*, **19**, 100 (1873).
van der Pijl, *Rev. Opt.*, **8**, 155 (1929).
van der Tweel, *Doc. Ophthalmol.*, **18**, 287 (1964).
Van der Velden, *Physica*, **11**, 179 (1944).
van Heel, *Dioptric Rev.*, **37**, 496 (1935); *JOSA*, **36**, 237 (1946).
van Heuven, *Brit. J. Psychol.*, **17**, 127 (1926); *Acta Ophthalmol.*, **15**, 109 (1937).
Van Hof, *Vis. Res.*, **6**, 89 (1966).
Vaughn and Hull, *Nature*, **206**, 720 (1965).
Veniar, *J. Psychol.*, **26**, 461 (1948).

Verhoeff, *AMA Arch. Ophthalmol.*, **13**, 151 (1935); *Am. J. Ophthalmol.*, **48**, 339 (1959).
Vermeulen and de Boer, *CIE, 1948.*
Vernon, *Med. Res. Council Rept.*, **148**, 1930; *Brit. J. Psychol.*, **25**, 186 (1934); **28**, 1, 115 (1937).
Verwey, *Arch. Neerl. Physiol.*, **3**, 76 (1918).
Vieth, *Ann. Physik*, **58**, 233 (1818).
Vilter, *Compt. Rend. Soc. Biol.*, **148**, 220 (1954).
Volkmann, *Wagners Handworterb. Physiol.*, 1846; *Graefe's Arch. Ophthalmol.*, **5**, 1 (1859).
Volkmann (F. C.), *JOSA*, **52**, 571 (1962).
von Bahr, *Acta Ophthalmol.*, **23**, 1 (1945).
von Frisch, *Bull. Animal Behavior*, **1**, 5 (1947); *Experientia*, **5**, 142 (1949).
von Kries, *Z. Physiol.*, **8**, 694 (1895); *Z. Sinnesphysiol.*, **9**, 81 (1895); **44**, 165 (1909); *Naturwiss.*, **11**, 461 (1923).
von Kries and Eyster, *Z. Sinnesphysiol.*, **41**, 373 (1906).
Vos, *JOSA*, **50**, 785 (1960); *Inst. Perception Soesterberg Rept.* **IZF 1961–11**; *Ophthalmologica*, **145**, 442 (1963).
Vos and Boogaard, *JOSA*, **53**, 869 (1963).
Vos and Bouman, *JOSA*, **54**, 95 (1964).

Wadworth, *Trans. Am. Ophthalmol. Soc.*, *1876*, p. 342.
Wagner (J.), *Z. Psych.*, **80**, 1 (1918).
Wagner (R)., *Z. Biol.*, **100**, 421 (1941).
Wald, *JOSA*, **42**, 171 (1952); *Licht Technik*, **No. 6**, 1953.
Wald and Griffin, *JOSA*, **37**, 321 (1947).
Waldram, *Trans. IES*, **10**, 147 (1945).
Walker, *J. Aviation Med.*, **12**, 218 (1941).
Walls, *The Vertebrate Eye*, Cranbrook Press, Bloomfield Hills, Mich., 1942; *JOSA*, **33**, 487 (1943); *Opt. J. Rev. Optom.*, **85**, 33 (1948); *Am. J. Optom.*, **28**, 55, 115, 173 (also as Monogr. 117) (1951); **31**, 329 (1954).
Walls and Mathews, *New Means of Studying Color Blindness*, U. Calif. Publ. Psychol. **7**:1, Berkeley, 1952.
Warden, Brown, and Ross, *J. Exptl. Psychol.*, **35**, 37 (1945).
Washburn, *Proc. Natl. Acad. Sci. U.S.*, **19**, 773 (1933).
Washer and Rosberry, *JOSA*, **41**, 597 (1951).
Waterman, *Science*, **111**, 252 (1950).
Watrasiewicz, *JOSA*, **56**, 499 (1966).
Weale, *CIE, 1955.*
Weaver, *Psychol. Bull.*, **28**, 211 (1931).
Weber, *Wagner Handworterb. Physiol.*, **3**, 588 pp. (1846).
Weekers, *Acta Ophthalmol.*, **110**, 43 (1945).
Weigel, *Licht*, **5**, 15, 43, 71, 211, 279 (1935).
Weinstein and Arnulf, *Commun. Lab. Inst. Opt.*, **2**, 1 (1946).
Wentworth, *Am. J. Ophthalmol.*, **14**, 1118 (1931).

Wenzel, Z. *Sinnesphysiol.*, **100**, 298 (1926).
Werner, *Psych. Monatschr.*, **49**, No. 218 (1937); *Am. J. Physiol.*, **51**, 489 (1938).
Wertheim, *Graefe's Arch. Ophthalmol.*, **33**, 137 (1887); Z. *Sinnesphysiol.*, **7**, 172 (1894).
Wertheimer, Z. *Psychol.*, **61**, 161 (1912).
Westheimer, *AMA Arch. Ophthalmol.*, **52**, 710, 932 (1954); *Opt. Acta*, **2**, 151 (1955); *JOSA*, **49**, 504 (1959); *JOSA*, **53**, 86 (1963).
Westheimer and Campbell, *JOSA*, **52**, 1040 (1962).
Westheimer and Conover, *J. Exptl. Psychol.*, **47**, 283 (1954).
Westheimer and Tanzman, *JOSA*, **46**, 116 (1956).
Weston, *Trans. IES, Sept.* 1949; **18** (1953).
Weston and Taylor, *Med. Res. Council Rept.*, 1935.
Weymouth, Anderson, and Averill, *Am. J. Physiol.*, **63**, 410 (1923); **64**, 561 (1923); *J. Comp. Physiol.*, **5**, 147 (1925).
Weymouth, Hines, Raff, and Wheeler, *Am. J. Ophthalmol.*, **11**, 947 (1928).
Wheatstone, *Phil. Trans.*, **2**, 371 (1838); *Trans. Microscop. Soc.*, **1**, 99 (1853).
White and Ford, *JOSA*, **50**, 909 (1960).
Whiteside, *J. Physiol. (London)*, **118**, 65P (1952).
Whitteridge, *Quart. J. Exptl. Physiol.*, **44**, 385 (1959).
Wilbrand, *Die Erholungsausdehnung des Gesichtfeldes*, Berlin, 1896; *Die Theorie des Sehens*, Munich, 1913.
Wilcox, *J. Gen. Physiol.*, **15**, 405 (1936); *Proc. Natl. Acad. Sci. U.S.*, **18**, 47 (1932).
Wilcox and Purdy, *Brit. J. Psychol.*, **23**, 233 (1933).
Willmer and Wright, *Nature*, **156**, 119 (1945).
Wirth, *Neue Psych. Studies*, **60**, 269 (1930).
Wirth and Zetterström, *Brit. J. Ophthalmol.*, **38**, 257 (1954).
Witkin, *J. Exptl. Psychol.*, **40**, 91 (1950).
Witte, *Physik.* Z., **19**, 142 (1919).
Wolf, *Proc. Natl. Acad. Sci. U.S.*, **32**, 219 (1946).
Wolf and Gardiner, *JOSA*, **53**, 1437 (1963).
Wolf and Morandi, *JOSA*, **52**, 806 (1962).
Wolf and Zigler, *JOSA*, **40**, 211 (1950); **53**, 1199 (1963).
Wolfe and Eisen, *JOSA*, **43**, 914 (1953).
Wood and Bitterman, *Am. J. Psychol.*, **63**, 584 (1950).
Woodburne, *Am. J. Psychol.*, **46**, 273 (1934).
Woodworth, *Experimental Psychology*, New York, 1938.
Wright, *Proc. Phys. Soc. (London)*, **B44**, 289 (1951).
Wright and Granit, *Brit. J. Ophthalmol.*, *Suppl.* **9**, 80 pp. (1938).
Wülfing, Z. *Biol.*, **29**, 199 (1892).
Wundt, *Phil. Studies*, **14**, 1 (1898).

Yarbus, *Biofizika*, **1**, 435 (1956); **2**, 698, 703 (1957).
Young (C. A.), *Phil. Mag.*, **43**, 343 (1872).

Young (T.), *Phil. Trans. Roy. Soc. London,* **91**, 43 (1801); **92**, 12, 387 (1802).
Yourevitch, *Compt. Rend.*, **188**, 137 (1929).

Zanen, *Bull. Soc. Belge Ophtalmol.*, **No. 90** (1948).
Zangwill, *Proc. Intern. Congr. Psychol., Edinburgh,* 1948.
Zeeman, *Arch. Augenheilk.*, **100**, 1 (1929).
Zegers, *J. Psychol.*, **26**, 477 (1948).
Zollner, *Ann. Physik,* **110**, 500 (1860); **117**, 477 (1862).
Zwaan, *Inst. Perception Soesterberg Rept.* **IZF 1958–10** (1958).
Zwanenburg, thesis, Amsterdam, 1915.

Subject Index

Author Index